Agricultural Valuations

Agricultural Valuations: A Practical Guide has long been the standard text for students and professionals working on agricultural valuations. Taking a practical approach, it covers all the relevant techniques and legislation necessary to correctly value farms, assess farm rents, carry out arbitrations, inventories and records of condition, including valuation clauses on sales of farms, livestock, soils, management agreements, valuation in court proceedings and a glossary of useful information.

In this fifth edition, Gwyn Williams's original text is taken on by Jeremy Moody and Nick Millard, renowned experts in the field, bringing the book right up to date to reflect recent changes in the rural economy, including development, diversification and renewable energy and specialist valuations and reference to all the latest legislation. Clear and accessible to students and professionals alike, readers will find *Agricultural Valuations* an invaluable guide to best practice in agricultural valuations.

Jeremy Moody is Secretary and Adviser to the Central Association of Agricultural Valuers (CAAV), Vice Chairman of the European Valuation Standards Board (EVSB), an independent adviser, Visiting Professor of Rural Land Management and Policy at the RAU and an honorary doctor of Harper Adams University. Working on farm structures, land occupation and use, taxation and tenancy issues throughout the United Kingdom, he was closely involved with the Agricultural Tenancies Act 1995 and is a member of the Tenancy Reform Industry Group (TRIG). Involved with the practical operation of agricultural policy since the 1980s with milk quota, the MacSharry reforms, the Single and Basic Payment Schemes and now post-Brexit policies, he liaises with governments and others in all parts of the UK. With work on matters from natural capital and soils to planning policy and the UK's recent Electronic Communications Code governing masts and cables, he has addressed conferences abroad from Portugal and Kosovo to Turkey and China.

Nick Millard began his professional career in a West Country market practice and joined Bruton Knowles in 1989 where he became Chairman of Partners in 2005. Having spent much of his career with the firm in estate management, latterly he concentrated on strategic property advice for clients in the public and private sectors. He has combined his commercial career with academic work and currently lectures at the Royal Agricultural University and Henley Business School at the University of Reading and is a consultant to Michelmore Hughes Stags. He is a member and former chairman of the CAAV Valuation, Compensation and Taxation Committee and a former delegate to TEGoVA. He has been a member of the RICS Tax Policy Panel and a lecturer and Visiting Fellow at the University of Plymouth. He is currently a Visiting Research Fellow at the University of Exeter, where he has worked with the Centre for Rural Policy Research. He has contributed to research for Defra, the devolved administrations and government agencies on a variety of land tenure and rural economy issues. He is a past president of the CAAV.

Agricultural Valuations
A Practical Guide

Fifth Edition

Jeremy Moody and Nick Millard

LONDON AND NEW YORK

Fifth edition published 2021
by Routledge
2 Park Square, Milton Park, Abingdon, Oxon, OX14 4RN

and by Routledge
52 Vanderbilt Avenue, New York, NY 10017

Routledge is an imprint of the Taylor & Francis Group, an informa business

© 2021 Jeremy Moody and Nick Millard

The right of Jeremy Moody and Nick Millard to be identified as authors of this work has been asserted by them in accordance with sections 77 and 78 of the Copyright, Designs and Patents Act 1988.

All rights reserved. No part of this book may be reprinted or reproduced or utilised in any form or by any electronic, mechanical, or other means, now known or hereafter invented, including photocopying and recording, or in any information storage or retrieval system, without permission in writing from the publishers.

Trademark notice: Product or corporate names may be trademarks or registered trademarks, and are used only for identification and explanation without intent to infringe.

First edition published by Estates Gazette 1984

Fourth edition published by Routledge 2008

British Library Cataloguing-in-Publication Data
A catalogue record for this book is available from the British Library
Library of Congress Cataloging-in-Publication Data
Names: Moody, Jeremy, author. | Williams, R. G. (R. Gwyn)
  Agricultural valuations.
Title: Agricultural valuations : a practical guide / Jeremy Moody and
  Nick Millard.
Description: Fifth edition. | Abingdon, Oxon ; New York, NY :
  Routledge, 2021. | Revised edition of: Agricultural valuations : a
  practical guide / R.G. Williams. 4th edition. 2008. |
  Includes bibliographical references and index.
Identifiers: LCCN 2020039898 (print) | LCCN 2020039899 (ebook) |
  ISBN 9781138678040 (hardback) | ISBN 9781138678057 (paperback) |
  ISBN 9781315559162 (ebook) | ISBN 9781317194293 (adobe pdf) |
  ISBN 9781317194279 (mobi) | ISBN 9781317194286 (epub)
Subjects: LCSH: Farms—Valuation—Handbooks, manuals, etc. |
  Farms—Valuation—Great Britain—Handbooks, manuals, etc.
Classification: LCC HD1393 .W55 2021 (print) |
  LCC HD1393 (ebook) | DDC 333.33/520941—dc23
LC record available at https://lccn.loc.gov/2020039898
LC ebook record available at https://lccn.loc.gov/2020039899

ISBN: 978-1-138-67804-0 (hbk)
ISBN: 978-1-138-67805-7 (pbk)
ISBN: 978-1-315-55916-2 (ebk)

Typeset in Baskerville
by Apex CoVantage, LLC

Contents

Preface	viii
Foreword	x
1 Introduction and history	1

PART 1
Foundations   7

2 Agricultural land	9
3 Basic property law	14
4 Business structures for farming REMPARI	26
5 Agricultural support: from the EU to beyond Brexit	31
6 Farm accounting REMPARI	48

PART 2
Valuations   53

7 Professional issues	55
8 Introduction to valuation standards and bases of value	65
9 Undertaking a valuation	77
10 Valuation of farm property with vacant possession	87
11 Valuation of let property	113
12 Valuations for insurance	131
13 Woodland and sporting	134

14	Diversification	144
15	Development	153
16	Livestock, machinery, growing crops and produce	187
17	Environmental valuations	202

PART 3
Valuations for taxation, compulsory purchase, utilities and communications — 213

18	Valuations for capital taxes	215
19	Agricultural stocktaking for income tax	232
20	Valuations for business rates and council tax	237
21	Compensation for compulsory purchase	247
22	Cables and pipes for electricity, gas, water and sewerage	313
23	Masts and cables for communications	333

PART 4
Agricultural tenancies — 341

24	Introduction to agricultural tenancies	343
25	Issues during a tenancy requiring valuation	345
26	Rent reviews under the Agricultural Holdings Act 1986	348
27	Rent reviews for farm business tenancies	374
28	Rent reviews for Scottish agricultural tenancies	379
29	End of tenancy: tenant's claims	391
30	Compensation for disturbance	412
31	End of tenancy: landlord's claims	416
32	Valuing the tenancy	422
33	Valuations for other agreements	430

PART 5
Professional practice   435

34	Farm agency	437
35	Dispute resolution and expert witness work	456
36	Future skills and services	467
37	Advice to young practitioners	475

Appendices
A	*Calendar of dates for farming agreements and other purposes*	481
B	*Headings for a valuation report for agricultural property: the valuation report*	483
C	*Schedule of statutes and cases*	488
	Index	494

Preface

Gwyn Williams was a tenant farmer's son from Montgomeryshire, born in 1930. Left physically unsuited for farming by osteomyelitis, he became an agricultural valuer and auctioneer. Starting as an assistant to Kenneth Glenny, then also CAAV Secretary, he moved to south Herefordshire and Ross on Wye in 1954 where he practised until he died in 2018, working from the town and Ross Livestock Market as well as farming. Greatly respected and liked, he achieved many awards including the CAAV's Kenneth Glenny Prize for writing the first edition of this book, with a further three editions then following establishing it as a point of reference for many students and those progressing in the profession.

That project, taken on by two authors, now reaches its fifth edition with a scope showing how the work of an agricultural valuer has broadened with the widening range of economic activity on farms and in the countryside. It is published on the threshold of the different countries of the United Kingdom introducing their own agricultural policies on leaving the European Union with an independence not possible since 1972: a decision-making moment not seen since 1947 but now taking account of devolution. With potential post-Brexit changes in trade, the developing response to climate change and environmental issues and unfolding technologies, these offer challenge but also opportunity and are likely to drive major change in which the professional skills of agricultural valuers will be in great demand. Building on the work of Gwyn Williams, the authors offer this book to support them in helping clients identify and deliver the most apt answers.

In taking this book forward at this time, we pay not only homage to Gwyn Williams but also tribute to those who supported our careers. More generally, we express our appreciation of the fellowship of agricultural valuers in all parts of the United Kingdom without whose good spirit, practicality, knowledge and humour a task like this would be much harder. We thank Sarah Anderton and Flossy Freeman-Inglis for bringing fresh eyes in reviewing drafts of this text as it developed and set on record the tolerance of our editor. Finally, we thank our wives, Tricia

Preface ix

Moody and Laraine Millard, for their support in all our varied work that has made a text such as this possible. The flaws and omissions in this text are, though, ours, but it is nonetheless offered to stand the profession, and so farming and the countryside, in as good a stead as we can in the coming years.

Jeremy Moody   Nick Millard

Foreword

Williams has long been an authoritative text on the subject of agricultural valuations, but this latest edition, published in 2020 at a time of great potential change and with a political will and government majority seemingly able to deliver such change, not only updates earlier editions but opens a window onto the post-Brexit world of 2021 and beyond. Law is taken up to the Agriculture Act 2020 and with reference also to the delayed Environment Bill. Case law is up to date and its application explained. As such, it is a book for those embarking on their career, seasoned practitioners and anyone wishing to learn exactly what is required to perform such a task.

The art of the agricultural valuer mystifies many. It is understood by some but misunderstood by more. This book provides a clear understanding of what undertaking a valuation of rural property actually involves and what skills are needed to arrive at the valuation itself.

The book explains the breadth of knowledge required to undertake agricultural valuations, far more than likely to be needed for other types of valuation. Methods may be similar, but their application and variety of valuation types require a working comprehension of all facets of property occupation in addition to an understanding of peripheral matters such as residential and commercial tenancies, environmental matters, waste, invasive species and soils, to name but a few.

The late Gwyn Williams's original text has been expanded to include advice on other practical issues relating to valuations of all types and also matters which all those working as a rural-based professional would be likely to encounter. Jeremy Moody and Nick Millard have together produced a book which is both readable and understandable. Each author is a noted professional with a full understanding of the subject and the result is something which should not only be standard reading for any student but be in the armoury of every agricultural valuer in the country.

David Brooks, BSc, FRICS, FAAV, REV
President, CAAV 2019–20
Colchester, Essex

1 Introduction and history

1.1 Agricultural valuations have been needed as long as there have been transactions in and taxation of farmland. The oldest recorded lease in England was granted soon after 700 with references in Anglo-Saxon literature to those managing land in the decades before and after that. Much of the Domesday Book of 1086 can be seen as a register of the values of agricultural properties for taxation purposes, prepared with the skills of preceding centuries of Anglo-Saxon administration. Changes in land occupation and taxation have continued as key drivers of rural valuation work.

1.2 The rise of agricultural tenancies saw an increasing need for valuations, whether of premiums to take tenancies or the rack rents that replaced that approach in the early nineteenth century. The last hundred or so years have seen, first, a swing from a rural world that was almost wholly tenanted to one in which ownership and farming use became steadily more aligned in the mid-twentieth century and then, undoing that, a growing division between the ownership, occupation and use of farmland in recent decades with the variety of ways in which this can be achieved.

1.3 Transactions and borrowing – Across the centuries, farmland with its fixed equipment, housing and other attributes is an asset to be valued whether as a whole farm or as parts. That might be for a proposed sale or purchase. Equally, the collateral offered by agricultural land can offer access to competitively priced finance and this too requires valuation. Valuations may also be needed to recognise transfers of value between connected parties. The knowledge of value and factors affecting it may inform and influence other business and family decisions.

1.4 Landlord and tenant – The eighteenth-century focus on agricultural improvement and the needs of the French wars saw the emergence of a market-based answer to the problem of encouraging a tenant to invest in the land for the longer term, with fertility and improvement. Tenant right grew as a complex of local customs responding to the farming needs and practices of each area. Those customs of the country with provisions for hold over, early entry and compensation for work improving fertility, applying in an overwhelmingly tenanted landscape, created the initial core of the profession in agricultural valuations. The corollary was the need to assess dilapidations, the tenant's compensation to the landlord for shortcomings in his management of the property. The assessments, first done by respected farmers, were increasingly handled by skilled local practitioners with some of today's practices founded in the eighteenth century. They then formed local groups to discuss and refine practice – such as the present Suffolk Association of Agricultural Valuers founded in 1847. Those groups are the roots of the Central Association of Agricultural Valuers (CAAV), founded by them as a national body for the profession in 1910.

1.5 The custom of tenant right was progressively overtaken by statute law for England and Wales with a parallel process unfolding for Scotland. The present provision for an agricultural tenant's fixtures was introduced in 1851. A voluntary basis for tenant right provided in 1875 was made mandatory in 1883. The successive Agricultural Holdings Acts from 1908 onwards provided occasions for valuation from tenant's improvements to game damage, as well as the formalisation of the Evesham Custom to support tenants' larger and commercial investments in fruit bushes. Custom itself was formally abolished by the Agriculture Act 1947, leaving the matter to statute and contract. The post-war security given to agricultural tenancies in Great Britain by the Agricultural Holdings Acts was accompanied by statutory rent review provisions which became ever more detailed with successive Acts.

1.6 By contrast, in Ireland the tenancy system which had seen the similar emergence of Ulster tenant right was, after further political pressure over land tenure, dismantled from the late nineteenth century, creating a landscape of small owner-occupiers. However, both parts of the island are now looking at reviving the use of tenancies rather than the seasonal right to farm offered by conacre.

1.7 With the 1995 reversion in England and Wales to the much greater freedom of contract for Farm Business Tenancies (FBTs), the essential problem of recognising the continuing value of beneficial improvements was handled by statutory provisions for the compensation of improvements, with tenant right now addressed as routine improvements.

1.8 Since the late 1970s, there has been increasing experimentation with alternatives to tenancies for a landholder to engage with a farmer to secure the farming of land. Contract farming, partnerships, joint ventures and other arrangements have all required valuation support.

1.9 Other uses – Continuing development pressure for housing and other uses creates opportunities for value while the economic incentives for diversification outside agriculture and the release of assets from farming have seen a growing need to consider not only residential lettings but also business and other uses with new legal regimes, issues, opportunities and risks. This has been illustrated by the explosion of activity with renewable energy production since 2010.

1.10 Development control does of course do exactly that, limiting access to a consequently greater value, and, with the historic bias of the system against new development in open country, there have been specific provisions for agriculture and forestry to be able to have new dwellings, usually subject to agricultural occupancy conditions, and buildings. A widening range of permitted development rights now requires understanding and appraisal for the value they might bring.

1.11 Statutory acquisition – Statute law has imposed on the countryside in other ways to require agricultural valuations. Turnpike roads, canals and the railways all drove the need to consider the value of the land taken and the effects on retained land in the compulsory purchase code first pulled together as general legislation in 1845. That has then been applied to the later needs for highways, water, electricity, gas and other compulsory acquisitions of land or rights from reservoirs to new towns, so much of which are necessarily on or across rural land. As the current HS2 railway scheme shows, these all raise issues for the agricultural valuer. The 2019 UK government's emphasis on infrastructure is likely to add to the volume of compulsory purchase work.

1.12 Alongside that, there is a separate regime for communications apparatus from Napoleonic-era semaphore stations to today's masts and fibre optic cables, most recently revised in 2017 as part of the government's strategy for broadband with as yet unresolved valuation issues.

1.13 Taxation – Governments have always needed to raise money and land has usually been an easy target. Rating applied to agricultural land from its mediaeval origins until substantially exempted in 1929 while some buildings and uses are liable and so need assessments. Taxes on inheritance began with Legacy Duty in 1796 and then Succession Duty in 1853 – both paid by the heir. They were overlaid with Estate Duty in 1894 and then Capital Transfer Tax (CTT) in 1975, relabelled as Inheritance Tax (IHT) in 1986.

1.14 Both Estate Duty and then CTT/IHT have had important but incomplete reliefs for agricultural property. Valuations have been needed under the various forms of Capital Gains Tax, Stamp Duty now replaced by the differing transactions taxes around the United Kingdom, VAT and even the Annual Tax on Enveloped Dwellings (ATED). Since at least the Second World War, stocktaking has been relevant to the tax assessment of farmers under Income Tax or Corporation Tax.

1.15 Other farming property – While other farming property such as livestock and machinery often have to be valued (whether for sale with the farm on their own or for other reasons including divorce), this can be especially critical when required by state disease control policy. This was seen sharply in the 2001 outbreak of food-and-mouth disease across much of England, Wales and south-western Scotland, as in the earlier 1967 outbreak, the culling for BSE and then TB (now rarely requiring valuations in England but continuing in Wales)

1.16 Government intervention in agriculture and the countryside – Valuers had had roles in assisting the effects of the last century's regimes and arrangements for two world wars while the support policies that operated from the Wheat Act 1932, the various marketing boards and then post-war deficiency payments framed the marketplace. Conservation measures saw the identification of Sites (or Areas) of Special Scientific Interest (and so management agreements) from 1949 and the creation of national parks.

1.17 The UK's membership of the European Union (EU) from 1973 to 2020 meant that the Common Agricultural Policy (CAP) was a powerful influence on farming. While intervention payments again did not need direct consideration, the developing policies of restraint with milk quotas, set aside, livestock premium rights, status for Arable Area Aid and then entitlements to the Single and Basic Payment Schemes all variously created identifiable assets, restrictions or income streams that directly required work. Professional ingenuity rapidly made milk quotas transferable, creating markets for their sale and lease, with the lessons learnt from that applied to later developments in the CAP.

1.18 Evolving policy has then made not only environmental restrictions significant but also created opportunities for value, starting with Environmentally Sensitive Areas and then the developing variety of schemes in each part of the United Kingdom with detailed rules and interactions.

1.19 The future – Looking ahead with the United Kingdom leaving the EU, issues of resource management and climate change seem likely to become more significant with a renewed attention to soils and water quality as well as flood management and mitigation with restrictions and opportunities both affecting or creating value. Energy efficiency is likely to bear more heavily in the consideration of property. There is discussion of the potential for new markets in environmental services as a means to manage issues or offset other development.

1.20 Throughout, formal valuations and valuation advice with accompanying services have been needed to aid decision making by owners, farmers, businesses, lenders and others. From purchase and tenancy through investment and business actions to partnership dissolutions and divorce, agricultural valuations have been required as an objective appraisal.

1.21 Dispute resolution has been a longstanding skill for senior and experienced agricultural valuers. Tenancy statutes historically focused on arbitration, but we now also see expert determination and mediation. Building on the more ordinary but vital work of negotiation between parties, these should all be means to give practical answers in a timely and cost-effective way for businesses to move on with their work, with application across the whole rural economy.

1.22 What is clear is the continuing relevance of agricultural valuations recognising and managing value within the United Kingdom's market economy with new price signals driving change and adaption. They need to be undertaken by those who understand the three quarters of the country that is farmed together with the people that do that, bringing the skills of practical appraisal and judgment to deliver the required professional support for all parties needing these services. Those underlying practical skills were recorded a thousand years ago in the Anglo-Saxon model of the management of a rural manor in *Gerefa* and Walter of Henley's land management lectures at Oxford University in the 1270s. This book follows those in a long line, and directly on Gwyn Williams's work, to assist those resolving agricultural and rural issues in the twenty-first century.

Part 1
Foundations

2 Agricultural land

2.1 Land

2.1.1 Land (typically defined in the United Kingdom to include the buildings on it) is very distinctive as an asset as its character comes from both its nature and its physical location. Especially for farming, it can be seen not only as a fixed physical asset but also as the working material of the business, in that sense akin to plant and machinery. It may have buildings and fixed equipment or simply be bare. The legal framework for land varies between the parts of the United Kingdom.

2.1.2 All these points and any other factors may affect the value of rural property within markets that express the supply and demand for it and such attributes.

2.2 Character of the land

2.2.1 The nature or character of any individual parcel of land or farm will be the accumulation of a wide range of possible points of interest. The list offered here can only be illustrative and not exhaustive, while the balance of factors will vary between cases, over time and between markets:
- underlying geology
- type and depth of soil
- height above sea level, slope and aspect
- climate
- size and layout of fields
- boundaries with fences, walls, ditches, etc
- drainage
- contamination
- access
- proximity to markets
- fixed equipment
- dwellings
- buildings
- water supply

- vulnerability to flooding or drought
- other services (including for renewables, grid connection possibilities)
- public access
- exposure to and risks from intrusion, rabbits, deer, game damage, etc
- sporting opportunities
- state of repair, maintenance and management.

2.2.2 There may be specific legal issues binding the land such as:
- obligations (such as to repair the chancel of the local church) or opportunities (first refusals) under the legal title
- restrictive covenants (real burdens in Scotland) and, potentially, commitments for conservation management
- reservation to others of an access route or other rights such as sporting or drainage over the land
- easements (servitudes in Scotland)
- rights taken by utilities.

2.3 Location and designations

2.3.1 The place of any parcel of land in the market will depend on its location. In the modern world that is not only a function of its accessibility or otherwise, its position vis-à-vis neighbours and other such factors but also how it is affected by official designations.

2.3.2 In particular, the market for farmland, especially for smaller blocks, is usually very much locally driven as few will want to buy a block of land that is not conveniently placed for their operation and so will not compete for land further away, segmenting the market. Thus, location may largely drive the market of potential buyers, save for larger freestanding units.

2.3.3 One particular consideration for the value of farmland, more than for many assets, is the potential for neighbours or others nearby to see convenient land not only for itself but for the additional value it might bring to their own businesses, whether by better supporting overheads or using facilities such as buildings or irrigation. Each might be a special purchaser but, where there are several such possible buyers, they may make a market.

2.3.4 More generally, relevant issues can include:
- the country within the United Kingdom, each having its own legislation and approaches
- agricultural designations, including:
  - whether in the Less Favoured Area (LFA) and so Severely Disadvantaged Area (SDA) (generally upland except for the Scilly Isles) or Disadvantaged Area (less material now). This can be relevant to access to rates of payments, such as the Scottish LFA Support Scheme (LFASS) and its headage payments
  - direct payment region with the differing rates for area payments (currently Basic Payment)

- relevance and impact of cross compliance, greening and equivalent provisions of state agricultural policies
- inclusion in a Nitrate Vulnerable Zone
- environmental and landscape designations:
  - National Parks (and areas with related designations such as the Norfolk Broads)
  - Areas of Outstanding National Beauty (AONBs)
  - Conservation Areas
  - designations of buildings, structures, parks, etc. as listed for their historical, architectural or other interest
  - Natura 2000 sites
  - Ramsar sites
  - National Nature Reserves
  - Sites of Special Scientific Interest (SSSIs), called Areas of Special Scientific Interest (ASSIs) in Northern Ireland and the Isle of Man
- development control:
  - is it in a Green Belt?
  - is it brownfield land?
  - is it with in the village envelope or other settlement boundary?
  - how does the land sit within the local planning policies?
  - the permitted development rights available and useful
  - have permitted development rights been withdrawn (under Article 4 or equivalent powers in Scotland or Northern Ireland)?
  - is it subject to planning restrictions because of issues such as a high-pressure gas main?
- public rights of way and open access provisions
- areas with particular costs, such as an Internal Drainage Board or other area with a drainage rate.

2.4 Buildings and fixtures

2.4.1 The basic principle is that anything fixed to land belongs with it and so passes with it when ownership changes or, potentially, when it is let.

2.4.2 While that may appear obvious for substantial items like major buildings, road, drains or hedges, there are often items that have been added by an occupier or which can appear to be chattels (moveables in Scotland). Sometimes these are clearly removeable without damaging the land and so likely to remain personal property. In other cases, they may have needed to be fixed to land to be useful (as with the calibration of a bulk milk tank). In yet further cases, they may have become part of the land. The issue was summarised by the court in *Holland v Hodgson*:

> Thus blocks of stone placed one on the top of another without any mortar or cement for the purpose of forming a dry stone wall would become part of the land, though the same stones, if deposited in a

builder's yard and for convenience sake stacked on the top of each other in the form of a wall, would remain chattels.

2.4.3 These situations have been considered in a number of cases over many years with the key tests being the degree and the object of fixing an item to the property. In *Elitestone v Morris*, the House of Lords decided that a prefabricated bungalow had become part of the land on which it only rested, contrasting it with a greenhouse that could be regularly moved, with this review:

> Many different tests have been suggested, such as whether the object which has been fixed to the property has been so fixed for the better enjoyment of the object as a chattel, or whether it has been fixed with a view to effecting a permanent improvement of the freehold. This and similar tests are useful when one is considering an object such as a tapestry, which may or may not be fixed to a house so as to become part of the freehold: see *Leigh v. Taylor* [1902] AC 157. These tests are less useful when one is considering the house itself. In the case of the house the answer is as much a matter of common sense as precise analysis. A house which is constructed in such a way so as to be removable, whether as a unit, or in sections, may well remain a chattel, even though it is connected temporarily to mains services such as water and electricity. But a house which is constructed in such a way that it cannot be removed at all, save by destruction, cannot have been intended to remain as a chattel. It must have been intended to form part of the realty.

2.4.4 These issues are relevant when considering what is to be valued for the sale, mortgage or other purpose for a property. Recent disputes concern whether fish in lakes or solar panels are part of land. Additional points arise for tenancies.

2.4.5 Tenant's fixtures – That common law position has been overlaid by statutory intervention whereby the tenant, here particularly an agricultural tenant, is given a statutory procedure for removing some fixtures – "unfixing" them.

2.4.6 Under what is now s.10 of the Agricultural Holdings Act 1986, s.8 of the Agricultural Tenancies Act 1995 for farm business tenancies and s.18 of the Agricultural Holdings (Scotland) Act 1991 and in slightly differing terms, the agricultural tenant has the right to remove buildings and many fixtures but must first give the landlord the chance to buy the item in question at its value to an incoming tenant. This does not apply to the various forms of Limited Duration Tenancy in Scotland.

2.4.7 For Northern Ireland, s.17 of the Landlord and Tenant Amendment (Ireland) Act 1860 provides the right to a tenant to remove many fixtures where this can be done without substantial damage to the freehold or the fixture at any time up to two months after the end of tenancy, compensating the landlord for any damage.

2.4.8 Crops, etc – More fundamentally and a key reason for the existence of separate agricultural tenancy legislation, the rule that what is fixed to land, belongs with land applies to crops, fertilisers and other basic aspects of farm work, revealing land not only to be premises but here more akin to plant and machinery used in production. Without some form of statutory intervention, a seed would, once sown, belong to the landlord with the tenant having no right in it.

2.4.9 The tenancy agreement allows the tenant to harvest and have the benefit of work during the tenancy, but agricultural tenancy statutes in Great Britain intervene in two broad ways:
- providing a structure (often referred to as "tenant right") at the end of the tenancy for compensating the outgoing tenant for what are then still growing crops, applied fertilisers and improvements to the soil, harvested crops (such as perhaps silage in the clamp), hefted hill sheep and some other works that must be left so that value is paid for what is left behind, encouraging the tenant to maintain the farm to the end
- disregarding at a rent review the generality of fixed equipment provided by the tenant (limited in Scotland to compensatable improvements) so that the tenant does not pay rent on the benefit of works the tenant has made.

3 Basic property law

3.1 Introduction

3.1.1 Agricultural valuation work requires a knowledge of land law to understand the rights and obligations relevant to a property. That law is a mix of basic common law and the growing volume of specific statutes. That combination then interacts with other regimes of law, such as those for subsidy policy and for development control.

3.1.2 Devolution in the United Kingdom – Basic land law is very similar in England, Wales and Northern Ireland, but it differs a little more in Scotland.

3.1.3 England and Wales have essentially had the same statute law until very recently, but with devolution some aspects of law relating to agricultural valuations are now beginning to diverge as the Welsh Parliament (Senedd) enacts its own legislation, varying or replacing existing acts and substituting Land Transactions Tax for Stamp Duty Land Tax.

3.1.4 Northern Ireland has had a different legislative history. While largely having the same common law as England regarding land, Ireland had a Parliament in Dublin until 1801 when legislation passed to Westminster before moving to Stormont in 1921. Much nineteenth-century legislation remains in place, as for conveyancing and the basic framework for all tenancy law (the Landlord and Tenant Law Amendment (Ireland) Act 1860 – "Deasy's Act").

3.1.5 Scottish law, including its land law, has developed slightly separately. It was a key term of the 1707 Act of Union that Scotland retained its own law and courts. In addition to its earlier history (including the Leases Act 1449 with its definition of what is a lease found to catch arrangements that might be licences elsewhere in the UK), more specific Scottish legislation was made in the second half of the twentieth century (such as the Agricultural Holdings (Scotland) Act 1991). Since its creation in 1999, the Scottish Parliament has been passing statutes within its wide range of competence (including the Land Reform (Scotland) Act 2016 and introducing the Land and Buildings Transactions Tax).

3.1.6 Taxation – Not only are land tenure arrangements and business agreements important in themselves and their associated rights and obligations, opportunities and restrictions, but they also have consequences in the treatment of the parties under both capital and revenue taxes. In this, a critical distinction lies in whether the taxpayer is using the land in business or, more passively, as an investment. Landlords' interests and rental income are generally treated differently from an owner-occupier's interest and business income. There are both general rules which apply irrespective of the nature of the land use and specific rules for agricultural use. These are discussed in Chapter 18.

3.1.7 Tenancies and valuation – Whatever the tenancy, the agricultural valuer may need to consider values for:
- the rent when the agreement is granted, whether in submitting or reviewing rent tenders with their justifications or negotiating it, taking into account the property, the proposed terms of the lease and the market
- the rent if it is later reviewed, when the valuer could be acting for either party or determining a dispute between them, similarly having regard to all circumstances
- the impact of the tenancy on the value of the land, whether for negotiation, security for lending, taxation purposes, compulsory purchase or other reasons
- end of tenancy claims for compensation between the parties for improvements and dilapidations
- very rarely and, if at all, more likely in commercial and residential lettings, the value and effect of any premium paid.
- the tenancy itself.

3.2 Introduction to England and Wales

3.2.1 The basic form of ownership is of freehold land (more formally, held from the Crown by tenancy in fee simple absolute in possession). The land may be owned by a single individual, individuals jointly or in common, a company or trust or in other ways.

3.2.2 In principle, the owner of land owns the subsoil and everything below it to the centre of the earth and all the airspace above. However, not only does practicality impose some limits on the real value of those dimensions but statute law has limited that in a variety of ways including the reservation of:
- gold and silver to the Crown, including treasure trove
- coal, oil and gas to the Crown
- rights for overflying by aircraft.

3.2.3 Statutory powers for authorised bodies to be able to compulsorily acquire land or rights in land intrude on the quality of freehold ownership.

3.2.4 The use of land is not solely at the owner's discretion but limited by, for example, the development control regime, public rights of way, the protection of buildings and sites of historic or scientific importance and other regimes as well as the risk of nuisance from others and the need to avoid causing nuisance.

3.2.5 The property may also benefit from or be regulated by other rights and obligations, such as may be created by:
- restrictive covenants, binding the land to benefit other land or vice versa
- easements, whether taking rights over neighbouring land or giving rights over the owned land as for water pipes, drainage and other reasons
- impositions under statutory powers as by utilities for water, electricity and gas regimes or for communications apparatus under the Electronic Communications Code
- the reservation or alienation of minerals, sporting or other distinct rights by a previous owner
- limitations under development control policies, most obviously agricultural occupancy conditions for a planning permission for a dwelling, limiting who may live in it, but also other limitations as may have been required under a planning agreement or with reduced permitted development rights in AONBs and National Parks and occasionally their full withdrawal under an Article 4 direction.

3.2.6 The land may be subject to mortgage where it has been used as security for a loan, whether to buy it or for other purposes. If the owner, as borrower, fails to meet the payments due, the lender can then take possession of the land pledged as the security for the loan.

3.2.7 Land referred to as a "common" will have an owner but be subject to the rights by others ("commoners"), often for defined powers to graze but with no ownership or larger management powers over the land.

3.2.8 Easements – An easement is a legally protected right over or use of someone else's property that might typically be used for a private right of way across a field to a house or for a water supply or drainage pipe. It is subject to the terms on which it is granted.

3.2.9 Legally, an easement requires:
- a dominant tenement or landholding that benefits from the right
- a subservient tenement over which the right applies
- that those two properties are adjacent.

It can then bind the legal title of the affected property and so benefit future owners of the dominant tenement.

3.2.10 Agreements Between landowners and farmers – Owners may have many reasons for owning farmland but not all wish or have the skills to farm it themselves without the help of others. Accordingly, there are a variety of means by which an owner may have an agreement with a farmer regarding the farming of the land. These include:

- arrangements in respect of the use of or access to the land, such as:
  - a tenancy, with some 35% of the agricultural area of England let on tenancies
  - a grant of rights such as profit of pasturage
  - a licence for access, such as for grazing, mowing or taking a standing crop
- business arrangements between farmers having one of those relationships with land:
  - a partnership or company as a means of business collaboration
  - instructing a farmer or other business as a contractor to provide services to the owner's own business
  - a variety of arrangements loosely referred to as joint ventures and share farming.

3.2.11 These differ in the legal nature of the rights granted, rules provided by statute especially for tenancies, and the actual agreements themselves. They will have differing consequences in terms of tax treatment, eligibility for subsidies and support schemes based on the land, livestock regimes and other issues, but each can be practical means to a desired end. The agricultural valuer will often be required to advise on appropriate structures and agreements.

3.2.12 Clear sighted analysis is important when negotiating, preparing, using and identifying such agreements since the label used will not commonly matter much. It is the practical effects of the benefits and obligations and how the parties really behave that will have far more weight should there be a dispute. Thus, a "partnership" or a "licence" may yet prove to be a tenancy while a "share farming arrangement" may, on inspection, prove instead to be a tenancy, a partnership or an employment contract.

3.3 Tenancies in England and Wales

3.3.1 The hallmarks of a tenancy have been clarified by case law, most recently, the House of Lords decision in *Street v Mountford* which found that a tenancy requires:
- a landlord and a tenant, who have
- an agreement to allow exclusive occupation
- of a defined property
- for a definite period whether for a fixed term or a repeating period, such as from year to year
- usually in return for a consideration.

3.3.2 That grant of exclusive occupation gives the tenant all rights in the land for the duration of the tenancy save what has been reserved in the agreement or is limited by statute. Thus, a tenant can assign or sublet the tenancy unless (as is usual) the agreement or statute prevents that – though the tenant cannot pass on any more rights than he or she has. Such issues explain the importance of discussing these issues first and then putting a practical agreement in place to record what has been agreed between the owner and the tenant.

3.3.3 Agricultural tenancies – The first critical point is whether the owner retains legal possession of the land or not. A tenancy gives possession of the land to the tenant who has the right to quiet enjoyment of what has been let to him, subject only to the terms of the agreement and any rules provided by statute. That may often arise by formal grant, sometimes written, sometimes oral, but can simply be found to have been created by the way the parties have behaved, requiring care in managing arrangements so that they do not lapse into something that was not intended.

3.3.4 This section considers farmland lettings (which are then considered in more detail in later chapters), but the underlying points also arise when considering the lettings of dwellings (with the effect of housing tenancy legislation) and for commercial uses (with business tenancy legislation, save in Scotland). For uses that are not necessarily agricultural, such as equestrian ones, it will be a matter of fact to determine if they are in, say, England and Wales, a farm business tenancy, a common law tenancy or a business tenancy.

3.3.5 Agricultural law intervenes to provide two main tenancy structures for farmland in England and Wales:

- the Agricultural Holdings Act 1986 which, with some subsequent amendments, consolidated the law as it had developed to that date, providing much detailed legislation for tenancies in place before September 1995 and in particular:
    - protecting the tenant's occupation beyond the agreed term of the tenancy by only providing very specific grounds for the landlord to enforce recovery of the land
    - the opportunity for all tenancies from before July 1984 to give rise to two successions being granted to qualifying family members
    - statutory provisions for the review of rent and a default position for repairs
    - end of tenancy compensation between the parties for improvements and dilapidations.
- the Agricultural Tenancies Act 1995, amended in 2006, which introduced the Farm Business Tenancy for almost all new lettings since September 1995. This provides much greater freedom of contract for the parties to qualifying tenancies, so that:
    - the security of tenure is just that granted by the agreement
    - tenancies let for more than two years can though continue from year to year until terminated on 12 months' notice

- the rent is to be a market rent unless otherwise agreed
- there are rules for the compensation for qualifying tenant's improvements but dilapidations are left to the tenancy agreement.

More detailed aspects of both codes will be considered throughout this book.

3.3.6 A theme common to the application of agricultural tenancy law in England, Wales and Scotland is that it applies to agricultural holdings. That concept, with its expectation of the tenant using the land for agriculture as a trade or business, is defined on the basis of it being in the words of s.1 of the 1986 Act with similar words in the Scottish 1991 Act:

> the aggregate of the land (whether agricultural land or not) comprised in a contract of tenancy which is a contract for an agricultural tenancy,

with "agricultural land" defined as:

> land used for agriculture which is so used for the purposes of a trade or business.

Farm business tenancies must satisfy the business condition of s.1 of the 1995 Act:

> that all or part of the land comprised in the tenancy is farmed for the purposes of a trade or business

The importance of that emphasis on business use was shown in the Scottish Land Court decision in *Fyffe v Esslemont* where it was found in a decision seen as relevant across the United Kingdom that such farming as was being conducted on the land was by others for their business and so the tenant lost the protection of the 1991 Act.

3.3.7 Business tenancies – A series of statutes govern aspects of the regulations of commercial lettings (other than for agriculture) for issues from tenant's improvements to guarantors. These will be relevant to commercial lettings of such property as buildings on farms or of land for largely non-agricultural purposes, such as equestrian businesses.

3.3.8 The statute most commonly referred to is the Landlord and Tenant Act 1954, Part II of which provides a regime for many business tenants to have a statutory right to renew their leases subject to the grounds for repossession by the landlord. There are arrangements for contracting out of this right, now under the Regulatory Reform (Business Tenancies) (England and Wales) Order 2003 and using an exchange of notices (easiest if done some weeks ahead of the lease) and this procedure will be adopted as a matter of course for many smaller and rural business lettings.

3.3.9 Residential tenancies – There is a separate body of statute law for residential lettings, now based on the Housing Act 1988 as amended. Most lettings of dwellings on farms and estates will now be Assured Shorthold Tenancies with substantial freedom of contract as to the agreements but within increasing regulation of the let residential sector overall.

3.3.10 However, this sector is becoming more regulated by tightening legislation. In particular, England is now expected to legislate to remove the cornerstone of the shorthold regime since 1980, the ability of the landlord to serve a notice terminating the lease without giving a ground to be challenged.

3.3.11 Wales has now enacted its own statute replacing existing legislation with law for contracts in the Renting Homes (Wales) Act 2016. At the time of writing, much of this Act was still to be commenced.

3.4 Other arrangements for farmland

3.4.1 Profits of pasturage – It is possible to grant a profit giving a more limited access to farmland only for the period, usually short, for which it is granted. In England and Wales, this will most commonly be a profit of pasturage giving a farmer a right to take grass whether, as the case may be, by grazing or for cutting and taking for conservation. The owner retains occupation of the land and is simply selling the access to the grass.

3.4.2 Licences – Often used very loosely as a word, a licence is simply a permission that could be given by the owner or tenant a third party to do something that would otherwise be a trespass. In *Thomas v Sorrell*, the court ruled:

> A dispensation or licence properly passeth no interest, nor alters or transfers property in any thing, but only makes an action lawful which, without it, had been unlawful.

Unlike a tenancy which grants everything that is not reserved by the owner, a licence only gives authority for what is expressly licenced.

3.4.3 Such a licence might range from the simple access needed by a contractor for hedge cutting to that for a stubble-to-stubble contractor or a farming partner. The more comprehensive the licence, the more it can be found to have the hallmarks of a tenancy, as may easily arise with the facts of arable cropping in the interests of a third party business. Such issues require care and they also have interactions with other matters such as tax and subsidy regimes.

3.4.4 Grazing agreements offer one very common subject for licences, often referred to as "grass keep", "grass lets" or, in the north, "grass parks". Recognising that a licence is simply a permission, with no larger framework of law, it will be important to be clear about what is granted. The CAAV model grazing licence is expressly prepared so that, if properly followed, the owner remains the farmer of the grass, producing and managing it, with the licensee simply having a right of access to graze and take the grass. This has been considered in *Charnley*, a Tribunal decision on an Inheritance Tax case where the taxpayer's access to both Agricultural and Business reliefs was upheld after evidence from the grazing licensee that the deceased was "farming his land using my stock" and was "the boots on the ground". The positive husbandry shown here made the difference between this and the Inheritance Tax decision in *McCall*.

3.4.5 A larger right to grazing and management could have differing tax, subsidy and other consequences. It will also typically be important not to effect an unintended change of who is the keeper of the livestock for movement control purposes, with recording, reporting and ear-tagging liabilities.

3.5 Northern Ireland

3.5.1 The province, with essentially the same common law as England and Wales, shares much of the previous analysis. As regards statute law, a general framework for tenancies was laid out by the Landlord and Tenant Law (Amendment) Act (Ireland) 1860 (Deasey's Act), now overlaid for business and residential tenancies by other provisions.

3.5.2 Agricultural lettings – The important difference is that the historic structure of agricultural tenancies was dismantled over a century ago by the Irish Land Acts enabling tenants to buy out their landlords. No replacement provisions were made. That means Northern Ireland has very little current experience of agricultural tenancies while subsequent legislative repeals mean that there is now no legal impediment to reviving the practice. That can be done with the freedom of contract largely allowed by Deasey's Act. With no further statutory provision, topics such as compensation for tenant's improvements need to be considered in the tenancy agreement.

3.5.3 Conacre – As Ireland became a land of owner-occupiers but with little turnover of land by sale, arrangements between owners and farmers came to be handled by agreements, often very informal, called "conacre". Conacre was originally the grant of a right to farm, usually for arable cropping, akin to a profit with the conacre taker able to remove his crop on payment of the fee. In practice, it has shifted its shape into a means for seasonal grazing arrangements now applying to over almost 30% of the farmland in the province. Often with the practical effect of a broad but short-term grant of grazing rights, it has not been found to provide a framework for continuity or investment.

3.5.4 Business tenancies – These are governed by the Business Tenancies (Northern Ireland) Order 1996 with more prescriptive regulation of renewal than in England and Wales and no opportunity to contract out.

3.5.5 Residential tenancies – Northern Ireland has its own legislation for these lettings, now mainly in the Private Tenancies Order 2006.

3.6 Scotland

3.6.1 Since land reform legislation in 2003, the ownership of all land in Scotland is now simply freehold.

3.6.2 Statutory powers for authorised bodies to be able to compulsorily acquire land or rights in land intrude on the quality of freehold ownership.

3.6.3 The use of land is not solely at the owner's discretion but limited by, for example, the development control regime, public rights of way, the protection of buildings and sites of historic or scientific importance and other regimes as well as the risk of nuisance from others and the need to avoid causing nuisance.

3.6.4 The property may also benefit from or be regulated by other rights and obligations, such as may be created by:
- real burdens, binding the land to benefit other land or vice versa
- servitudes, whether taking rights over neighbouring land or giving rights over the owned land as for water pipes, drainage and other reasons
- impositions under statutory powers as by utilities for water, electricity and gas regimes or for communications apparatus under the Electronic Communications Code
- the reservation or alienation of minerals, sporting or other distinct rights by a previous owner
- limitations under development control policies, most obviously agricultural occupancy conditions for a planning permission for a dwelling, limiting who may live in it, but also other limitations as may have been required under a planning agreement or with reduced permitted development rights in AONBs and National Parks and occasionally their full withdrawal.

3.6.5 The land may be subject to mortgage where it has been used as security for a loan, whether to buy it or for other purposes. If the owner, as borrower, fails to meet the payments due, the lender can then take possession of the land pledged as the security for the loan.

3.6.6 Beyond these issues, Scotland's land reform agenda has a growing policy expectation of community interest in, engagement with and sometimes acquisition of land and also with its uses. That is expressed in the community right to buy and more widely by the role of the Scottish Land Commission (including the Tenant Farming Commissioner) and the land rights and responsibilities statement under the Land Reform (Scotland) Act 2016.

3.6.7 Tenancies – While the analysis of *Street v Mountford* is relevant, Scotland has its own land law. The Leases Act 1449 sets out the definition for a lease in Scotland which, rendered into modern English, provides that

> It is ordained for the safety and favour of the poor people that labour the ground that they and all others that have taken or shall take lands in time that come from lords and have terms and years thereof, that suppose the lords sell or alienate these lands, that the tenants shall remain with their leases until the end of their terms, no matter into whose hands that ever the lands come to, for the same rent as they took them for before.

3.6.8 Providing the tenant's real property right in a lease, it requires that a lease must have a fixed duration. It has thus been found that an agreement for an unspecified farming rotation was not a lease (*Stirrat v Whyte*) while it was considered in *Brador Properties v British Telecommunications plc* that in Scotland the definition of a lease in Scotland was wider than that in England, so that in Scotland it may be enough that the alleged tenant receives the right to certain uses only of the lands, as opposed to the entire control of them.

3.6.9 Agricultural tenancies – Scotland has its own agricultural holdings statutes with:

- the Agricultural Holdings (Scotland) Act 1991 which governs all tenancies from before 2003, protecting the tenant's security of tenure which can run indefinitely for generations under Scottish rules for succession. It also makes provisions for rent review, fixed equipment and end of tenancy claims.
- the Agricultural Holdings (Amendment) (Scotland) Act 2003 which, supplementing and amending the 1991 Act rather than creating a completely new regime, introduced the options of:
  - a Short Limited Duration Tenancy (SLDT) for up to five years which cannot be followed by another one to the same tenant
  - a Limited Duration Tenancy (LDT) which can now operate for fixed terms of any length for 10 years or longer subject to lengthy procedures for termination as well as powers for tenants to diversify and to buy the landlord's interest if it is to be offered on the market
- the 2003 Act also carries forward the provision for grazing tenancies of less a year which can be repeated without creating further rights provided there is a clear break in the grazier's access to the land between them
- the Land Reform (Scotland) Act 2016 which introduced:
  - the Modern Limited Duration Tenancy (MLDT) available from November 2017 in place of LDTs, very similar to the LDT but allowing a break clause (not otherwise allowed under these laws) on the fifth anniversary where it was let to a new entrant

- a long-term Repairing Tenancy where the burden on fixed equipment can be carried by the tenant with the aim of improving run-down unit (but yet to be introduced at the time of writing)
- the Small Landholders (Scotland) Acts 1886 to 1931 also govern some tenancies.

3.6.10 The north-western crofting counties also have properties subject to separate crofting legislation.

3.6.11 Other arrangements – Scottish law, as interpreted in cases such as *Brador Properties*, acts to treat as leases many arrangements that would be licences elsewhere in the United Kingdom. As seasonal grazing or mowing tenancies are allowed, now by the 2003 Act, that resolves many issues about such access to pasture land; the CAAV equivalent to its model grazing licence for England is for a seasonal tenancy in Scotland. However, with this provision being just for grazing or mowing, this issue calls for care more generally, particularly when considering seasonal cropping (as for potatoes) or share farming agreements.

3.6.12 In Scotland, a partnership has its own legal personality as a firm and so can be the tenant of a lease in its own right. As a result, there was a history of using limited partnerships (with the landlord as a limited partner) to provide fixed-term leases, but this was ended by the 2003 Act.

3.6.13 Business tenancies – Scotland has virtually no specific law governing business tenancies, which therefore largely operate with freedom of contract under the general law.

3.6.14 Residential tenancies – Scotland has its own legislation for residential tenancies, now in the Private Housing (Tenancies) (Scotland) Act 2016 which introduced a new Private Residential Tenancy regime from December 2017 that replaced the Short Assured Tenancy and removed the landlord's ability to serve a no-fault notice to terminate a lease.

3.6.15 Servitudes – A servitude (an easement elsewhere in the United Kingdom) is a legally protected right over or to use someone else's property that might typically be used for a private right of way across a field to a house or for a water supply or drainage pipe. It is subject to the terms on which it is granted.

3.6.16 Legally, a servitude requires:
- a dominant tenement or landholding that benefits from the right
- a subservient tenement over which the right applies
- that those two properties are adjacent.

It can then bind the legal title of the affected property and so benefit future owners of the dominant tenement.

Further reading

An Agricultural Valuer's Guide to Business Tenancies in England and Wales (CAAV, 2012)
An Agricultural Valuer's Guide to Residential Tenancies in England and Wales (CAAV, 2013)

Arable Farming with Contractors (CAAV, 2020)
Contract Farming for Breeding Livestock Enterprises (CAAV, 2020)
Grazing Arrangements (CAAV, 2013)
The Letting and Management of Residential Property in Wales (CAAV, 2017)
Rural Workers Dwellings: Planning Control in the United Kingdom (CAAV, 2014)
Tenancies, Conacre and Licences: Arrangements for the Occupation of Agricultural Land in Northern Ireland (CAAV, 2015)
2015 Update on Residential Tenancy Regulations in England, Wales, Scotland and Northern Ireland (CAAV, 2015)

4 Business structures for farming

4.1 Farming business structures

4.1.1 Farming businesses in the United Kingdom are typically family concerns, not involving third parties and structured as sole traders or partnerships of only family members. Relatively few operate as companies, though these will be more evident among larger and more commercial businesses.

4.1.2 A sole trader is where the business is in the name of a self-employed individual who is the business with its opportunities and risks for profit and loss. The business is funded by the sole trader's resources, supplemented as necessary by borrowing. Others in the business will be employees of the sole trader.

4.1.3 A partnership, discussed further later on, is an association of individuals sharing the risks and profits of a business, otherwise operating in the style of a sole trader.

4.1.4 A company, also discussed further later on, is a separate legal and taxable entity, distinct from its owners who are shareholders while its directors, appointed by shareholders, will be responsible for its management. In a family business, family members might fill all these roles with shareholdings potentially varying between family members with their transfer being a means to manage succession in the business. With parties so closely connected it should be noted that money moving between the company and its shareholders, directors and other parties will be taxable as that is a transfer of value between different entities.

4.2 Business arrangements between farmers

4.2.1 The increasing interest in the ways in which owners and farmers may work together builds on the experience of seeking alternative arrangements to using a tenancy in the days before 1995 when the 1986 Act was the only means to let. These arrangements variously see:
- parties coming together in one business, as through a partnership or by forming a company in which each has shares

- one party retaining the other as contractor to provide services, though the balance of practical responsibility between the parties may vary widely between agreements
- two businesses remaining independent in producing a common output (a broad definition of the loose phrase "share farming").

4.2.2 One common theme of these mutual farming arrangements is the level of practicality and trust they require from both parties, especially to handle matters where losses arise or circumstances prove adverse. The more serious the level of engagement, the more it should be seen as a medium- or longer-term relationship rather than a short-term commercial deal.

4.3 Partnerships

4.3.1 In agriculture, these are commonly found between members of farming families and only rarely involve a third party.

4.3.2 Partnerships are governed by the Partnership Act 1890 and the individual agreement, where one can be found. S1(1) of the 1890 Act states:

> Partnership is the relation which subsists between persons carrying on a business in common with a view of profit.

The hallmarks of a partnership are that more than one person, whether individuals or a legal person such as a company, jointly carry out a commercial enterprise, with the effect that they share the costs and so the profits and losses of the business.

4.3.3 As a commercial partnership, each partner carries liability for the obligations entered into by the other partners. That exposure to risk has compounded the reluctance of farmers and landowners to form partnerships outside the family. That is especially so where rights to land might be thought at risk with the courts anyway regularly considering such farming partnership disputes between family members.

4.3.4 The Limited Partnership Act 1907 does allow a partner to limit his liability to a stated financial amount but that also limits involvement in the business (creating the "sleeping partner").

4.3.5 In England, Wales and Northern Ireland, a partnership does not itself have a legal personality and so a lease to a partnership will be to the named individuals and will not be affected by any later changes in the membership of the partnership.

4.3.6 The Limited Liability Partnerships Act 2000 allows the LLP structure. Originally designed with large professional partnerships in mind, some farming businesses have taken up this model whose features:
- a partnership structure
- the limited liability of a limited company
- some public reporting of accounts.

This is most likely to be useful where there is the potential for substantial liability to others, whether financial or the risk of potential claims (as may be the case for a food producer). However, lenders to the LLP may usually seek personal guarantees, so bypassing one attraction of the LLP structure.

4.4 A company structure

4.4.1 The Victorian creation of the limited liability company has been very widely adopted throughout the economy, partly encouraged by tax legislation. It is not so commonly seen in general agriculture, though it is more usual among the largest and most commercial businesses. It allows a business with one or more shareholders to limit its liability to third parties, gives a vehicle allowing some separation between ownership (shareholders) and managers and rules for the payment of income through dividends to shareholders alongside wages to directors and staff. The price for limiting liability is a level of disclosure of accounts through Companies House records. Again, lenders may seek personal guarantees for loans, so overcoming the limited liability in respect of those borrowings.

4.4.2 It is most likely to be appropriate where the business can be seen to have continuing life of its own, warranting leaving money in it for future issues and a potential sale, rather than just as a vehicle for individual remuneration. Moving assets out of a company can be expensive in tax terms as the company pays tax on any gain on the disposal and then the recipient may also pay tax on the receipt of value. As a result, while many earlier farming companies included land, buildings and dwellings, it might now be more natural to incorporate only the farming business without the property.

4.4.3 Directors face particular issues over the deemed income for taxation of benefits in kind, especially over housing, for which this can be expensive. Companies owning houses worth over £500,000 are also subject to the Annual Tax on Enveloped Dwellings (ATED), though there are reliefs for dwellings qualifying as "farmhouses", occupied by qualifying employees, let out commercially or in a number of other categories.

4.4.4 A company structure has also been adopted, as an alternative to a partnership, in situations where two or three farmers wish to pool (and reduce) resources of labour and machinery to pursue a common farming business between them on a more efficient basis. Whether done as a partnership or a company, that might see each participant take responsibility for a particular part of the business.

4.5 Farming with contractors

4.5.1 The essential principle here is that one person (the farmer) retains another person (the contractor) to do something at his behest. It could be as simple as hedge cutting, but it could be more comprehensive. In practice, while this is commonly discussed in the form of stubble-to-stubble contracting of whole cropping operations, a nearly infinite number of arrangements are possible. Contractors are widely used for specific operations like silage making or spraying. While the greatest use of contracting arrangements has been with combinable crops for which they are often an apt business model, variations have been applied for use in dairy and livestock enterprises, again with a wide range from foot trimming services to whole unit management.

4.5.2 "Stubble-to-stubble" contracting has though been a successful model for combinable cropping, reflecting the business reality of many situations where a farmer, whether owner or tenant, can no longer justify the machinery or labour required to undertake operations to current standards and so retains someone else, a contractor or other farmer, able to do just that. The contract will be defined by the agreement which should also provide for remuneration. Much turns on the relationship between the farmer and the contractor with a farm plan, a budget and regularly recorded management meetings to appraise progress and discuss issues.

4.5.3 That whole enterprise contracting model has been developed for other livestock operations, notably for breeding livestock businesses with income from progeny and produce.

4.5.4 There are various bases for remuneration ranging from a standard payment for the services provided by the contractor to the contractor retaining all profit from an enterprise over and above a return to the instructor. Many will see some division of that remaining surplus between the parties, sometimes with the contractor taking a higher share of a first slice and then lower shares above certain thresholds where income may be more the result of fortunate conditions than the contractor's skill. As noted later on, the contract farming structure of New Zealand's "share milking" arrangements can involve payment with or in respect of a rising share of the dairy herd itself. In part, these may be influenced by tax considerations as well as practicality.

4.5.5 The business is that of the farmer who pays for the services of the contractor. That makes the point that for crops, the crops are the farmer's crops, not the contractor's crops, even if the contractor is retained to handle the sale from his grain stores – though it may be better to use pooled grain storage. They may agree a transfer of ownership of produce once harvested but, up to that point, the contractor is at risk from the financial failure of the farmer.

4.5.6 That also poses a particular issue for high value cropping where the specialist grower is likely to have a relationship and understanding with the purchaser of the crop who may be interested in the management of the land but not in the landowner – those facts may rather point to using a tenancy for such cropping.

4.6 "Share farming"

4.6.1 This phrase has been in general use for some 35 years but still lacks very much specific meaning. In broad terms, the phrase is applied to business arrangements which see two businesses, remaining separate, whose endeavours result in a common output whose gross proceeds they then share on a pre-agreed basis. This distinguishes this concept from:
- a partnership in which the partners in a combined business divide net income (or losses) and share liability between them
- contracting where one party provides services on instruction directly to the other's business
- employment where one party is simply employed by the other.

4.6.2 The original thinking behind the discussion of share farming was based on observation of dairying in New Zealand where formal structures for share milking developed in the late nineteenth and early twentieth centuries and were then given formal statutory frameworks from the 1930s. These are, however, essentially structured as contract farming arrangements and models based on them which can include reward as a share of the dairy herd are better considered as a contract for services.

4.6.3 The model most discussed in the United Kingdom is more of share land management, having a structure with:
- a land-providing business, managing and maintaining the land and buildings
- a farming operator using the land on a licence and running the farming business, each carrying their own costs and then dividing the farming sales between them on a pre-agreed basis.

The success of this model is very sensitive to the split of the gross income between the two businesses and assessing that depends on an understanding of the likely costs involved in each business. It will be a matter of individual facts as to how this works in taxation terms for the land provider while it may be that the farming operator would be the better Basic Payment claimant.

4.6.4 Other variations see a more intermingled mix of farming operations carried out by both businesses dividing the output. That can see a structure using a single company co-owned by the two participants (see 4.4.4 earlier).

Further reading

Arable Farming with Contractors (CAAV, 2020)
Contracting Farming for Breeding Livestock Enterprises (CAAV, 2020)
Grazing Arrangements (CAAV, 2013)
Tenancies, Conacre and Licences: Arrangements for Occupying Agricultural Land in Northern Ireland (CAAV, 2015)

5 Agricultural support
From the EU to beyond Brexit

Caution – This chapter offers an overview of this topic for which the legislation and administration is complex. At the time of writing, financial support policies for agriculture are on the common basis of the European Union's (EU) legislation for its Common Agricultural Policy. However, the application of these policies already differs between parts of the United Kingdom and can change over time. Further change is expected to follow with the United Kingdom's departure from the EU with the probability of greater divergence in support policies between the parts of the United Kingdom, though the inherited policies are likely to be relevant for some time.

Specific care should be taken over this in the light of circumstances at any given time and reference made to the then current legislation for the relevant part of the United Kingdom and official guidance.

5.1 Introduction and background

5.1.1 The policies and legislation of the EU, including its Common Agricultural Policy (CAP), have been a major influence on agriculture in the United Kingdom since it joined what was then the EEC in 1973. That changed as the countries in the United Kingdom became responsible for agricultural policy on leaving the European Union (Brexit). This chapter offers a general guide to the policies that the UK has inherited on Brexit and the issues of professional practice that they raise and then reviews prospective policy developments and some of the issues that might arise for professional practice.

5.1.2 The basic policy framework – The EU's Common External Tariff with its schedule of rates offers significant protection to parts of the EU marketplace, notably for livestock and dairy products, albeit with approved tariff free access for third countries by tariff rate quotas (TRQs) or other agreements. Behind that tariff barrier, a range of policies:
- initially supported production with intervention support and specific payments which have then been substantially converted into direct area payments
- provided money jointly with member states for rural development generally, including agri-environment schemes,

with effects on farm businesses, farm structures and values.

32 *Foundations*

5.1.3 Post-Brexit changes to that triple structure of external border controls (a UK issue), internal support and rural development funding (both handled at country level within the UK) will influence markets, decisions and values. While the October 2019 Withdrawal Agreement transferred power over direct payments to farmers to the UK for 2020, significant change is only practical after that.

5.1.4 The evolution of EU policy – The CAP initially acted to support production (and so it assumed incomes) by guaranteeing prices for major products and acting as a buyer (intervening) where prices fell below those set. In time, this led to stored volumes of production that could not be sold ("beef mountains", "wine lakes", etc) and proved to be an open-ended financial commitment for the EU. In 1984, the rising cost of intervention in the dairy sector (and the associated "butter mountains" being dumped on world markets) saw the introduction of milk quotas (considered in previous editions) which finally lapsed in 2015. Direct support through intervention prices (and export subsidies) was substantially replaced from 1993 by area payments and a revised structure of livestock headage payments which capped expenditure and met the requirements of the Uruguay World Trade talks.

5.1.5 The payments to farmers under that regime were consolidated into area payments with the creation of the Single Payment Scheme in 2005, now replaced by the Basic Payment Scheme from 2015. In principle and following World Trade Organisation rules, that reform broke the direct link between support payments and production decisions, as qualifying claimants with eligible land can be paid on the basis of land management without production ("decoupling") using payment entitlements allocated to and transferable between them. In the UK, only Scotland has continued any form of payment coupled to production decisions (the Suckler Beef Support Schemes and Sheep Upland Support Scheme).

5.1.6 With the scale of the devolution of agricultural matters within the CAP to Scotland, Wales and Northern Ireland, each took their own choices, differing from those in England, as to the introduction of the Single Payment Scheme (SPS) in 2005 with:

- England moving over the years from 2005 to 2012 to standard area payment rates in each of three payment regions
- Wales and Scotland assessing individual payments from each farmer's historic patterns of payment
- Northern Ireland basing its payments on a mix of an underlying area payment with a substantial element based on historic payments.

Payments were administered through a structure of entitlements (rights to payments expressed in area terms) which had to be matched against equivalent areas of qualifying land. Scheme rules imposed Statutory Management Requirements and cross compliance obligations, breaches of which attracted penalties.

5.1.7 In 2015, that was replaced by the more complex Basic Payment Scheme (BPS) with its implementation in each part of the United Kingdom starting from the already divergent positions of the four territories and diverging further. England carried SPS entitlements with the standard area payment values forward as BPS entitlements. Wales (with one payment region) and Scotland (with three) moved from historic payments values to reach standard area payment values in 2019; the policy in Northern Ireland, with one region, was to complete that in 2021, but this has since been paused.

5.1.8 The various CAP schemes are administered through the Integrated Administration and Control System (IACS), which comprises registers of claimants, land parcels and other information with a structure of inspections to check compliance.

5.2 Direct payments – The Inherited Basic Payment Scheme (BPS) and allied payments

> *Caution* – *This is a very basic sketch of the BPS as the UK inherited it from the EU, ahead of any developments by each of the countries of the UK after Brexit.*

5.2.1 The main EU regulations provide a common framework for the Basic Payment Scheme and allied payments which took effect from 2015 and were initially carried forward after Brexit under the Withdrawal Act.

5.2.2 Each territory of the United Kingdom made its own choices about implementing those regulations with their complexity, using a mixture of regulation and administrative guidance. That again created complexities for claimants with land in more than one part of the United Kingdom.

5.2.3 The basics – The **Basic Payment** is at the centre of the present CAP payments regime (itself subject to EU discussions about revision after 2020). It is available:
- to eligible claimants
- on the basis of eligible land
- where it can be matched against the claimant's entitlements to the payments
- but subject to penalties and reductions for breaches of the eligibility, cross compliance conditions, Statutory Management Requirements and other regime rules.

5.2.4 It is then supplemented by the **Greening** element. With 30% of the total direct payments budget, this is paid where three rules are met:
- on pasture land, the rules for retention of overall national area of "permanent" pasture areas are met

- on arable land and depending on its scale, the claimant:
  - meets the requirements of the crop diversification rules
  - delivers any required Ecological Focus Areas (EFAs).

In addition to the general rules for penalties and reductions, failures to meet these rules will see reductions in greening payments.

5.2.5 Those who qualify as **young farmers** have been able to claim a 25% top-up, capped by area, to their Basic Payment claims for up to five years. That is a personal payment that cannot be transferred with entitlements.

5.2.6 Scotland has then used some of its direct payments budget to operate **livestock support schemes** for beef across Scotland (in part, at least, to support volumes for the slaughterhouses) and sheep in the most remote areas of rough grazing. No other part of the UK offers coupled payments.

5.2.7 Wales makes a higher payment on the first 54 hectares of a valid claim (the "**redistributive payment**").

5.2.8 Payments are administered by the payment agency in each UK territory: the Rural Payments Agency (RPA) in England, Rural Payments Wales (RPW), the Scottish Government Rural Payments and Inspections Department (SGRPID) and the Northern Irish Executive Department for Agriculture, Environment and Rural Affairs (DAERA).

5.2.9 Annual claims are to be made by 15 May in each year if they are to be paid in full.

5.2.10 Payments are made in the period from the following 1 December to 30 June, though Northern Ireland has made substantial advance payments in October.

5.2.11 Claimants are subject to inspection, increasingly done remotely, and can suffer penalties or reductions where a compliance failure is found.

5.2.12 There are also circumvention provisions that can exclude claimants found to have artificially created the basis for their claim.

5.2.13 Eligibility for BPS – The EU regulations define who can qualify as a "farmer" as someone with a holding and who "exercises an agricultural activity". That is in turn defined as any of the following:

- farming as conventionally understood
- maintaining farmland so that it can be readily grazed or cultivated
- carrying out a minimum prescribed activity on land that, by its natural condition, is ready for grazing or cultivation without intervention. England and Northern Ireland decided they had no such land while Scotland has prescribed particular rules (including a minimum stocking rate) for two of its three entire payment regions deemed to fall within the definition. Wales has recognised only saltmarsh and coastal dunes as being in this category.

5.2.14 To be eligible for payment, such a "farmer" has had to be an "active farmer". This has been a misleading label which by no means requires active farming but most of its meaning is now spent save in Scotland. It now means,

> where more than half the land is naturally kept ready for grazing or cultivation, meeting the minimum prescribed activity on it. This has no force in England and Northern Ireland (which confusingly uses the phrase "active farmer" in a different context).

All These rules are accompanied by a developing structure of guidance and interpretation. Reference should be made to the guidance current at the time, especially as this is likely to see progressive changes after Brexit.

5.2.15 Eligible land – Claims can only be made in respect of eligible land recording on the Land Parcel Information System (LPIS). This is land that is predominantly used for agricultural activity throughout the calendar year of the claim. Any non-agricultural activity must not significantly hamper that. Again, there is particular guidance as to how this is understood in each country. Payment may not be made on other land.

5.2.16 Land at disposal – A qualifying farmer can only claim on eligible land that at his "disposal". This has been tested in the CJEU decision in *Landkreis Bad Dürkheim* which found that it requires the claimant to have control over the agricultural activity on the land and to have the power to make autonomous decisions in that.

5.2.17 This requirement has very significant consequences when considering agreements for farming land between owners and farmers as the nature and terms of the agreement will govern which of the two parties is the proper BPS claimant on that land. The usual position in the United Kingdom will be that owners and tenants in occupation will qualify while landlords, graziers and contractors will not. Where an agreement expressed as a licence allows the licensee a level of control that meets these tests, that arrangement may in reality be a tenancy.

5.2.18 In Northern Ireland, this requirement has been used since 2015 to exclude almost all owners of land subject to a conacre agreement with the expectation that the conacre user will generally be the eligible claimant, having that land at his disposal. While "active farmer" means something else (see earlier description) in EU rules, Northern Ireland calls this its "active farmer" approach.

5.2.19 Dual use – As different CAP schemes have required differing relationships with land, it is legally possible for different parties to make claims in respect of the same land parcel under different schemes, provided each genuinely meets the rules of the scheme used for their application. Thus, at present agri-environment agreements require the claimant to have management control of the land. Management control is interpreted to have a different meaning to what is require for land to be at the BPS claimant's disposal. In England and Northern Ireland, that has generally made it possible for a landlord to be the agri-environment agreement claimant while the tenant is the BPS claimant on the same land, provided that the necessary provisions are made in the tenancy agreement to show management control. Wales no longer accepts that this can arise.

5.2.20 What might an ineligible claimant do? – Where a potential claimant proves on consideration to be ineligible, the options include:
- accepting that fact
- resolving the difficulty if that is reasonably possible (bearing in mind the prohibition of artificiality)
- making the land available, probably on a tenancy, to someone who can qualify.

Where the affected person has entitlements, they may be sold or leased to another claimant, rather than lost through inactivity.

5.2.21 Entitlements – Each claim is possible to the extent that the claimant has a number of entitlements to payment that matches the area of eligible land at his disposal. Entitlements were allocated or confirmed on the basis of 2015 claims. New ones can be claimed each year by qualifying new entrants from the National Reserves. Otherwise, entitlements can be transferred by sale or lease between farmers.

5.2.22 Entitlements can only be matched against land in the payment area for which they are allocated and cannot be moved between them. These areas are:
- England – Upland SDA moorland
- England – Upland non-moorland SDA
- England – non-SDA
- Scotland – Region 1 – based on the areas that were not rough grazing
- Scotland – Region 2 – based on rough grazing land
- Scotland – Region 3 – unimproved rough grazing in the Less Favoured Areas
- Wales
- Northern Ireland.

5.2.23 Use or lose rule – Under BPS, all entitlements must be used in at least one of every two years. Any not so used will be lost to the national reserves without compensation, but after Brexit this might be relaxed. Those affected may wish to consider selling or leasing out vulnerable entitlements.

5.2.24 Payment rates on entitlements – The payment area in which the entitlements are allocated and used will govern the payment rate for the claim, to date as an EU scheme set in euros. These rates are based on the available money for each area and the total area claimed.

5.2.25 In England, there are standard payment rates, payable on any claim in each English area which for 2020 were (excluding the greening element) to be:

- £162.77 per ha for non-SDA (Severely Disadvantaged Areas)
- £161.56 per ha for upland SDA, other than moorland
- £44.33 per ha for upland SDA moorland.

5.2.26 Since 2015, the other parts of the UK have been phasing individual payment values to standard rates, achieving this in 2019 for Wales (estimated at €125.65/ha) and Scotland (estimated as €165.65/ha for Region 1, €36.16 for Region 2 and €10.48 for Region 3 but with adjustments to be made). Northern Ireland was to have completed this process in 2021 but paused for 2020.

5.2.27 Welsh redistributive payment – Wales, operating a single payment area and so the same underlying standard BPS payment rate across its whole area, adopted the Redistributive Payment option whereby it pays a higher rate on the first 54 hectares of any claim, so moving money from larger claims to smaller claims. In 2019 that additional rate was €103.24/ha.

5.3 Greening

5.3.1 Part of the initial presentation of the Basic Payment Scheme was that it would make a larger environmental contribution through the creation of an additional "greening" element. This has used 30% of the direct payments budget as a supplementary payment to BPS claimants, requiring many, essentially on arable land, to undertake particular land management measures, albeit ones that are sometimes more disruptive of some businesses than environmentally beneficial. Most farmers will see the Greening payment as an intrinsic part of the Basic Payment. They is no longer applies in England from 2021.

5.3.2 So far as claimants have "permanent grassland" (defined here as land that has been in grass without an intervening crop for five years) this simply provides that if, when assessed nationally, the proportion of such grassland falls by more than 5% from an initial threshold, then farmers can be required to return some of that converted land to grass.

5.3.3 The rules for "arable" land (as defined and including temporary grass) are triggered by the area of such land a claimant has in the year such that, with detailed exemptions:
- those with more than 10 ha of "arable" land must have at least two crops
- those with more than 30 ha of "arable" land must have at least three crops
- those with more than 15 ha of "arable" land must have 5% of that land in options available for Ecological Focus Areas. Each part of the United Kingdom sets its own options from among those allowed by the EU regulation or which the Commission agreed were equivalent to them.

Based on areas reported on the claim form, these apply for the individual claim year, though some EFA measures can run into the following year.

5.3.4 There are specific penalties for failures to comply with the greening rules.

5.3.5 In all parts of the UK the greening element has been paid at a standard rate in each payment area, irrespective of the claimant's actual BPS rate. For 2020 claims in England the rates were:
- £70.45 per ha for non-SDA land
- £69.59 per ha for upland SDA, other than moorland
- £19.09 per ha for upland SDA moorland

For England and Northern Ireland, these are now to be merged with the main area payments rates which are then to be reduced.

5.4 The rural development regulation and agri-environment schemes

5.4.1 A much smaller sum of money is made available through the Rural Development Regulation. Despite its title, England and Wales use most of this money for agri-environment schemes while Scotland currently uses a large part of its money for the Less Favoured Areas Support Scheme (LFASS) of headage payments for beef and sheep farmers, varied by type of LFA and being reduced as required by the EU. However, many other measures are possible.

5.4.2 Agri-environment schemes – In each part of the United Kingdom and, to date, within EU rules, these provide a locally determined framework and options for agri-environment agreements. These are contracts between the claimant and the state, usually for five years, jointly funded by the EU and the member state, by which specified land management options are to be delivered. Under EU rules, standard payment rates are based on an overall assessment of income foregone by the option with an additional margin. EU rules now require the agreements to start from 1 January.

5.4.3 The United Kingdom has seen a variety of schemes over the last 30 years, starting with Environmentally Sensitive Areas. Those currently to be found are:
- in England:
  - Countryside Stewardship
  - existing 10-year Higher Level Stewardship agreements
- in Wales: Glastir
- in Scotland: Agri-Environment Climate agreements
- in Northern Ireland: the Environmental Farming Scheme.

5.4.4 The Treasury has agreed that, after Brexit, it will meet the element of funding for existing agreements that has been paid by the EU, so providing continuity for agreement holders.

5.5 Penalties and reductions

5.5.1 Where a claimant under any of the CAP schemes, including an agri-environment scheme agreement as well as area payments, is found to have breached the cross-compliance rules there is an escalating structure of financial penalties.

5.5.2 A more serious scale of penalties – even exclusion for up for four years – applies where land used for the claim has been found to be ineligible for that claimant.

5.6 CAP schemes and valuation

5.6.1 The often complex issues and administrative weaknesses of the CAP schemes make a large call on professional support for those affected, from assisting with or preparing claims and applications understanding and advising on the impact they and their rules have on plans being considered by clients.

5.6.2 Entitlements – As tradeable assets, entitlements have a value which, in the marketplace, may be agreed as a unit value per hectare but may also be understood as a multiple of the payment value.

5.6.3 These multiples are generally low, typically around 0.8 to 1.2 year's payment reflecting the relatively plentiful supply of entitlements and the scarcity of new land against which to use them. Any Young Farmer top-up that has been received by the vendor is not relevant to this since, as a personal payment, that cannot be transferred although a qualifying purchaser may be able to claim that top-up in their own right, depending to their circumstances. However, the area cap on this payment and the rules for claimants limit this as a factor in the market.

5.6.4 The BPS rules allow qualifying new entrants the option of claiming fresh entitlements from the national reserves, altering the supply of and demand for entitlements.

5.6.5 Where land is sold or leased with entitlements for a single price or rent, apportionment of that value for the capital or rental value of those entitlements may be required as that should not be subject to Stamp Duty Land Tax (Land and Buildings Transactions Tax in Scotland; Land Transactions Tax in Wales).

5.6.6 For Capital Gains Tax on the disposal of entitlements, the base value for calculating the gain taxable gain will be:
- nil, if they were originally allocated to the vendor or subsequently from the national reserve
- the purchase price or probate value, if they were bought or inherited.

Outside England, BPS entitlements were created in 2015. In England, the previous SPS entitlements were rolled forward to be BPS entitlements. They remained the same asset, albeit giving access to the new payments, and so their earlier history is still relevant. As an example, entitlements bought in, say, 2008 will still have that price as their base value.

5.6.7 BPS and land – As it is the combination of eligible land and entitlements in the hands of a qualifying claimant in any one year that unlocks the payment, this can be expected to have an effect on land transactions.

5.6.8 This is most direct and obvious for short-term rental values as for farm business tenancies in England and Wales, short limited duration tenancies in Scotland and conacre or tenancies in Northern Ireland. These agreements both provide access to the land that enables a claim that could otherwise not be made and substitute for the claim that the person providing the land could often otherwise make.

5.6.9 The Scottish Court of Session determined in *Morrison-Low v Paterson* that the access to area payments offered by a 1991 Act tenancy is relevant to a rent review under the statutory provisions of that Act. The same logic essentially applies to other tenancies in England and Wales under the terms of the 1986 and 1995 Acts.

5.6.10 The relative scale of the payment and the many other factors involved in sale values of farmland make it much harder to discern the effect of the subsidy scheme in the capital value of land. One effect of area payments appears to be to act as a restraint on activity in the land market as they provide an incentive for claimants to remain in occupation of the land that enables the payment. The potential withdrawal of payments after Brexit might, all other things being equal, have market effects in reducing or removing income that land can yield, removing that incentive to remain in occupation so adding to land on the sale or rental markets and so far as it might, in combination with other factors, more generally affect confidence among those in the market.

5.6.11 The inherited CAP schemes and changes in occupation of land – There can be several important issues where the occupation of land changes, whether on sale or a change of tenant. In this it is important note that:
- a sale is an agreed action in which the old owner and the new owner have a contractual relationship and are able to agree its terms and so cover these issues
- the letting of land by the owner who was previously farming it is similar to that
- however, where one tenant succeeds another they have no contractual relationship. The outgoer's contractual relationship is the tenancy agreement with the landlord, often agreed long previously, while the incomer will have the new tenancy agreement with the landlord.
- where a tenant leaves on a notice to quit or to terminate for the owner to farm or take other actions, it is only the tenancy agreement that provides an agreed basis for what happens
- where a tenant surrenders the tenancy by agreement, that agreement can cover all necessary terms to ease the change of occupation.

5.6.12 Points to consider for BPS issues include:
- where the change of occupation is after 15 May within the calendar claim year, it is likely that the outgoer will have made a claim and will wish to ensure that the land remains eligible with cross compliance and his chosen greening obligations, so that he is fully paid
- where the change is before 15 May, the incomer, as prospective claimant, will want to be clear that there no pre-existing breaches that could lead to penalties or reductions.
- where there might, ahead of the 15 May claim date, be a reason for doubt as to who will be claiming BPS on what land, clarity is needed to avoid a problem over dual claims. This might arise, for example, where there is a change of occupation in late April or early May, as with a 13 May tenancy in the north-east of England or where there is some holdover or early entry across the 15 May date.
- whether the number of entitlements equivalent to the area of land in question are to be transferred
- ensuring access for the new occupier to the necessary records and data including land parcel registration, eligible areas and the relevant parts of previous claim forms. These are not automatically available from the payment agency.

5.6.13 Where the change in occupation happens during the period of an agri-environment agreement, the outgoing occupier as signatory to the agreement is reliant on the new occupier accepting and continuing the obligations with risk of repayment being required if that does not happen. Where the change of occupation is by a contract between the two parties then this matter can be handled within the contract. An outgoing tenant can thus be more exposed, but if the tenancy could have been expected to have ended in the agreement period then landlord's consent would have been needed. The new occupier needs a full transfer of relevant information.

5.7 Prospects after Brexit

5.7.1 On 23 June 2016 the United Kingdom voted to leave the European Union. After an extended period of negotiation, irresolution in the UK and further negotiation, a revised Withdrawal Agreement was concluded in October 2019 and, with the outcome of the December 2019 General Election, the UK left the EU on 31 January 2020. The combination of the re-enactment in UK law of much EU legislation by the European Union (Withdrawal) Act to ensure legal continuity and the Transition Period (sometimes also called an "implementation period"), which sees no change in rules until at least January 2021, means that little initially changes on the ground. That Transition Period is to be used to settle the UK's future trade and other arrangements with the EU, a process that will reflect the tension since the 1950s between the defence of UK sovereignty and the desire for free trade with Europe.

5.7.2 As agriculture has been so entwined with the EU during the UK's membership, withdrawal could alter much, especially through the larger economic forces provided by changes to trading relationships and the more direct changes in domestic agricultural policy after leaving the CAP.

5.7.3 While international trade is a matter for the United Kingdom, domestic support policy is essentially for each of the four governments, already with differing regimes and now likely to diverge further in both approach and timing. The Chancellor of the Exchequer has confirmed the overall level of farm support for the life of the 2019 Parliament (and consistent with that given under the CAP in 2019). However, it will be for each government to choose how to spend its share of the money.

5.7.4 Trade – International trade is an important factor for the prices of agricultural produce and inputs. The UK, with a self-sufficiency ratio around 60%, is a significant net importer of food and its agricultural trade is much more concentrated on the EU than is the rest of the economy. Some 70% of UK agricultural exports go to the EU, but larger volumes of food imports come from the EU. Lamb is an example of a significant export sector with almost all the nearly 40% of UK production that is exported going to the EU. Most other sectors see significant imports. That trade has been conducted within the mutual rules of the Single Market under which member states recognise each other's produce as meeting shared standards. The UK's intention, on withdrawing from that Single Market, is to negotiate a new trading relationship.

5.7.5 Overall, the EU's Common External Tariff, with its schedule of taxes on imports to the EU, offers a high level of protection to EU agricultural produce, notably for livestock and its products. UK agriculture will be affected by any changed relationship with the EU's Customs Union and new trade agreements negotiated with third countries. That means that issues can arise over time from both:
- the question of whether UK trade with the EU becomes subject to tariffs and regulatory friction
- the arrangements for future new trading agreements between the UK and other countries, especially where they are agricultural producers.

It is not simply a matter of tariff charges but also the arrangements for regulatory recognition between the UK and the EU of the standards for production and warranting compliance with them. Where there is more trade friction, it may benefit farm sectors where much produce is imported, such as dairy and horticulture.

5.7.6 The Single Market has also seen much movement of EU citizens to work in UK farming, notably dairy, horticulture, fruit, pigs and poultry, and its processing trades. Those already in the UK before Brexit have the right to stay, but there may be restrictions on future movement to the UK. Where this adversely affects a business, it may add to the pressures to adopt further automation and the use of robotics.

5.7.7 The outcomes of these issues will have their effects on the markets and processors for agricultural produce and also the supply of inputs.

5.7.8 Future policies? – Agricultural policy is much more than payments to farmers, but may, over time, come to cover a number of areas including:
- a focus on encouraging farming's business ability to compete, with measures from research to support for innovation and productivity
- agri-environment concerns, appearing increasingly to focus on soils, water and climate change issues
- resilience in the face of risks
- animal and plant health and welfare
- approaches to regulation.

5.7.9 That could see a range of approaches including:
- agri-food strategies
- policies for land tenure and business structures with opportunities for new entrants, progress and retirement
- concerns about economic activity in sparsely populated areas
- insurance schemes and other economic tools to manage risk.

These would still have to comply with international commitments, such as the rules of WTO membership which limit the ability of countries to use support policy to distort production and trade.

5.7.10 While there are relatively few costs for most claimants associated with the Basic Payment Scheme, the new payments are more likely to require costs to be incurred or income foregone. So far as that approach reduces the net financial benefit for claimants from within the total payments available, it may encourage more land onto the sale and rental markets by reducing the financial benefit from occupying land.

5.7.11 First steps – The European Union (Withdrawal) Act has ensured that existing EU legislation is carried forward, giving legislative continuity in the UK and the basis from which new policies can develop in each part of the UK. That has then been reinforced for 2020 by the UK-wide Direct Payments for Farmers (Legislative Continuity) Act 2020.

5.7.12 All parts of the UK have held consultations on future policy approaches. More directly, the Agriculture Act will provide the framework of powers for that policy management and development, particularly for England over an agricultural "transition period" running to 2027. It also includes initial supporting provisions for Wales and Northern Ireland with their own legislation to follow. Scotland has its own Agriculture (Retained EU Law and Data) (Scotland) Act. Policy development and implementation would then accelerate in all parts of the United Kingdom.

5.7.13 England – While the Agriculture Act 2020 provides powers to manage agricultural policy, the uses to which those powers would be put was the subject of the *Health and Harmony* consultation in early 2018 with the following general policy proposals made in September 2018. They focus on:

- using public money to buy public goods, that is, diverting money currently used for BPS to fund land management for goals that society wants but which are not paid for in the marketplace. With the 25 Year Environment Plan, these are largely environmental issues with climate change and biodiversity with concern also about animal welfare, flooding and other issues.
- improving farming productivity and so its efficiency and competitiveness with a range of measures including schemes reusing other money released from the BPS and policy changes for tenancies, skills, innovation, investment and technology.

5.7.14 With some operational changes to the inherited CAP such as the removal of greening, there would be a seven-year transition as the new regime is phased with what is currently the Basic Payment finally removed in 2027. This is to start with an initial 5% reduction in 2021 with more taken from larger claimants, but the path beyond that is not known. At some stage, Basic Payment would no longer be linked to a requirement to farm ("de-linking"), at which point it might be renamed. Some claimants may have the opportunity to take remaining BPS payments as a single lump sum.

5.7.15 The main destination for the money withdrawn from the BPS is intended to be environmental agreements for land management. Existing schemes would broadly continue until the Environmental Land Management Scheme is progressively introduced as the main vehicle for the "public money for public goods" policy offering agreements to farmers. It could be that it might pay sufficient for farmers to see options under this as a potentially profitable business enterprise alongside, supporting or instead of other production enterprises – so developing a market approach, rather than a compensation basis, for the commercial evolution of farming and land management. Where it is offered in this way, such profit is likely to be much less than the net margin from Basic Payment.

5.7.16 For the first time since the Second World War, supporting farm incomes will not be an official policy objective. Instead, this is a move towards markets with farmers to make their own business choices. In recognising that, an early use for some of the redeployed Basic Payment money would be to fund schemes to improve farming productivity.

5.7.17 Basic Payment (with few associated costs) forms some 15% of turnover for many broadacre combinable cropping and grazing farms, less for higher value enterprises such as dairy and horticulture but a higher fraction for extensive grazing. Subsidy forms a higher proportion of assessed profit with the estimated 2019 figure for the Total Income from Farming (TIFF) of £3.9 billion including £3.3 billion of subsidy. That masks great differences between high performing and poorly performing farmers in each sector, with some of the latter losing money before subsidy. However, those proportions show that a 10% fall in produce prices would typically have the same effect as a halving of subsidy income.

5.7.18 With that background and combined with any changes that may come from trade and other issues, the redirection of BPS money can be expected to have a major effect on farm economics driving reappraisal and new decisions by many, though others may just absorb the loss.

5.7.19 This process is likely to influence costs, not only rents, that have been funded with the benefit of BPS income as well as choices about farm structures, operations, enterprises, diversification and off-farm income. With the interaction between these pressures and the seven-year transition period with the life cycles of individuals and families in farming, this may also lead to consideration of whether some or all land is sold, let out or managed on another farming arrangement, so offering opportunities for others to start or develop their businesses. All this will touch on values and valuations and drive a need for advice and action to facilitate farming families understanding their situation, coming to decisions and then acting on them. Government statements have endorsed the role of advice; some of that advice will come from agricultural valuers.

5.7.20 Wales – The Welsh government has consulted on broadly similar themes in two papers, *Brexit and our Land* and *Sustainable Farming and our Land*, developing policies for the replacement of BPS with a Sustainable Farming Scheme (SFS). It acquires general powers to manage policy until the end of 2024 under a schedule to the Agriculture Act 2020 but is to bring forward its own legislation in the Welsh Assembly after the 2021 elections, making 2023 or 2024 the start of transition to the new scheme.

5.7.21 The SFS would again completely replace BPS with the path for that changeover still to be determined.

5.7.22 A sustainability review is to be a precondition for entry to an SFS agreement, considering both business and environmental sustainability. The SFS could then offer an agreement with mandatory provisions and such further options as the farmer takes up, with the resulting payments for land management forming part of farm income. The consultation papers have seen advice as an integral component of this process, saying

> advice should be seen as an investment in the capacity of farmers and farms rather than a cost.
> (*Sustainable Farming and our Land*, 1.48)

5.7.23 Capital funding would be available for productivity improvements on and off the farm.

5.7.24 Similar issues arise for agricultural valuers as in England.

5.7.25 Scotland – With the consultation paper, *Stability and Simplicity*, looking to defer major change until 2024, the Scottish government has wanted to remain closely aligned with the EU's CAP. The Agriculture (Retained EU Law and Data) (Scotland) Act provides the "housekeeping" powers to manage that policy with the ability for ministers to bring forward regulations to modify (that is, "simplify" or "improve" rather than change) the main inherited CAP legislation (which it allows to be rolled forward). New primary legislation would be required for any substantive change in policy, as perhaps might be needed for its use to assist climate change mitigation and adaptation, whether in accordance with future CAP changes by the EU or not.

5.7.26 Scotland had, as required by the EU, made sharp reductions in its Less Favoured Area Support Scheme (LFASS) payments to beef and sheep farmers but, outside EU rules, is now reversing that.

5.7.27 With an already complex model used in 2015 for adopting BPS in Scotland, some operational changes are expected to ease that. Capping payments to larger claimants is under consideration.

5.7.28 That looks to leave the present BPS essentially in place until at least 2024 with its stimulus to existing claimants to remain in occupation of the land that supports their subsidy claims.

5.7.29 Northern Ireland – While without a government during most of the pre-Brexit period, a stakeholder engagement was held in 2018 on a Northern Ireland Future Agricultural Policy Framework. That considered initial measures and then policy after 2021, focusing on productivity, resilience, environmental sustainability and an improved supply chain. Money would be moved to assist those objectives but, Northern Ireland having the highest area payment rate in the UK, without necessarily removing some form of underpinning area payment altogether. Since then there has been some discussion of potential livestock payments. While a new Executive was formed in January 2020, no decisions had been taken at the time of writing about the shape of post-Brexit policy in the province.

5.7.30 Northern Ireland acquires general powers to manage policy under a schedule to the Agriculture Act 2020 but will then need to develop its own legislation, which might perhaps not come forward until after the May 2022 Assembly elections.

5.7.31 With the importance to the province of both trade with Great Britain and across the Irish border, the operation of the Northern Irish Protocol in the Withdrawal Agreement could have a particular effect on farm economics and therefore on values in Northern Ireland. Subject to the negotiations with the EU, there would be an open border between the two parts of Ireland, "unfettered access" for Northern Ireland trade to Great Britain but potential costs and complexities for trade from Great Britain to Northern Ireland. Northern Ireland remains within the EU's Single Market so far as it relates to goods, including agricultural produce, and therefore its regulation on such matters as standards and labelling.

5.7.32 The more there is economic change, the more there may be a need to consider the renewed use of tenancies or other farming agreements as means of managing that change. Other needs to improve productivity and meet environmental expectations and climate change objectives are likely to be relevant.

5.8 Future retrospect?

5.8.1 Looking back on this period from the early 2030s may give a clearer view of what was really happening and the aptness or otherwise of the different responses coming forward as they are tested in each part of the United Kingdom. With economics as a more important driver than support, it might be that the sum total of the individual decisions made over the early and mid-2020s will see renewed change in farm structures, not necessarily resulting in larger units, and change in who is farming the land.

5.8.2 It will be seen how far financial pressure and the assistance offered have improved overall productivity and how far new export markets have been won to support higher margins. The trend could be for a shift from commodity production, though still with good commodity farmers controlling costs, to more specialist or added value production with more diverse businesses. Less productive areas, arable as well as grass, might tend to see more environmental use of land, albeit generally compatible with a more extensive approach to agriculture whether as an adjunct to other sources of income, from on or off the farm, or by specialist farmers. More high value produce might come from within buildings. Much of that complex pattern of decision making, finding the better answers for the people in each situation, will turn on how values move and the review, advice and support of agricultural valuers.

Further Reading

Beyond Brexit: The UK's New Agricultural Policies (CAAV, 2020)

6 Farm accounting

6.1 Introduction

6.1.1 There will be a number of occasions when it will be important for an agricultural valuer to be able to review a set of farming (or, indeed, other business) accounts, whether for a specific enterprise proposal, to appraise a business or provide advice. This may become particularly pertinent in helping clients review their businesses as economic circumstances and support policies change after Brexit.

6.1.2 The common distinctive feature of farming accounts lies in the closely related points that:
- established businesses will often have strong balance sheets (especially where land is valued on a current basis) with less debt than many conventional businesses
- they will show a low return on capital employed.

6.2 Farm management accounting

6.2.1 The essential insight for farm management accounting is to separate consideration of each individual enterprise (such as wheat or suckler cows) from the consideration of the overall business. In doing that, the key concepts become:
- enterprise sa1083+89+5les – the output of, say, wheat, milk or lamb
- the direct costs in producing that (such as seeds, fertilisers and sprays). These are enterprise-specific and vary with the scale of the enterprise and so are commonly called "variable costs".
- the difference between sales and variable costs is the gross margin, the money earned for each unit of the enterprise (such as an acre of wheat or a suckler cow)
- other costs (such as labour and machinery) which could be spread across enterprises and do not vary so readily with scale are seen as overheads and here commonly called "fixed costs"
- what is left after those is the net margin or the return to the farmer on his labour, management and investment.

This structure has offered a powerful and simplifying analytical tool in considering management advice to a business and to compare the production efficiency of businesses. While it may in practice be much harder to vary variable costs than it is to change fixed costs, this terminology is driven by the way this model operates as the size of an enterprise changes.

6.2.2 As an example, winter wheat grown for feed might have this set out, using an acre as the unit of output, as:

Output – sales – 3.6t/acre @ £130/t		£468
Less Variable Costs		
Seed	£20	
Fertiliser	£74	
Sprays	£89	
Sundries	£10	£193
Gross Margin		£275

6.2.3 That is the margin that, on those assumptions, each acre of winter wheat would provide to help fund the overall business in paying for its overheads and providing a return to the farmer for his time, skill and investment. If the yield was higher or the price worse, the sales figure would change. As noted, variable costs might be harder to alter.

6.2.4 This model readily allows sensitivity testing of the outcome for differing yields and prices of product and so aid consideration of forward decision for the farm's enterprises.

6.2.5 A 650-acre cereals farm with some rented land might then have a gross margin and overheads of:

Total Gross Margin,	
simply assuming all is winter feed wheat	£178,750
Less Fixed Costs	
Paid Labour	£30
Machinery running costs	
and depreciation, contracts	£34
Rent	£38
Interest	£10
Building repairs, depreciation	£10
Sundries (fees, utilities, etc)	£51
	£173
Over 650 acres at £173/acre, Fixed Costs are	£142,379
Net Margin	£36,371

6.2.6 That analysis can not only assist decisions about moving land between enterprises but also about taking on or giving up additional land with the relationship with the farm's overheads. It thus offers a well-recognised basis for laying out budgets for tenders when bidding for a tenancy or seeking to justify an enterprise to a lender.

6.2.7 That structure of accounting is very much focussed on the agricultural operations and decision making of the business. There will be also other income and costs, sometimes marginal and sometimes very substantial, associated with other activities, commonly including agri-environment commitments, diversification and cottage lettings.

6.3 Financial accounts

6.3.1 While farm management accounting may assist the operation of the business and its decisions, financial accounts are also required to make tax returns and for lenders as well as providing a retrospective snapshot of the business itself. Detailed rules for this are laid down by Financial Reporting Standards (now, for farms, usually FRS 102) which must be followed and are endorsed by the tax authorities.

6.3.2 These will see both a profit and loss account and a balance sheet, each prepared according to the required conventions. They will cover an accounting period, conventionally a 12-month period running to a balance sheet date, for farming that may often be 31 March or 30 September.

6.3.3 The profit and loss account will consider the business as a whole and have the basic structure of:
- Income – recording sales of crops, animals and produce, as well as subsidies and other income and rents, all summed up as turnover
- Costs – covering all the inputs for the business including depreciation for past capital spending as it benefits the business in that year
- the resulting profit or loss.

6.3.4 For costs, a critical distinction in this is between current or revenue spending and capital spending. Current costs are those incurred for use in the year; capital spending has a longer-term benefit and so might be spread over the accounts for several years on an agreed basis, reflecting the expected life of the asset in question. This can call for care when distinguishing between works of repair (revenue) and improvement (capital). Capital spending goes to the balance sheet and is then depreciated to give an annual cost for the profit and loss account.

6.3.5 Costs may have been incurred in one year to support income to be made in the next year. An arable farmer with a 30 September year end might well already have spent money on establishing some of the next year's crops. That spending needs to be carried forward into next year's accounts by recognising it as stock in the stocktaking. That cost can then be carried forward into the next year's accounts to be set against income from the crop that has just been planted. Stocktaking, work for the agricultural valuer, is considered in more detail in Chapter 19.

6.3.6 That links the profit and loss to the balance sheet, which records the assets and liabilities of the business as at the end of the accounting period, comparing them to the previous year end (and so the start of the present period). The balance sheet will record:
- the fixed assets – most obviously farmland but also other assets enduring over the years. Owned farmland may often be included here at its historic cost to the business rather than its current value. That may mean many farming businesses have significantly stronger resources than the formal balance sheet will show.
- the current assets including:
  - stocks – based on costs already incurred to be set against future income
  - debtors – those people who owe the business money
  - cash – the money in the bank accounts
- creditors falling due within a year – people to whom the business owes money to be repaid in the near future, often bills that simply happen to be outstanding at the balance sheet date
- longer term creditors – such as mortgages – that are a liability to be paid over time.

6.3.7 The object of the balance sheet is to add up all the assets of the business at the end of the year, deduct all the liabilities and show what is left as the reserves, which may be in property, stocks, money owed to the business or cash.

6.4 Using the accounts

Various tools can be used to use the accounts to describe the health or vulnerability of the business. The choice will vary with the type of business. These include:
- measures of the liquidity of the business, how well it can meet demands for payment:
  - the Current Assets ratio – current assets/current liabilities to show how well the business can meet its short-term commitments
  - the Quick Assets ratio – a sharper version of that which shows the ratio between current assets less stock to current liabilities; the stocks are already committed to current production
  - the Net Working Capital Ratio – the ratio between net current assets (current assets less current liabilities) and total assets
- measures of the business' profitability:
  - Return on Assets ratio – the ratio between net income and average total assets (having averaged the opening and closing balance sheet totals)
  - profit margin – net income as a share of sales

- perhaps more usually relevant for more diversified businesses measures relating turnover to assets
- measures of indebtedness such as:
  - gearing – debt as a percentage of the business' assets

6.5 Tax accounts

6.5.1 Tax law requires some adjustments to be made to accounts for the purposes of the tax return. The major issues for faming accounts are:
- the replacement of depreciation with capital allowances.
- where the herd basis election has been made by the farmer the stock-taking should include production animals (such as breeding ewes or dairy cows) which the accountant will remove from the tax accounts (see Chapter 19).

Such items are work for the accountant, but it is useful for the valuer to understand the process and, in working with the accountant, to ensure that nothing is missed between them.

6.5.2 The accountant may then also apply averaging for Income Tax. This is the opportunity, recognising the potential volatility of many farm accounts, for the profits (but not losses) to be averaged over two or five years, reducing the exposure of business to higher tax rates in unusually good years. Averaging is available only to farmers, market gardeners and creative writers and artists.

Further reading

Reviewing a Business: An Introduction for Agricultural Valuer's (CAAV, 2019)

Part 2
Valuations

7 Professional issues

7.1 Introduction

7.1.1 Aside from the technical challenges which arise from working as a rural valuer, there are a range of professional issues which the valuer may encounter, whether on a regular or occasional basis. This chapter explores some of those issues but with two distinct caveats:
- the list of dos and don'ts sometimes reproduced on such topics runs the risks both of sounding rather too pompous and of being undone by the author in his subjective approach, leaving out something the reader may feel is fundamental; and
- in a rapidly developing technological environment the nature of the challenges facing professionals is changing at a much faster pace than for previous generations, although that should not obscure the fact that the core principles of our work remain substantially the same.

7.1.2 Indeed, there are some watchwords which have remained relevant for professionals from the quill pen through to the laptop and beyond, and for the author these focus on:
- the need to treat other people with respect and equity
- the need to be honest in one's professional work and
- the need to be efficient and, increasingly in the modern environment, cost effective (which is not the same as cheap).

7.2.2 The first two should be self-evident to anyone wishing to join any profession or indeed pursue any career with genuine success. This chapter therefore considers some of the issues around effectiveness and efficiency learnt as much if not more from failings on the author's part than any particular expertise.

7.2 Clarity and understanding

7.2.1 Other chapters will consider the need to confirm instructions when undertaking valuations, but this applies equally to all professional work undertaken on behalf of clients. This is necessary to ensure compliance with statutory (e.g. money laundering) or professional (e.g. Red Book and CAAV Bylaws) requirements but more particularly to ensure that the agent understands what the client wants, and the client appreciates what the agent will provide. This extends beyond the question of the task involved to issues of timing of delivery, cost and, where negotiation with another party is involved, the parameters within which the agent should negotiate.

7.2.2 This latter issue was captured in the Form of Appointment which valuers traditionally asked their clients to sign when undertaking a tenant right valuation. Practice was that valuers would not engage in negotiations unless they knew their opposite number had an Appointment Form signed. That approach subsisted longer in some parts of the country than others but has now been superseded by letters of instruction or similar arrangements. Latterly, the apparent unwillingness of some infrastructure developers to empower their appointed valuers with the autonomy to negotiate and particularly agree terms has become increasingly frustrating for both affected landowners and their agents.

7.2.3 Securing confirmation of instructions may sound relatively straightforward, but it can be more than simply attesting the identity of the client. It can be particularly challenging where projects extend beyond the original instruction or where they arise as part of a longer-term relationship between client and agent. A valuation required by a client for whom the valuer is undertaking day to day management work will still need a separate instruction. Similarly, new instructions will be required when a valuation for a client contemplating a purchase migrates into negotiating that purchase, which will then bring new requirements including money laundering checks.

7.2.4 The temptation when dealing with these compliance issues is to make the process as streamlined as possible. This has much to recommend it, but it is important that forms of instruction should not be so generic that they bear no relevance to the specifics of the case involved. In the same context whilst the parties will wish to ensure the exercise is as swift and simple as possible, it is important that it is clear to both that they are making a commitment to each other. Some agents are still apologetic when asking their clients to confirm instructions, but they should not be; they are asking the client to pay for a service, the costs of which may run into thousands of pounds. There are few other activities where one would make such a commitment without a clear contract explaining what is involved.

7.3 Client and project management

7.3.1 Securing instructions is not the end of the project management process. It is important that throughout any case the valuer should keep clients aware of progress and likely timescales. There are elements of work where this is a formal requirement, in agency activity for example; but where this is not the case it is still important for clients to be kept apprised of developments, or where relevant the lack thereof and the likely timescale for progress. One well known industrialist who became head of a large property institution greeted all its retained agents with the caution that the only thing he did not want were surprises; on reflection, that seems an appropriate dictum for all client management.

7.3.2 Timing is an important element in this. Clients will often anticipate that matters can be delivered sooner than is likely to be the case; after all, for many their transaction is likely to be the only one they are involved in at the time whereas the valuer may have twenty or more cases underway. It is important to try to set realistic timescales at the outset and to identify potential impediments to progress. A development sale may depend on the grant of planning permission and that, in turn, will depend on the work pressures on a planning officer who may now work part time in a much-reduced planning department. Identifying those constraints is important not least in ensuring that the client's professional team try to make it as easy as possible for the planning officer to recommend approval of that application, rather than complaining about the perceived inadequacies of local government.

7.3.3 In more complex cases, it is helpful to set out both the timescale and the critical path in the work for the client: thus, in negotiating the surrender of the tenancy of a farm for a landlord when notices would be served, when agreements would be signed and if all that goes to plan when possession would be secured. These are the cases where the valuer's ability to communicate and, indeed, empathise with people comes to the fore. The valuer's duty is to the client, in this case the landlord, but the tenant, who may choose not to be represented by an agent, is the person most affected by the transaction. As an aside, there are very strong arguments for the tenant to be represented, not just for equity but to make the task easier for all concerned. The good landlord's valuer will make sure that he keeps the tenant informed on progress as well, within the bounds of client confidentiality, if not for common decency then because if the tenant feels informed, valued and content the transaction will proceed more swiftly and easier.

7.3.4 This goes to the heart of the issue of the treatment of others which is more critical to the rural valuer's profession than any other. The rural valuer, acting as managing agent or as an advisor on a specific issue, sits between the parties involved and whether the proposed transaction is completed is partly in the hands of the parties and partly down to their agents. There will be some circumstances where settlement can never be reached, no matter how hard the agents try to bring their respective clients to an agreement. However, there will be other instances where an extra effort of diplomacy, and often an almost saintly ration of patience, will bring parties to a successful conclusion; managing such situations can be amongst the most fulfilling parts of the job, no matter how small the transaction may be.

7.3.5 Unfortunately, this may not always be the perception of the role of agents amongst other members of the rural economy. Whatever one may feel about the land tenure debate in Scotland, and the perhaps not entirely coincidental implosion of the tenanted sector there, the fact that there was felt to be a need for a code of practice for agents would suggest that there are areas where some are falling short of the standards to which we should aspire. Even if this was no more than a political prejudice, and it might be difficult to argue that case, then clearly the perception was of a lack of empathy at the very least. Further regulatory intervention will only make business operation more expensive and, equally as important, interacting and negotiating with others more constrained and less creative. That is a very practical argument for making sure that we do not attract or warrant more adverse comment.

7.4 Conflicts of interest

7.4.1 Again, reference will be made to conflicts of interest in later chapters. This is another area which has been the focus of considerable concern in recent years, not least since the financial crisis.

7.4.2 Clients are naturally anxious that their agent should be free to act for them without feeling at all inhibited by their role acting for another party. In some instances, the conflict can be very straightforward, between two different parties in the same transaction, for example. However, in other cases, it is rather less apparent and valuers will have to consider the risk of future potential conflicts as well as those less evident from existing work. Thus, a valuer approached by a landowner asking them to assist in promoting land for development will need to think both whether they are involved with any existing sites which may be in competition for limited allocations and whether they have established relationships with any other owners who have not yet come forward but whose instructions they may prefer to accept or would be embarrassed to turn down. There are similar difficulties in some family arrangements. In cases of succession under the Agricultural Holdings Act, for example, the position may change so that one potential successor becomes more favoured than another. The valuer, who may well have been instructed by the incumbent tenant in the first instance, may find themselves in the middle of an internecine struggle between potential successors and a direct conflict of interest.

7.4.3 The mechanics of checking for a conflict becomes more complex with the size of the firm; in larger firms, a valuer may not even know all colleagues, let alone whether they might be acting for a client, which creates a potential conflict. The use of document management and client relationship management software can enable far easier searching for potential conflicts, providing all colleagues use the system. One difficulty may arise when valuers have had preliminary discussions with clients which have not yet matured into an instruction, but a potential conflict arises in the

Professional issues 59

interim. No system is flawless, and the important thing is to notify both clients as soon as possible if such a conflict does arise so that they are aware of it and can consider it. Unless some satisfactory and regulatorily acceptable arrangement can be reached, the valuer might do best service by withdrawing from both instructions.

7.5 Data and record keeping

7.5.1 Data are at the core of the valuer's practice, whether evidence of transactions, property inspection notes or records of meetings or conversations. Storing these data so that they are safe, readily traceable and easily recovered is a challenge for all professionals, particularly where speed of communication and the potential for electronic access to data means clients expect a far more rapid reply to most enquiries. There is a danger that the pressure for more rapid response is undermining the quality of some advice to clients and, as set out earlier, it is important that expectations are managed as far as possible.

7.5.2 For some, nothing will beat hard copy records, notwithstanding the significant costs involved in printing and storing paper files; others now operate with the minimum paper possible, making notes on tablets or laptops and scanning in all other documents. Both regimes have their advantages and disadvantages, but both are far better than the approach which many valuers, and most likely other professionals, now adopt unwittingly, by filing data received in paper form whilst leaving email and other electronic communications stored in a parallel universe in virtual form only.

7.5.3 This can leave acute gaps in some files, which may not be a problem to the person involved in the case who was party to and thus more likely to recall a relevant email exchange, but it will be completely unknown to a colleague picking up the file in his or her absence leading to all sorts of complications.

7.5.4 Thus, unless filing is wholly paperless, requiring all information however collected to be scanned into a document management system, important emails should be printed off for paper files. Where the system is paperless it is important that communication and report writing takes place on media which is connected to that paperless system. An acceptance of an offer sent in haste from a home mobile or laptop is just as inaccessible as an unfiled paper copy of the same offer.

7.5.5 Whilst filing and file administration may seem rather too mundane for the qualified valuer the reduction in administrative support in most firms means that valuers are more likely to be involved in managing their own files, particularly where they work remotely part of the time. Firms adopt various approaches. Some RICS and external quality assurance auditors are particularly keen for valuation files to be maintained thematically (e.g. "instructions", "inspection", "workings", "draft report", "final report", etc), which can be a particularly good approach where large or complex cases are involved. Otherwise there is nothing to better chronological filing, and it only takes one or two occasions of having to read back through an entire file to find a letter filed out of date order for valuers to become disciples of that approach.

7.5.6 One slightly disconcerting aspect of electronic filing is the pace of change of the technologies involved. Many offices may have a repository of floppy discs, whether five inch or three and a half, but far fewer will now have a machine with the relevant drive. What may have seemed an exciting space-saving initiative at the time might now look rather less well-advised now if an historic record is required.

7.6 Standardisation and systemisation

7.6.1 Standardisation is an important element of practice across the professions; in the simplest example, this may be the use of a constant style in correspondence across the firm. For valuers, this is increasingly likely to extend to the use of templates for valuation and other professional reports, most commonly internally designed and generated, and the use of externally produced templates for licences, tenancies and contracting agreements.

7.6.2 There is great virtue in adopting a standardised approach, not least more efficient use of time and a consistent approach across the practice, providing that the standard template is kept up to date and colleagues comply with that approach. One of the difficulties is in finding a template suitable for all purposes, a commercial property valuation template, commonly driven by the requirements of a high street bank at highly competitive fee rates, is unlikely to be suitable for a valuation of a farm for Capital Gains Tax purposes. That said, it should not be beyond the wit of most property consultancies to develop a series of reporting templates to suit their principal activities.

7.6.3 Indeed, in most cases it is not the technical challenge but, more likely, the cultural one which may lead to standardised approaches being ignored or cannibalised to the extent that they become both unintelligible and non-compliant with either internal standards or external regulation. There is something of the individual about most valuers, but there are some, even in the rural profession, where this approaches prima donna status. Anyone involved in management of a firm will have endured the lecture from a sector specialist, self-appointed or otherwise, explaining that whilst templates are, of course, a good thing, they cannot be applied to his specialist area or his level of skill. There will, of course, be exceptions (a proof of evidence, for example, will of itself be specific to the particulars of the case), but it should still hold to the core elements required where valuations are involved and match the requirements of the Woolf Reforms to court procedures. Otherwise, where there are to be departures from the standard approach, these should be as limited as possible and approved on a case-specific basis, not simply for a few senior, or vocal, individuals.

7.6.4 There are some areas, however, where it is particularly important to challenge a template and make sure that it does apply to the circumstances of the case. This is especially relevant where drafting Farm Business Tenancies or other occupational agreements. The CAAV and other organisations, as well as some firms of solicitors and some institutions, produce comprehensive templates for tenancy agreements, which save the busy valuer from having to research the relevant legislation and current practice every time they draft a tenancy agreement. However, improperly (which most commonly means lazily) used these can lead to a lack of imagination in the overall agreement reached between the parties.

7.6.5 Templates, for example, naturally, include a bar on alienation and constraints on use; that can in turn be echoed by an unwillingness amongst some agents to consider a deviation from either approach where some changes might be better suited to both parties. At a simpler level, most templates suggest rent payments on even periodic, typically quarterly or half-yearly bases. Dairy farmers aside, few farm businesses now have symmetrical cash flow and, in some cases (soft fruit production for example), receipts may be highly concentrated in the year. Again, some imagination on that front may suit both parties far better than the traditional approach. This is not to advocate wholesale abandonment of the templates but simply to hope that future generations of valuers will think twice before shoe-horning a letting of a block of land for maize production for AD feedstock into the last agreement they did, which happened to be for some hill grazing.

7.6.6 Standardisation does not simply apply to the documentation used and produced but also the approach one takes to undertaking the professional task at hand, often referred to as "systemisation". Commonly this may be generated in the first instance by the adoption of particular time recording, document management or other support systems, and in many firms, led by those involved in public sector or utility work, this will be formalised in some form of Quality Assurance system.

7.6.7 In summary there is a question of judgement here. A standardised approach, as commonly expressed through a template, should be at the very least a guide, and, where it is fit for purpose, a substantial aid in drafting the relevant document. However, there is a serious risk that they encourage valuers, particularly those at the beginning of their career, simply to fill in the gaps. The valuer must ensure that they actually think about the task at hand when preparing the report, valuation of agreement involved.

7.6.8 Again, for those of us, possibly including the authors, who regard themselves as working outside of normal parameters, this may seem like a constraint on their individual professional approach, and those involved in

implementing standardisation will encounter some resistance. However, in most cases, standardisation offers considerable support for professionals with the particular merits of:
- relieving the professional from time involved in administrative activity;
- providing confidence for clients of a consistent approach (particularly relevant to larger clients in both public and private sectors);
- ensuring that the various elements of the task are documented and completed effectively; and in turn
- where followed, providing a robust response to any professional indemnity or other complaints.

7.7 Fees and time management

7.7.1 There may have been a time where clients were content to pay fees which were ill-defined in advance and vague in their construction. Indeed, current valuers may be intrigued by older fee accounts which range from multiple pages of detailed description of the activity, often trawled from a painstaking review of the file, to a single line 'to professional fees' and an amount. Counterintuitively, the latter abbreviated form was just as common for large fees as very small ones.

7.7.2 Nowadays, clients will expect, and are entitled to, a clear explanation of the basis for fee charging at the point of instruction and where possible some indication of the likely amount. The abandonment of the fee scales which provided the framework for pricing for much professional work in the past means that fees for non-transactional work are, as shown by the annual CAAV Fees Survey, predominantly either time based or on a fixed fee basis.

7.7.3 That, in turn, will require some form of time recording whether, at its simplest, either manual recording of time on files, a home-designed spreadsheet system or a more sophisticated proprietary time recording system. Many valuers find occasional jest, as do clients, in complaining about another profession and its approach to time recording in one-tenth of an hour intervals, but this discipline provides far greater robustness in explaining fees to those who may be paying them, whether clients or third parties. The CAAV guidance note offers some further guidance on the merits of time recording and the Fees Survey provides an annual insight into both fee levels and the basis of charging, which can be of considerable assistance when establishing fee structures with clients.

7.7.4 It is beyond the scope of this publication to consider the merits of different forms of time recording, but one underlying consideration should be that, whilst sufficiently rigorous to provide the discipline referred to earlier, the system should serve the business, not dominate it. Most systems can be managed to deliver that delicate balance, but it is important that firms involve the professionals as well as practice managers in the procurement of such systems, and not just the long-established fee earners but also the younger valuers who often have a better understanding of some of the applications involved and, if nothing else, will have to put up with the system for longer in the future.

7.8 Utility and advice

7.8.1 The practice of the rural valuer is a complex mix of legal and practical knowledge, market interpretation, technical application, negotiation and occasionally social and emotional support to clients and others, all consolidated into appraising the (sometimes perverse) facts facing the valuer. It requires academic skills, but it is not an academic exercise and, occasionally and perhaps increasingly, the need for compliance, whether imposed by statute or professional institution, and commercial pressures seem to be conspiring so that valuers produce answers which fall short of proper advice.

7.8.2 There is an apparently subtle distinction here which may not apply in every case but is fundamental where it does. The RICS Red Book, for example, explains in considerable detail what constitutes a valuation and, in turn, Market Value. Valuers, particularly in the early stages of their career, will follow those dictates assiduously to produce a highly professional and hopefully accurate valuation. That tells the client what the sale price might be if a purchaser reflects the market; it is specifically not an assessment of what the client will actually receive if considering selling the property. That may seem obvious to us but less so to a farming family, who may not have sold any property for many generations.

7.8.3 At the simplest level, that transaction will incur costs, but what about the incidence of taxation, specifically excluded from consideration in the Red Book valuation?

7.8.4 Some valuers may feel that this is something for the client's accountant or solicitor, but what if they do not consult them until after the decision to sell is made and the property is on the market or, worse still, heads of terms agreed. A client selling property to reduce debt will get a nasty shock when the proceeds of the sale, which was carefully crafted to generate sufficient funds to clear a loan or meet an Inheritance Tax bill, is reduced by Capital Gains Tax at 20% or 28%.

7.8.5 All valuers will provide a valuation, but the good valuer will then draw attention to the risk of the impacts of taxation and other costs. That may mean simply a recommendation in the letter accompanying the valuation to seek advice on these issues before making any decision. In other cases, it will help prompt teams of professionals to work together in modelling the impact of any decision for clients and providing legal, accounting and valuation advice to create a bespoke solution.

7.8.6 The valuation is the technical answer, but it can be of limited utility to the client. The added value which the best professionals, rural valuers or otherwise, will bring to the party is to turn that technical answer into advice specific to the client's situation.

7.8.7 Gwyn Williams in writing the previous editions of this book understood that implicitly as the role of the professional rural valuer as would most of his contemporaries in reading the publication. Sometimes it can be lost on those of us responsible for training the next generation of valuers, whatever our role, that our best work only begins once we have delivered the technical advice.

Further reading

Notes for Probationers on Office Systems (CAAV, 2010)
RICS Valuation Global Standards & UK Supplement (The Red Book) (2020)
Scottish Land Commission Tenant Farming Codes of Practice (Scottish Land Commission, 2015 to date)

8 Introduction to valuation standards and bases of value

8.1 What is valuation?

8.1.1 Valuation is the appraisal of the value that can reasonably be expected for something for a specific purpose in its context. In agricultural terms, that may be a particular piece of land or rights in or over it, a building, livestock or machinery. It can concern the relationships between people with different interests in the land, such as landlord and tenant, and offer a means to handle issues between them such as the crops that one has grown in the other's land or issues of rent or compensation.

8.1.2 That requires an understanding of:
- the physical nature of the item in question: what are its qualities and deficiencies, its potential and its risks?
- its legal nature: what is the quality of the rights in the item held by its owner? What can be passed to someone else? Do other people have claims on it? Are there restrictions on it?
- the marketplace for what has to be valued: what do people want? What do they not want? Why? What is the balance of supply and demand?

8.1.3 That appraisal of the physical and legal nature of an item, such as a block of farmland, with an appreciation of what is known about it, what might reasonably be assumed and recognising what cannot be known, is necessary to define what it is that must be valued.

8.1.4 Values, as they are found and as they change, are the essential mechanisms of a market economy, driven by supply and demand, reporting what it is expected that could be paid at a given date for an item. That value will change over time: perhaps rising if more people find that item attractive and so compete with each other for it or falling if more feel it has too many risks or is no longer useful. That may reflect not only perceived income earning opportunities but also other personal desires, including for agricultural assets such issues as amenity, sporting, privacy as well as a common concern not to miss rare opportunities in a land market constrained by location. These changing price signals then drive change throughout the economy, assisting economic change and so informing other decisions, the allocation of capital and business energy. Those dynamic processes of a market economy, as values move over time between assets, regions and sectors with the associated finance, all turn on valuations informing all parties involved.

8.1.5 It is fundamental in that that value is not a measure of cost. An individual may be able to create or acquire something advantageously or, for their own reasons or miscalculation, do so for more cost than it is worth. Equally, the value of, say, land or machinery may have changed since it was last acquired. The valuer is usually required to find a value as between notional sellers and typical buyers, each of whom will have their own views and access to funds in judging what they might wish to pay.

8.1.6 Equally, value is not necessarily the actual price paid in any transition which may reflect particular aspects of the transaction or the actual parties. The task is to find what figure can be advised as the one that can reasonably be expected to be achieved for the item.

8.2 Evidence and analysis

8.2.1 The work of valuation is thus about understanding a property and its place in the market with the interaction between buyers and sellers with motives, concerns and information, and facilitated by access to finance.

8.2.2 Using that understanding will require gathering evidence:
- to appraise and report clearly on the property
- as to current behaviour in the marketplace, especially as revealed in recent transactions
- as to relevant physical, legal, financial and market developments and then applying professional skill and judgment to that evidence to arrive at an opinion as to value.

8.2.3 Much of this, especially for agricultural valuations, will rely on the ability to compare transactions, identifying and justifying appropriate adjustments to compare differing farms or other properties with each other. This may see consideration of matters such as:
- differences in the date of the transaction
- differences in size
- differing levels of fixed equipment with buildings, drainage, house or cottages
- differences in the character of the land
- differences in services, access and similar points
- presence or absence of legal restrictions or development opportunities
- differences in access to support or schemes
- differences in other factors considered by the market.

8.3 Valuation standards

8.3.1 Over the years, professional discussion, court cases and statutes have developed structured understandings of the rules for appraising valuations. The general rules, definitions and assumptions for this are brought together in valuation standards. The standards most likely to be met are:

- International Valuation Standards (IVS) prepared by the International Valuation Standards Council (IVSC) for all assets and liabilities, including financial instruments (and adopted by the RICS and incorporated in its Valuation – Profession Standards, the "Red Book")
- European Valuation Standards (EVS) prepared by The European Group of Valuers' Associations (TEGoVA) and focussed specifically on real property, that is, land and buildings.

8.3.2 Other recognised Valuation Standards include the US Appraisal Institute's Uniform Standards of Professional Appraisal Practice (USPAP). With generally little substantive difference between the texts, these provide standard and recognised rules so that valuers, clients and others who may rely on the value that is being determined are all working in the same terms.

8.3.3 While there may be occasions when statute (such as tax law) imposes some specific points, such as a definition, Valuation Standards provide a common framework generally focusing on:

- the recognition, development and understanding of "market value" as the key concept in appraising value – the default basis of value
- the use of assumptions about unverified matters or special assumptions about hypothetical situations (such as forced sale value)
- definitions and understandings of alternative bases of value which may be required for particular purposes or in some situations (such as one of the forms of fair value, marriage value and worth or investment value)
- the importance of the valuer having the necessary skills and knowledge to value the item in question and compliance with professional ethics on matters such as integrity, objectivity and conflicts of interest
- the requirements of the valuation process in inspecting, gathering information and undertaking the assessment
- the approach to reporting the value determined, giving a single figure as at a specific date with the supporting appraisal, assumptions, qualifications and perhaps advice.

8.3.4 Such a structure is especially important for property and so agricultural issues. US dollars, for example, may be bought and sold in large volumes in liquid markets with one dollar indistinguishable from another. They can be sold in any volume required, with no marketing period and low transaction costs. As a result, their price is easy to find. However, land and buildings exist in more limited markets, are fixed in their location, are often harder to subdivide or aggregate and are useful only to a more limited pool of buyers. While all US dollars may be taken to be the same with no more knowledge needed, there are many more issues of interest about any specific parcel of land and points on which knowledge may necessarily be imperfect or unknown, requiring judgments. Valuation Standards provide the framework for that process.

8.3.5 When valuing for financial accounts, that is the difference between homogenous assets with widely reported prices (described for accountancy purposes as a Level 1 input) and individual properties, such as farmland, that require judgment (and so described as Level 3 inputs).

8.3.6 What follows are the larger definitions of the bases of value; there are then different approaches or methods commonly used to assess the value of an asset. The most generally used are:

- the *comparative method* – reviewing values paid for other similar items, making adjustments for differences in quality, time or other factors. The great majority of agricultural valuations follow this approach. For a single asset like a three bedroomed house, the property to be valued might be compared directly with recent sales of other three bedroomed houses in the same area and that might be considered by the same buyers. A farm might require closer analysis of other relevant sales to consider different qualities of land, scale or condition of fixed equipment, houses and other factors. This analysis is considered further in Chapter 10 and elsewhere in this book.
- the *income method* – where an item gives rise to an income, such as rent, that can then be valued as a capital sum by using an interest rate assessed on a judgment of the life and security of the income and how it compares with other assets. Where the asset has no other qualities beyond its income that derived capital sum could be its market value. This method may often simply be another way of comparing transactions, but where very low rates of return are applied, as often for agricultural assets, the outcome can be very sensitive to small changes in yield, which result in large changes in valuations, so requiring caution as it may be hard to justify distinguishing between small differences in yield.

> *Example* – An asset with a rent of £10,000 would have a value of £500,000 if capitalised at 2% but £666,667 if capitalised at 1.5%. How confident might the valuer be as to which of the rates in that small range to select when the difference in outcomes is large?

8.3.7 In reality, both methods can be seen as exercises in comparison – one directly of capital values and one of yields – reflecting the differing markets for which they tend to be appropriate. The more the income approach is divorced from market observation, the more precarious it might be, especially with the low yields typically seen for rural property and, more generally, the ultra-low interest rates prevailing since the financial crisis of 2007–08.

8.3.8 With valuation issues coming before the courts, a number of decisions have commented on preferred approaches. A good example, addressing realities of actual and limited evidence, is offered in *Living Waters*

Christian Centres v Fetherstonhaugh, in which the Court of Appeal endorsed the approach taken by the judge in the lower court:

> First, to describe a particular property as a "comparable" in the context of a rent review arbitration, as in relation to any other property valuation issue, is to refer to another property that can be used as a reference point to give some indication of the value of the property in question. The closer the other property stands on the scale of characteristic proximity to the property in question the more helpful indicator it will be of the value of the property. Thus, if all material characteristics going to value are the same, one would expect the value of the property in question to be very close to that of the other property. In such a case much weight would be attached to it as "a comparable". There might, however, be other cases where a number of properties of similar characteristics had one or more characteristics materially different from the property in issue but could still be used as the basis for a general indication of the value of that property, because an experienced valuer would know what adjustment to make to the value of that property by reference to the material differences in characteristics in order to obtain a general indication of the value of the property in question. In such a case it would no doubt be appropriate to describe the other properties as imperfect or even poor comparables. That would simply mean that their value could not be transposed to that of the property in question, but it would not mean that they were evidentially irrelevant to that value. At the extreme end of the scale would be properties that were so different in material characteristics from the property to be valued that they had no evidential contribution to make and could therefore be treated as irrelevant. Accordingly, the term "comparable" has to be treated as wide enough to cover any other property that has any evidential contribution to make to the assessment of the value of the property in question, whether that contribution is substantial or only relatively small because of differences in material characteristics.
>
> Second, information as to the material characteristics of the other properties, said to be comparables, may be complete or relatively incomplete. The less complete it is, the less weight can be given to that property as a comparable. Special circumstances relating to the negotiation of the rent at a particular property may render it a poor comparable or not a comparable at all, whereas, if there were no knowledge of those circumstances, that property might appear to be the perfect comparable.
>
> Third, the fact that one arbitrator has attached to a particular comparable or comparable's weight that would not have been accorded to them by a court is not misconduct. It may be at its highest an error of fact but as such is not to be treated as within the appeal regime of the Arbitration Act 1979. Consistently with that

approach it is not of itself to be treated as misconduct unless it is the product of procedural irregularity or unfairness in the course of the arbitration or unless it leads to the consequence that the conclusion arrived at is one that no reasonable arbitrator could reach on that evidence.

Fourth, to describe a comparable as inadmissible is only if the evidence of the material characteristics is inadmissible, for example, because it is hearsay, but not merely because the comparable is a poor comparable or one that is arguably off the scale of comparability.

Fifth, for an arbitrator to rely on inadmissible evidence may be misconduct or error of law. Whether in any particular case the conduct is admissible or inadmissible depends on whether the parties have expressly or impliedly agreed to the relaxation of the strict rules of evidence.

8.3.9 As will be seen with the valuation of land with potential development uses, the courts and tribunals prefer to work from direct market comparison, only resorting to residual calculations (working backward from a potential development value allowing for cost and risk) when that market evidence is not usefully available.

8.3.10 Discounted cash flow (DCF) may be a technique for use where income and costs accrue over time, as for some development projects and some renewable energy schemes. In particular situations, other methods can work from cost, such as finding a depreciated replacement cost as may sometimes be relevant for specialist buildings, but this requires care if it is to find a value.

8.3.11 However the valuation is approached, the essential and final test is whether parties in the marketplace could really be expected in practice to pay a price at the level of the value that has been assessed. That emphasises the importance of soundly analysing good quality comparable evidence where it can be obtained. This final test may be a particular challenge for any valuation arrived at with an essentially theoretical approach.

8.3.12 This is particularly applicable to valuations of real property, given the usual individual nature of the properties and the markets concerned, especially at times of flux. The valuation of agricultural property is an art requiring care, experience of the specific market, research and the use of market evidence, objectivity and an appreciation of the assumptions required and judgement – in short, professional skills.

8.4 Market value

8.4.1 This is commonly the basis of value to be used, unless specifically instructed otherwise.

8.4.2 A standard definition (in this case EVS 2016's EVS1 with its focus on property) reads:

> The estimated amount for which the property should exchange on the date of valuation between a willing buyer and a willing seller in an arm's length transaction after proper marketing wherein the parties had each acted knowledgeably, prudently and without being under compulsion.

As that is drafted with a capital value, as for sale, in mind, there is an equivalent definition for market rent.

8.4.3 The purpose expressed by that definition is to find a value that is neutral as between buyer and seller and so, in a market with competing parties, it is an estimate of the amount that could reasonably be expected to be paid – the most probable price in the market conditions holding at the date of valuation. While the item in question might have different values for different individuals who may be in the market, its market value is the estimate of the price in the present market on assumptions that are deliberately neutral to achieve a standard basis of assessment for both buyers and sellers.

8.4.4 The definition brings together:

(i) *the result* that is to be found – the most probable price reasonably obtainable for the property in an arm's-length market transaction at the date of valuation on the assumptions of this definition (and so disregarding atypical assumptions). It is the best price reasonably obtainable by the seller and the most advantageous price reasonably obtainable by the buyer.

(ii) the real *property* that is to be valued – with its legal, physical, economic and other attributes and all its actual opportunities and difficulties so far as they are recognised by the marketplace. That can include the possible uses of the property that may be unlocked by changes affecting it, whether new development control permissions, relevant infrastructure, market developments or other possibilities and so market value may reflect any "hope value" that the market may place on such prospects.

(iii) the *transaction* assumed – the phrase "should exchange" points to the price at which the market would reasonably expect a hypothetical sale achieved by negotiation and completed on the date of valuation in accordance with all the other elements of the Market Value definition.

(iv) the *date* of valuation – a value is specific to the date for which it is assessed which may be before or at the date the report is prepared, certainly not later than that.

(v) the nature of the *hypothetical parties* as willing and competitive – the motivations and character of the actual owner is disregarded. Both seller and buyer are motivated to pursue the transaction, but neither is compelled to complete it. The price is the value that resolves those

72 *Valuations*

two positions. The parties are independent of each other, acting at arm's length.

(vi) the necessary *marketing* – the property is assumed to have been marketed in a way that brings it to the attention of an adequate number of available potential purchasers so as to achieve its disposal at the best price reasonably achievable.

(vii) the consideration by the parties – they are required to have acted:
- "*knowledgably*" and so be reasonably well informed about the nature and characteristics of the property, its actual and potential uses, the state of the market at the date of valuation and what could reasonably be expected as at that date.
- "*prudently*" and so each acting in their self-interest, seeking their best outcome for their respective positions in the transaction, as understood at that time.

8.4.5 In specific circumstances, the valuer may make assumptions or special assumptions:
- an *assumption* may have to be made to arrive at a value, as where assuming (or is instructed to assume) the most likely case for a matter of fact or circumstance which is not or cannot be known or reasonably ascertained. Some issues, such as the detail of the title to the land or the presence of asbestos, may be beyond the valuer's ability to check independently. In that case, the valuer might make and state an assumption on such a point (for example, that the property has good freehold title or that there is no asbestos) and recommend that the client have the facts established by those with the appropriate specialist skills.
- a *special assumption* is where the valuer assumes, usually on instruction, a fact or circumstance that is different from those that are verifiable at the date of valuation. The result will be the market value on that special assumption. It might be required to give an opinion as to the value of the property were it vacant, when it is actually let, if planning permission for a particular use were achieved or if there had to be a forced sale on a particular timetable.

8.5 Other bases of value

These should only be adopted where instructed but include
- *special value* as between specific parties, most obviously *marriage value* where putting two properties or interests in a property together results in a property that is more valuable than the two properties assessed separately. That might arise between neighbours (as perhaps one with a water supply and the other needing irrigation) or a landlord and tenant negotiating over either the surrender of a secure agricultural tenancy or the tenant buying the freehold, unlocking additional value between them.

- *fair value* – when considering a transaction between two identified parties with the real characters and motives. The phrase, fair value, is also used by accountants to mean something that will often be similar to market value.
- *worth* or *investment value*, which is the value of a property to an individual irrespective of its wider market value. The value to that person with particular criteria or needs may be higher or lower than the market value. It might reflect such particular requirements as for a minimum yield from an investment or a level of environmental benefit. Worth thus appears a useful basis for considering the assessment of natural capital values (see Chapter 17) for one party, enabling choices between options and informing the later work on the actual values of transactions between parties that change outcomes.

8.6 The need for a sense check

Whatever the basis of valuation and the approach or approaches used to determine the required value, the valuer should always step back and look at the draft answer and consider if it makes sense. Could someone be expected to pay that figure? That process of review may lead to reappraisal of the valuation with resulting adjustments made as needed to its analysis, reasoning and conclusion.

8.7 Difficult markets

8.7.1 These tasks can be particularly hard where markets are very thin (and so offering few or no immediately relevant comparables), frozen or volatile, calling for a level of consideration, judgment and explanation to assist the client.

8.7.2 This has been a significant issue for rural values four times in the first two decades of this century with the immediate impact of the foot-and-mouth disease outbreak in 2001, the financial crisis of 2007–08, the 2016 referendum and the Covid-19 pandemic of 2020. Typically, the first effect is for property markets to freeze with the volume of current transactions dropping away. That limits or removes the comparables that are available, while even recent evidence from before the crisis may be of uncertain relevance. This generally creates circumstances described as *valuation uncertainty*. The valuer's approach may vary with the specific circumstances but will need to ensure that the client is aware of the position and that any valuation is subject to that specific uncertainty. As well as the opinion that the valuer may have at the time, it may be useful to discuss a pre-crisis value and propose a re-valuation in some months' time.

8.7.3 Then, as either transactions (sometimes forced sales) and valuations (as for Inheritance Tax) have to be done and then the human need for trade reappears, so evidence begins again to accumulate, itself generating more transactions and more evidence with confidence returning. The issues might then be less of valuation uncertainty than of *market risk*, the risk of values being exposed to coming changes. Again, that might usefully be the subject of commentary in the valuation.

8.7.4 An issue in disease outbreaks can be limited access to the property to be valued. The circumstances of the Covid-19 pandemic meant that the interior of occupied property might not be available for inspection. That limitation was readily recognised for agricultural property by lenders but required care in the valuation process with cases ranging from 1,500 acres of land with a cottage that the valuer had seen three months ago (where it might be relatively unimportant) to a seven bedroomed house on five acres (where that lack of access might be critical). The point was, though, that the valuation, drawing on all other available evidence including that from past records and what might be provided remotely by the occupier, would be a valuation on limited information, not a desktop valuation.

8.8 Longer-term value?

8.8.1 While agricultural land prices in the United Kingdom have generally seen little volatility since the early 1990s, there is a larger concern, both in macro-economic terms and for major lenders, since the financial crisis of 2008 about potential volatility (both boom and bust) in property prices with its effects for economic management and the stability of lenders. This concern has widened from the housing market, periodically seen as a source of instability, to commercial property and so may have effects across all property sectors, perhaps especially for secured lending for which a property is valued at a given time to underpin a loan extended over many years, whether with regular payments reducing the sum borrowed or repayment left to the expiry of the loan period. With farming's low return on fixed capital, secured agricultural lending typically operates with loan-to-value ratios that are much lower than those in other sectors, often lending to more than 50% of the value. That insulates farming mortgages from such risks, as the property value would have to fall by half before the loan was not fully secured.

8.8.2 While this issue might be better seen as a matter of overall economic management rather than for individual properties, anxiety about exposure to risk, especially in a low inflation environment, has led to discussion of whether a longer run value can be identified for a property and whether some of the risk in understanding that should lie with the valuer as well as with the lender. This a different topic from the valuations, sometimes called future valuations, that might support a development project which, in effect, says what the value would be today as each milestone in the project is reached and so trigger the release for funds for the next stage.

8.8.3 The valuer cannot give reliable opinions of the value at any future date for which the circumstances and market behaviour could be different in ways unforeseeable now. Nonetheless, it may assist some clients if the commentary on the valuation in the valuation report might record where present market circumstances significantly diverge from long-run trends, looking over decades. For agricultural property that might be by reference to historic ranges of values and potential changes, such as the possible issues where land is increasingly exposed to flooding. For let commercial properties, that might refer to the long-run average yield for the relevant class of property. Other data might be relevant for the housing market, such as prices as a ratio or multiple of incomes. The underlying insight is that markets tend over time to revert to the mean. If a relationship between values, such as yield, has diverged markedly from that average, it may well, all other things being equal, revert to that long run yield at some point in the future.

8.8.4 That observation of overall markets then needs to be tempered by the circumstances of the individual property, whose type or location may be improving or weakening in the overall market as the economy changes. Indeed, the finance markets are beginning to perceive the risks of climate change for property, with flooding, mainly from rain and rivers but also from the sea, being the largest risk in the United Kingdom. Compliance with more demanding energy efficiency rules seems likely to become a larger issue over time for buildings. Such issues could lead the market to take a different view of affected properties. A valuation might also be vulnerable should long term relationships in the markets, such as the yield on money, have changed.

8.8.5 While such an approach may be useful for some clients as commentary, it will not be a measure of market value which is specific to its point in time.

8.8.6 Furthermore, individual investors and lenders may, as a risk management strategy, seek to form a view on their own criteria as to a long run value or a minimum secure value that may exist in all but cataclysmic conditions. If given clear instructions as to the desired assumptions, a valuer may be able to assist with those judgments which are not, at this stage, about market value but about worth.

8.8.7 With the observation that agricultural values have often seemed to run counter to the larger economic cycle, these considerations do not remove the point of drawing on experience of previous market cycles. They are rather to observe that each valuation is at a given point in the economy at which there are prevailing trends and prospects (not necessarily limited to the cycle) which may be vulnerable to change, at least some of that change being unforeseeable. It is right that, where appropriate, the valuer comments on this.

8.9 Further explanation

As that discussion shows, the valuer can be of further assistance to a client in offering an explanatory comment or review explaining the figure and, if instructed, the risks that might attend it.

8.10 Measurement

8.10.1 In assessing and reporting on a property for a valuation, it is important to be clear about the measurements used. That also matters when considering other transactions that may be comparable in helping find a value.

8.10.2 While rural property has traditionally been measured in acres, the law requires official documentation to be in metric terms (and so area in hectares and length in metres). For farmland it can be useful to know the useable area as well as the actual physical area, so excluding land that cannot be cropped or grazed. The requirements of current subsidy regimes also see eligible areas for land parcels which are recorded on the official Land Parcel Information System for each country of the UK.

8.10.3 Particular conventions may also be relevant to the assessment of buildings and dwellings.

8.10.4 In addition to national or sector specific conventions for measurement, the more general ones are:
- RICS Property Measurement (incorporating the International Property Measurement Standards)
- TEGoVA's European Code of Measurement.

Further reading

European Valuation Standards (The European Group of Valuers Associations (TEGoVA), 2016)

RICS Valuation Global Standards & UK Supplement (The Red Book) (RICS, 2020)

9 Undertaking a valuation

9.1 Introduction

9.1.1 For many rural valuers the joy of the work and, indeed, the attraction for first joining the profession was the prospect of being able to work outdoors and not being "tied to a desk". Consequently, many will enjoy visiting and inspecting farms and land and there is no doubt that effective inspection is at the core of every valuation. However, this is only part of the valuation process, and the rural valuer has to apply the same professionalism to the rest of the process to ensure that the eventual valuation report is fit for purpose and useful to the client and their other advisors.

9.1.2 The principles underlying the conduct of valuations have not changed at all since the first edition of this book was published over 30 years ago. The valuer should know the 5 "w's"
- where is the property?
- what is to be valued?
- who is the client?
- when is the valuation date?
- why is the property being valued?

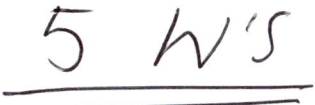

9.1.3 Those, in turn, will inform two important considerations – how will the value be calculated? How much will the fee be?

9.1.4 However, much to the chagrin of some valuers, the bureaucracy associated with the valuation process has become more formal, complex and expensive. Whilst this may occasionally appear routine and tedious, it is an essential part of the process, both in ensuring that the valuer and client understand what is required and what will be produced and in protecting the valuer should the valuation ultimately be challenged whether as part of court proceedings or on the grounds of professional negligence.

9.1.5 For those of us who embarked on our training in the twentieth century, the established senior valuers seemed to have either an uncanny knack or, depending on how elaborately and artistically they explained matters to trainees, an almost mystical ability to assess value. In a period of greater deference, their opinion carried weight even before they explained their analysis. Those valuers had great skill and would still get the right answer today. However, the modern valuer must combine this ability with a rigorous process capable of withstanding external audit from first instruction through to issuing the report and the fee invoice. Those valuers who are RICS Registered Valuers will be audited as part of that accreditation and firms with ISO 9001:14001 or similar quality assurance systems will be subject to internal and external audit as will those who are panel valuers for lending institutions. These inspections can seem an intrusion into already busy working schedules, but they are a precondition for instructions from some clients, and there is much to be said in working with auditors of such schemes to develop and improve systems.

9.2 Instructions

9.2.1 The first place that professional indemnity insurers will look when becoming involved in a case is the quality of the instruction letter on the file and evidence that it has been agreed by the client.

9.2.2 Whilst the precise nature of the instruction letter will turn on the requirements of the individual valuation, most practices now operate with a series of standard templates. That standardisation means that a degree of sense checking is required, not least as there are far more valuations of commercial and residential property than rural and if a single standard template is used it is likely to be designed for urban property.

9.2.3 As a farming client may look somewhat askance at an instruction letter which refers to common areas and multiple leasehold interests, there is a strong argument for making sure that there is a standard rural instruction template available in the practice for those undertaking rural valuations.

9.2.4 In some instances, instructions may come in writing from a lending institution or professional advisor, in which case the valuer must check that they can comply with the instructions before issuing the appropriate letter of confirmation with terms of business where those are not covered by an established contract. In other cases, the instruction may come directly from the client whether by phone, email or leaning on the rail at the store cattle ring. In that case, the valuer will still issue a letter accepting instructions but asking the client to sign and return a copy confirming their acceptance of the terms.

9.2.5 Guidance on what is required is set out both in the Red Book and, for Recognised European Valuers, of whom there are a number in the CAAV, TEGoVA's Blue Book. Valuers should refer to any updates, which in the former case are more frequent than swallows in summer, to ensure that their instruction letters are kept up to date. It may also be sensible to consult with professional indemnity insurers, perhaps when dealing with policy renewals, if nothing else to ensure their satisfaction with the process.

9.2.6 Whilst the precise terms will vary depending on the nature of the property and the instruction, the letter is likely, at the least, to include:

- the name of the client and their interest in the property – in some cases it may be appropriate to undertake identity and money laundering checks
- the name, location and a brief description of the property to be valued – although for portfolio valuations this may be rather a broad summary
- the purpose of the valuation which in turn will dictate:
  - the basis of the valuation generally by reference to the Red Book and
  - any Special Assumptions to be adopted
- the date of the valuation – or, in some cases such as valuations for Capital Gains Tax, multiple valuation dates
- the basis of tenure and occupation
- the nature of the report to be provided
- the basis and amount of the fee – whether hourly rates or a fixed fee and the treatment of expenses
- conflict of interest – confirmation that a conflict of interest check has been undertaken and that there is no conflict – or, if there is a potential conflict, the valuer's proposal to both clients involved in the potential conflict
- the identity of the valuer(s) involved in undertaking the valuation and their respective roles (e.g. Client Partner, Valuer)
- limitations on liability to the identified client
- confidentiality – and a bar on disclosure and publication, which whilst it cannot be enforced can provide a limit on liability to other parties
- reliance on sources of information
- reference to standard terms of engagement for valuations
- confirmation that the instruction is covered by the firm's complaints policy and where the client can find that policy.

9.2.7 In addition to the letter of instruction, most valuers also send out standard terms of reference or engagement for valuations which set out terms and conditions which will apply to all valuations. These include information which should otherwise be included in the letter confirming instructions and include, for example:
- limitation on inspection
- the fact that no form of survey, structural or otherwise has been undertaken and no services have been tested
- legal interest – and an assumption that there is good title for the property
- assumptions on planning and development
- clarification on the treatment of tax and expenses of sale or purchase – normally making no allowance for such costs
- assumptions on contamination – essentially that the property is assumed to be free of contamination and reserving the right to revisit the valuation should contamination be identified

- deleterious materials – an assumption that no deleterious materials have been used in the property
- environmental conditions – relevant in some areas where radon gas may be apparent and also in some areas where historic mining (or perhaps future fracking) may have an impact on ground conditions and stability
- notifiable diseases – particularly relevant in agricultural valuations, an assumption that the valuation is on a disease-free status

9.3 Information gathering

9.3.1 Information gathering takes place before, during and after inspection of the property.

9.3.2 In an ideal world, documentary evidence such as farm plans, tenancy agreements where relevant, planning permissions and environmental agreement papers can be gathered in advance of the inspection. However, often timescales are such that the information will not be available until, during or after the inspection. That does not preclude the valuer beginning to research comparable evidence and any other issues which may have a particular bearing on the property, planning policy or the activity of specific purchasers in the market for example.

9.3.3 It is important that the valuer prepares effectively for the inspection which is at the core of the valuation. Individual valuers will have their own approach to inspecting property and clearly new valuers will wish to adopt an approach which suits them. Whatever approach you adopt, the most important attributes are that it is easily repeatable, can be applied to a range of different properties and generates a comprehensive set of information. Indeed, occasionally, it is easy to forget that latter point, that the purpose of the inspection is to gather information on the specific property to enable you to assess its attraction to the market and the key features which will impact on value. In a thinly supplied market, or particularly where dealing with market uncertainty, as referred to in the previous chapter, it may seem that the relevant facts of a particular property will have a limited impact on value. However, there will be key features of the property, whether positive or negative, which will only be apparent from the inspection and which will have an impact on value. Intriguingly, where valuing for sale, the greatest challenge may be establishing which issues will be most relevant to different prospective purchasers; a neighbouring farmer, for example, is likely to be far more interested in the capacity of the farm buildings than a retiring stockbroker seeking a rural retirement idyll.

9.3.4 Some practices will have specific standardised inspection pro formas for their valuers to collect information, although relatively few rural valuers have moved to the laptop or tablet-based, tick-box approach now used by many specialist residential valuers. Whatever approach is taken to note taking, it will be a distinct advantage if those notes are intelligible to another valuer who may be looking at the valuation and the background notes and information in the future. The advent of lightweight highly portable digital cameras, whether in camera or phone format, has made recording the details of any property much easier; however, whilst a picture may be worth a thousand words, it can be incredibly silent without the relevant notes to back it up. In the same vein, whilst narrative notes are important, understanding can be significantly enhanced by simple annotated diagrams, particularly where farm buildings or the layout of the farmhouse is involved.

9.3.5 There will still be data to be gathered after the inspection to complete the full spectrum of information required to undertake the valuation. Where this is to be included in the report itself, copies of planning permissions or extracts from partnership agreements or tenancy agreements for example, the valuer needs to be careful to ensure that the information reproduced is accurate and properly validated and that permission has been granted for its publication. It is equally important to ensure that information relied upon in constructing the valuation is accurate and, where there are specific facts which have a marked effect on the valuation, that these are referred to and the reliance on them identified in the report. That is equally the case where a degree of informed speculation may be required, for example, in reflecting the possible hope value where valuing a property with the prospect but no certainty of development.

9.3.6 Information gathering does not always end when the report is issued. The most obvious example is where a valuation precedes a sale when it is important and occasionally challenging to record the outcome of the sale and, where very different from the valuation, the reasons why that might be.

9.4 Assessing the value

9.4.1 The mechanics of arising at a valuation will be explored in subsequent chapters. The point which bears repeating here is that the calculations, as well as being logical and checked for accuracy, should be set out so that they will be intelligible, whether to the author or a colleague, in the future.

9.4.2 Where the calculations involved are a relatively simple analysis of comparables used to value a block of accommodation land, this may not require significant explanation beyond setting out the comparables and adjustments made. Where the valuation is more complex (a rural portfolio or a large development site, for example), more comprehensive notes may be required. This is particularly the case where the valuer has relied on bespoke software to calculate values. Valuation programs are constantly changing and there is no guarantee that the version used will be familiar or available to a valuer revisiting the case in the future. Thus, detailed notes may well be required to support the printout from the program to explain why a particular discount rate was chosen, for example, or how the dates of likely receipts were forecast.

9.4.3 One particularly useful piece of information, commonly lacking from valuation files, is where the electronic versions of the report and valuations are kept; some masterly technical valuers may be less assiduous with their filing of hard or soft copy and it is extremely frustrating – and worse, limiting – to inherit a file full of spreadsheet printouts with no clue on which laptop or cloud the information might sit.

9.5 The valuation report

9.5.1 As with instruction letters, most firms will have standard forms of a valuation report which are designed to suit various property types and valuation purposes. As with all forms of a standardised document, it is important that these are kept up to date and fit for purpose. Whilst ultimately this is likely to be the responsibility of a senior member of the practice, it is not uncommon for the task of editing and updating to be delegated to a graduate or newly qualified surveyor on the simple dictum of their having been at college most recently and, therefore, being up to date on these things. Whilst this can be valuable experience for the member of staff concerned, audit bodies and clients will increasingly expect senior members of the practice to be responsible for such quality issues.

9.5.2 In larger firms responsibility for such generic documents may well be centralised in a quality assurance or technical department where people spend their time exclusively on such issues. This has the merit of efficiency and of allowing those involved to keep abreast of developments and hone their specialist skills in that area; but it requires sufficient financial turnover and profit to fund more non-fee earners and is vulnerable to the accusation that they are not practicing and so do not really appreciate what is involved.

9.5.3 Those responsible for practice management may also face the challenge from some valuers that their work and their skill is simply so specialised and so refined that it cannot be reduced to a generic template and that they should be left to do things as they always have. In some cases, there may be some force in the argument where very particular work is required, sometimes involving valuations of business activity as well as property assets. Then the answer may be for another experienced valuer to review the process and the style and content of the report; in many other cases, however, the unique specialism will have far more similarities with other valuations than differences. That may come as a shock to the specialist involved, and how best the practice manager can overcome this challenge is a subject beyond the scope of this publication and on which whole business degree courses are built.

9.5.4 Whatever the approach to standardisation it is important that valuation reports and the workings supporting the valuation are checked by another valuer before the report is issued. This is particularly the case now where many valuers will type their own reports rather than dictate them for typing by their secretary. For some reason, it seems much harder to spot errors in something one has typed oneself than in something typed by a third party. This is relatively straightforward where there is more than one valuer in a practice, although harder in a small practice during absences, but it is particularly difficult for a sole principal. However, most valuers in that situation have been able to reach reciprocal arrangements with other sole principals or valuers in small practices to check each other's valuations.

9.5.5 Different practices and clients will also have different approaches to signing valuation reports, whether the valuation is signed as if from a corporate entity or by an individual(s). How many signatories are required and who they should be will depend on the nature of the property and particularly the level of value involved, but there is an element of good practice in having more than one signatory. There is a certain frisson of excitement (and anxiety) when one becomes sufficiently senior to sign off the larger portfolio valuations, particularly when those are for secured lending. No matter how experienced the author of the valuation is, it is rather more than just a rubber stamp.

9.5.6 Example headings for a valuation report on agricultural property are set out in Appendix B.

9.6 Valuations for court proceedings

9.6.1 Courts presiding over disputes will often issue instructions for the parties in the dispute to appoint an expert witness to provide assistance of a technical nature where the court decides it is necessary. Disputes over valuation are a very common example of the use of expert witnesses and the expert is unique in the courtroom environment in that he is the only person entitled to give evidence based on his opinion using his specialist knowledge and skill. Further, in acting as a Court-appointed expert the valuer has a duty to the Court, not to an individual client.

9.6.2 While the basic approach to undertaking a valuation still applies, there are particular considerations when a valuer is instructed to prepare a valuation as an expert witness. The valuer acting as expert witness has a role in ensuring that the full facts are presented to the parties in such a way that, wherever possible, court proceedings can be avoided. Few courts will now hear such cases without first requiring the parties to attempt mediation and the valuer may play an important role in that process whether directly involved in the mediation or providing technical support and analysis of the potential settlement being discussed.

9.6.3 Where proceedings are involved the valuer's report and opinions will be severely tested and the valuer must be well prepared to present evidence effectively to the Court and be examined on it. It goes without saying that the valuer must be certain that of having the expertise necessary to fulfil such a role before accepting the instruction.

9.6.4 As an expert witness, the valuer will have to adhere to a collection of rules, protocols, practice statements and guidance notes in producing the report and will need to be sufficiently familiar with their content and the rapidly increasing body of relevant case law applicable to expert witnesses to be able to undertake the instruction with confidence.

9.6.5 In a civil court, the expert witness is governed by the Civil Procedure Rules (CPR), most particularly Part 35, issued by the Ministry of Justice. The CPR were brought about following the review of the procedures adopted in pursuing or defending a case through the civil courts by the Woolf Reforms of 1999. These rules have brought considerable clarity to an area which previously was by no means clear. They set out a procedural code with the overriding objective of enabling the court to deal with cases justly, including ensuring a consistent and objective approach by expert witnesses.

9.6.6 Part 35 is a template for both experts and instructing solicitors. It sets out the expert's overriding duty, which is to the court rather than to either party in the case, regardless of who has previously instructed him or may be paying his fees. It also stipulates what an instructing solicitor may or may not ask an expert to do and what to expect of an expert.

9.6.7 The Civil Justice Council's *Protocol for the Instruction of Experts to give Evidence in the Civil Courts* came into effect in 2005 and assists in the interpretation of CPR Part 35 and its associated Practice Direction. Those accepting instructions to act as an expert witness are required not only to be experts in their field but also to be familiar with CPR Part 35, its Practice Direction and the Civil Justice Council Protocol. Experts failing to adhere to this guidance may find themselves in considerable peril facing personal liability (see, for example, *Phillips v Symes*).

9.6.8 In addition to the Court protocols, the valuation expert is also supported by guidance from professional institutions. For members of the RICS, the fourth edition of the Guidance Note, *Surveyors acting as Expert Witnesses*, was published in 2014. It comprises both a practice statement for surveyors and an accompanying "Client Guide" which can be given to clients and explains how to instruct a surveyor as expert and the constraints and limitation on the expert surveyor, whether acting for one party or jointly between the parties.

9.6.9 Members of the CAAV will wish to refer to *Means of Dispute Resolution* (numbered publication 223) for broader guidance in this area, now supplemented by *Rural Arbitration in the United Kingdom*.

9.6.10 Most of the headings which would apply in a valuation report for any other purpose will be relevant to an expert witness report. However, it is even more important that in these cases the report should be easily

intelligible to a layman reading it as part of the proceedings; the report should:
- be addressed to the court and not to the party from whom the expert received his instructions
- distinguish where possible, between matters of plain fact, expert observations and inferences and conclusions drawn from those observations
- consider all matters relevant to the instruction whether prejudicial or not to the instructing client's case
- include:
  - a statement that the valuer understands his duty to the Court and has discharged that duty properly
  - a declaration of the valuer's suitability in terms of qualification expertise and experience to undertake the role of expert
- in terms of the technical content, include:
  - a clear summary of the elements of the case as they are relevant to the valuer's role, which does not require a full exposition of the dispute between the parties
  - a resumé of the expert's involvement in the case
  - the agreed facts and background to the case or assumed background where there is no agreement
  - the issues which arise
  - the expert's conclusions or findings with reasons and in terms of a valuation supporting comparable evidence

9.6.11 Acting as an expert witness can be a challenging but also fulfilling experience, whether in having one's opinion upheld in court, or perhaps even more rewarding in helping parties avoid the costs and anxiety of court proceedings. It is not an exercise to be taken lightly in terms of the emotional energy and sheer time that it will demand, not least sitting around on tenterhooks waiting to give evidence where something does go to court, but for those valuers who act regularly as expert witnesses it is a fascinating and ever-changing area of practice.

9.7 Final thoughts

9.7.1 The Red Book and other similar publications provide guidance to valuers on the processes and procedures to be adopted in undertaking a valuation. Similarly, the valuer's instructions, particularly where coming from a lending institution or the Court, will provide more detailed case-specific instructions. Previous editions of this book and other publications may also provide guidance for valuers, but beyond these various sources of guidance there are perhaps two key questions valuers should ask themselves in each case:
- can the reader, whether client or other advisor, find the answer easily from my report? And
- the sense check – does the value I have provided feel like the right value?

9.7.2 Those tests apply whether the valuer is advising on property strategies, landlord and tenant matters or property investments but above all, questions of value. The need for standardisation and compliance has never been more important for valuers and, properly observed and delivered, these are also critical for clients. However, as we move to a world where algorithms will make it easier for anyone to obtain an approximation of value (and those doubtful of that should consider the impact of websites such as RightMove and Zoopla on the far more segmented residential market), the real utility of a human valuer will be in providing case-specific information and conclusions which add value for the client.

9.7.3 Amongst all the confusion of the process the valuation profession should not lose sight of the fact that the opinion of a competent and respected valuer is, of itself, of value to the client. This is perhaps most relevant in emerging economies and, indeed, there is anecdotal evidence that a valuation from a UK-based professional practice is an important imprimatur in securing finance for some transactions. The combination of practicality, market awareness and intellectual robustness which have been hallmarks of rural valuers over many years will become even more critical to maintaining the value and utility of the profession.

Further reading

Means of Dispute Resolution (CAAV, 2015)
RICS Surveyors Acting as Expert Witnesses, Fourth Edition (RICS, 2014)

10 Valuation of farm property with vacant possession

10.1 Introduction

10.1.1 As previous editions have identified, the valuation of agricultural freeholds and their rental value is a skill only acquired after a number of years of experience, and the young valuer will have an advantage if he has worked on a farm, although that is rather less common than it was in the past and should not be seen as a barrier to new entrants. Again, as earlier editions have pointed out, there are many factors which impact on the productivity and profitability of any farm. However, times change and the extent to which these factors are reflected in value has been significantly affected by three issues: the shortage of farms coming to the market, increased demand from outside of agriculture and technological change.

10.1.2 In 1985, when the first edition of this book was published, records suggest that approximately 300,000 acres of farmland were traded in England; in 2015 that figure was approximately 120,000 acres, and that itself was an increase of some 16% on the levels of previous years (Savills 2016) and, for 2019, estimates suggest that less than 100,000 acres was traded openly. The 1985 figure already showed a marked reduction from the amount of land sold in the immediate post-Second World War period, which was over 650,000 acres. That lack of supply is exacerbated by much of the demand for land being location specific; whilst someone relocating will look outside of their immediate area, farmers, wishing to expand their existing holdings, will wish to purchase land convenient to their home base, not 30 miles away.

10.1.3 Whilst supply has contracted, the demand has been amplified by interest from prospective purchasers coming from outside of the traditional farming and landowning sectors. In principle this is nothing new. The Industrial Revolution produced generations of successful factory owners who chose to invest their wealth in landed estates. However, the level of interest has increased markedly, driven both by city bonuses and the marked increase in house prices which has seen those retiring or downsizing from London and other cities able to afford farms and smallholdings in the countryside. This effect has been most marked where the property for sale includes a dwelling, but once established such purchasers can also be active in the local accommodation land market seeking to add to their holdings. Similarly, whilst the impact has been most marked in the owner-occupier market, latterly there has been interest in investment property from high-net-worth individuals and speculative trusts particularly exploiting the opportunities created by institutional sales.

10.1.4 This dramatic reduction in the supply of land on offer has had two significant impacts; the average price of land sold has increased markedly over the last 40 years, from around £1,700 per acre to over £9,500 per acre with much of that growth in real terms coming since the turn of the century. At the same time, the difference in value between poor quality and good quality land and farms has become far narrower than was previously the case.

10.1.5 Over the same period, advances in farming technology, the development of larger machinery, greater use of fertilisers and chemicals and developments in livestock and crop breeding have given the modern farmer far greater capacity to change the nature of the land farmed and the output therefrom than preceding generations. Where a cereal farmer in the 1950s placed considerable store on the quality and ease of working of the soil, the modern farmer may, with the use of more powerful machinery and new techniques, create a seedbed from worse soils and do so faster and earlier in the season. In an under-supplied market, the capacity, whether actual or perceived, to improve the productivity of poorer land may lead the current generation to bid more strongly for that land than would their predecessors.

10.1.6 We may be on the cusp of a change in attitudes with the development of precision farming, anxieties about the effects of short-term, high impact farming and increasing interest in the health of the soil all pointing towards a more tailored approach, which again may emphasise the importance of land quality. However, the practicality of supply and demand has stood the test of time in explaining a wide variety of markets, and there is no sign of a return to 1980s-levels of land supply, even if the Brexit transition period may encourage some to sell. Beyond all this, the same sentiment exists amongst potential land buyers faced with some land for sale in their location as was the case 40 and perhaps 100 years ago: "we will never get another chance". In that light, whilst the comments in the remainder of this chapter draw heavily on the wisdom of previous editions, ultimately value is a function of the market, and in an under-supplied market logic is not always king.

10.1.7 In practical terms, farm valuations are usually based on Market Value (as defined in successive editions of the Red Book) and based on the comparable method. The application of this approach to various different circumstances is considered in greater detail in the following chapters. This chapter considers some of the specific factors which may influence value, and which should be considered in a valuation.

10.2 Quality of the soil

10.2.1 Setting aside intensive livestock units and even allowing for improved technology and inputs, the quality of the soil will be a fundamental influence on the productivity and profitability of agricultural land. With the very varied geology and topography of the United Kingdom, soils can be differ substantially within short distances. They range from the fertile silt fens, Chichester Plain, Lancashire mosses and the Black Isle to the acidic Lizard, the Culm Measures or peatbog with greensand, brickearth, Cotswolds brash and many others in between. The rich, deep soils of much of southern Herefordshire can be contrasted with thin, stony soils or free draining soils over gravel. Whilst technological advances have narrowed the margin between good and bad soils, although not the cost of working, the characteristics of good soil remain much as they have always been, and just as elusive. A good quality soil will be:
- deep
- free draining (but not so that it dries too quickly)
- on level land
- of reasonable pH and taking all that together able to grow a range of crops, particularly in the very best cases, high value vegetables and salads.

10.2.2 In the past, versatility, the capacity to grow both good arable and grass crops, would have been seen as key. That has become rather less so, and the nature of soils as well as topography and climate have led to an increasing polarisation of farm types in England, at least into a pastoral west and arable east. Whether the ambitions of DEFRA's *Health and Harmony* and wider policy direction will see a return to more mixed farming remains to be seen. On first impressions, the combination of weather and climate change, increasing specialisation leading to fewer farm businesses with the necessary skill set and the specialist investment needed by most farm systems suggests that is unlikely.

10.2.3 Land with clay soils, with finer particles and smaller air spaces, will be far from free draining and prone to water logging. That will make it difficult and risky to grow and, more particularly, lift later-harvested crops such as potatoes and sugar beets and lead to shorter grazing seasons in a wet year. That said, for the livestock farmer they also offer the prospect of good grass growth, unless and until a particularly dry year sees clay dry out very quickly. In the heaviest ground, autumn sown crops were a difficulty, but technological developments and earlier sowing have overcome that, albeit at the cost of a perennial weed problem in some areas. The rise of blackgrass has seen some farmers in badly affected areas delay planting until a more traditional date to allow the weed seeds to germinate and then be destroyed in establishing the new crop.

10.2.4 As might be expected, sandy soils, with coarser particles and larger airspaces, are the mirror to heavy clays. They dry out more quickly and, as also on thin soils, crops may "burn" due to the lack of moisture. Whilst generally easier to work, and available earlier in the spring, they are more hard-wearing on equipment with constant abrasion, although that happens with stony ground as well.

10.2.5 In an average year, if there is such a thing, sandy soils will be available to work later in the autumn and earlier in the spring, an advantage in both arable and pasture areas. They tend to support more autumn growth, although it is unlikely that overall vegetation growth will be as great as on heavier soils. They are likely to need irrigation to grow higher value crops and, with climate change, access to irrigation and the appropriate abstraction or storage licence may be an increasingly important element of farm value. Sandy soils may also blow and windbreaks will be required to prevent erosion. Recent innovations have included the planting of rows of apple trees as windbreaks with the added value of an annual cash crop over earlier conifer plantings.

10.2.6 Sandy soils, often more acidic, tend to be hungry, needing both more manure and lime, phosphates and potash which can leach, particularly on sloping ground. Soils can also tend to wash out with heavy rain, particularly some of the more dramatic weather events which seem to be a more regular occurrence. This has become an increasing problem as cultivation has moved higher up the hill and on steeper ground than was previously the case, again driven by the greater horsepower available in modern tractors and the conversion of pasture land to growing maize as part of a dairy regime. Soil runoff is a problem for farmers, losing their productive topsoil, and for householders and water authorities "downstream" when this leads to silting of watercourses and flooding.

10.2.7 Stony soils are also hungry and tend to burn, but in contrast to sandy soils there is little that additional organic or artificial manures will do to improve land where there is very little soil above the rock. Whilst generally only suitable for grazing, in the last four decades many have been ploughed sometimes for cereal growth and sometimes to improve pastures. Whilst stony soils are a common feature of uplands, where stone walls are a clear indicator of stone, these soils are not confined to the hills. That was demonstrated in a salutary lesson for one of the authors, arriving on a farm for what was hoped to be a final settlement of the Michaelmas rent review, to be greeted from a distance by the sound of some fairly outspoken commentary on the quality of the farm and the soil, all of which was attributed to the landlord rather than any higher authority. On reaching the tenant, the author was presented with two plough shares, in bits, broken by the landlord's stony ground. Discretion was the better part of valour, a new appointment was fixed as the tenant was "obviously busy" and the rent was subsequently settled.

10.2.8 The ideal soil is likely to be a deep loam which can grow a range of crops from grass to high value field scale vegetables. However, often these are alluvial loams in river valleys which can be at risk of flooding. Much of the increasing attention to the condition of soils is focused on the importance of a good soil organic matter component, able to assist both water retention and drainage, warmth and earlier germination and active soil ecosystems from earthworms to microorganisms as well as offering a significant means to sequester carbon.

10.2.9 Scientific analysis can provide a full picture of soil capacity and structure. Soil testing for a widening range of factors is increasingly undertaken and is, indeed, expected under DEFRA's Farming Rules for Water applying in England. Alongside that, information available from satellite, drone and other data capture and used to direct spray and fertiliser applications provides enhanced information for landowner, farmer, potential purchaser and agent. However, adoption rates of precision farming are still relatively low and, even when such data are more readily available, it will still be important for the agricultural valuer to be able to understand the nature of the soils from physical evidence available.

10.2.10 One of the best unscientific indicators of soil quality is the colour of the soils, with the red soils of parts of the south-west generally indicative of strong flexible soils, very dark soils suggesting peat capable of producing strong vegetative growth but suffering from all the problems of wet heavy soils, sandy soils with a colour to fit their name, as have chalk soils, and clay soils ranging from the browns of clay loam through to the greys of heavy clay which double your bodyweight with the soil attached to your boots if you venture to walk across them in a damp winter.

10.2.11 There are other natural clues. Vigorous hedge and tree growth suggest deep good quality soil and strong hedgerow trees; in the past, elm trees were a particular indication of that. The presence of ditches suggests the need for drainage, so clay rather than sand, chalk or limestone soils. In the wetter low-lying areas of Sedgemoor and the Fens, the management of ditches is an important element in maintaining a drained workable soil and, where there are livestock, acting as "wet hedges" to keep stock watered and confined to the right ground.

10.2.12 One of the challenges of assessing soil quality is the degree of variation both between and even within fields on a farm. Thus, an otherwise loam field might have clay patches which lie wet and need draining, which in turn may not be cost effective. The same problem may arise with springs, causing wet patches, which are found in all types of soils; again drainage will be the only effective solution. Former field boundaries marked on old maps may be a good indicator of where soils change or where springs arise. Indeed, in an era where satellite imaging and GPS seem to have replaced the paper map there is still much to be learnt from the previous generations of maps, whether hard copy or digitised. Thus, the early OS county series, tithe maps and private farm and estate plans can point the way to old farm pits with contamination risks, "osier beds" suggesting poor drainage and former buildings.

10.2.13 In a similar vein, soils overlying clay subsoils and strata will be waterlogged unless drained. Drainage is a challenge where areas are very low lying and thus there is little fall available to accommodate the drainage system. On the other hand, areas like Sedgemoor and particularly the Fens show what can be done by drainage engineering on what would now be known as a "landscape scale". In the latter case, drainage has converted low-lying land, in some cases below sea level, into some of the most productive arable land in the country. The Fens may seem a very homogenous landscape to the outsider, but in practice there are both organic peat fens, recognisable to a West Countryman used to Sedgemoor, but where the peat is now much diminished through oxidisation and wind blow, and silt fens that can support double cropping, a distinction important to both growing seasons and variety and productivity of cropping. They illustrate how a combination of natural and engineered factors come together to create a particular farming environment; the soils are inherently productive, but in this area of comparatively low rainfall, the presence of a fairly high water table, which might be a hindrance in the west, is an asset. That said, as observed earlier, in a period of climate change irrigation is becoming increasingly important to support continued production of high value crops.

10.2.14 The valuer should, as an introduction, study the Agricultural Land Classification maps. In England and Wales these maps grade the land in five categories, reflecting varying combinations of workability and fertility, with Grades I and II the best, Grade III, now split into two parts to cope with the predominance of land in this classification, ranging from quite good arable through to grassland. Grade IV is limited grassland and Grade V is usually rough grazing.

10.2.15 The maps are produced on a small scale dealing with a fairly broad brush on limited sampling and serve to illustrate general characteristics of land in the vicinity rather than the specific nature and quality of an individual farm or field. Whilst that provides a useful guide it is worth noting that many farmers will report multiple soil types in individual fields; consequently, some small areas of very good, or indeed very bad, land are not indicated as they should be.

10.2.16 In Scotland, the land classification produced by Macaulay Institute in the 1960s offers a more detailed classification of land based on land use capability. The system divides land into seven classes, with again Class 1 as the most productive arable land and Class 7 as rough grazing. Four of the classes are further subdivided, to produce a very detailed system with 13 different classifications.

10.2.17 There may be occasions where other land classification information may be useful and more detailed maps from the UK Soil Observatory are available. Site-specific information is increasingly available through the use of precision farming techniques. Indeed, the valuer of the future may have an embarrassment of data available, and that may tempt some into some very detailed quantitative analysis. However, the challenge will remain the same, working out what the market **will** bid, not what it **should** bid.

10.3 The farmstead

10.3.1 Dwellings – One of the broader changes over the 35 years since the publication of the first edition has been the attention to the value of rural dwellings. Houses were often seen as simply part of the fixed equipment of the farm, not a separate source of value and so, when surplus and in poor repair, even substantial historic houses were demolished within living memory. The limitations of the Rent Acts saw let cottages sold as soon as they fell vacant. That has all changed. A key factor in that has been the increasing influence of "non-farming" buyers on the farmland market. This is particularly the case in areas attractive to commuters but is by no means confined to those areas. Thus, the existence of a good quality residence has an important bearing on farm values, particularly with the general increase in residential values. That used to be reflected in a tendency to lot farmhouses and some immediate land for the residential market. However, the increase in the number of purchasers capable of buying a whole farm has seen something of a reversal of that trend and, whilst this may still be appropriate for some properties, it is by far less prevalent than was previously the case.

10.3.2 Whether the post-Brexit and post-Covid market sees a combination of more property on offer and a "dash for the countryside" which might, in turn, see more lotting remains to be seen. Whatever the case, the imperative for valuers will still be to identify which approach best suits the market and optimises the receipt, which will differ for each individual property and can vary over time as conditions change.

10.3.3 A general trend towards lower labour requirements has seen less of a demand for farm cottages to accommodate farm labour, whilst employees' families may prefer to live in the local village with better access to services. However, additional dwellings on a farm will have both utility and value. If not used to house labour or a second generation, they can provide income through letting whether long-term or in the right area for holiday use. Whilst the condition of farm cottages has not always been a paramount consideration when valuing a complete farm, there are increased obligations to provide habitable homes. Safety and energy efficiency regulations mean that, whilst there may be a lesser direct impact on value, a cottage in good condition, with a good EPC rating and meeting Minimum Energy Efficiency Requirements (MEES) rating will save on refurbishment costs, if nothing else. That pressure on dwellings seems only likely to grow as governments pursue measures in response to climate change.

10.3.4 Farm buildings – It would seem obvious that a good set of well-designed buildings in a good location will add value to a farm. That will often be the case, but particularly where a farm is more suited to the residential rather than the commercial farm market buildings may be a hindrance on value as much as a help.

10.3.5 Choices here will differ. Whilst the commercial farmer will look to a set of modern buildings which support the principal agricultural activities on the farm, of which more will be mentioned later, the lifestyle buyer may be more attracted by a courtyard of traditional buildings, whether those enhance the environment for the house or provide opportunities to house their own business or provide lettings for holiday or employment use. They are more likely to be put off by an extensive set of modern farm buildings if they intrude too much on the enjoyment of the house.

10.3.6 In either case, the limited supply of farms coming to the market means that the full value of the buildings may not be reflected in the relative values of well and poorly equipped farms. That has been the case for many years and indeed in many cases now it is the potential which buildings offer for alternative use which may be of greater relevance to their value or, perhaps more properly, their contribution to the value of the holding. Thus, obsolete buildings which might have been demolished a generation ago are now seen as an opportunity for additional income from what in the past would have been seen as a redundant asset.

10.3.7 Modern buildings – That is not universally the case; specialist buildings such as poultry and pig units can be hard to convert to alternative uses and certainly do not appeal to every type of buyer. They may have utility and value whilst they are fit for purpose and there is a profitable business to be run, perhaps ancillary to the main farm activity. However, the economics of intensive livestock production are volatile, with very tight margins per unit of production and thus require very considerable expertise, management and standards of husbandry. Such buildings tend to become obsolete in a relatively short timescale, particularly with changing welfare and food safety standards and thus the prospect is of their having little value after a relatively short economic life. If nothing else that demands care when planning such buildings to ensure that they are sited in a way which will have the least impact on residential properties.

10.3.8 Modern dairy buildings, especially if well designed with adequate housing for cows, feed and slurry, are an asset on a dairy farm. An effective parlour and dairy setup is important with adequate space for a fast parlour, sufficient bulk milk storage and a good circulation system for cows. However, in value terms, these are all "nice to have" rather than "need to have". Indeed, fashions change routinely in milking parlours from the abreast parlours still in operation right through to robots; and now with an increasing trend towards the latter amongst dairy herds of all sizes the dynamics of designing a dairy unit will change. Again, given the shortage of farms for sale the absence of specialist buildings may not be a bar to a dairy farmer purchasing a farm, but the investment required to create a modern dairy unit can be very significant and hence well-equipped dairy units can trade at a premium.

10.3.9 It used to be said that the most popular and useful farm buildings are portal-framed general purpose buildings which have a variety of uses; indeed that was reflected by some landlords and their agents who would favour such buildings as tenant's improvements, or even fund them, but would baulk at granting the same status to "specialist buildings", such as grain stores. However, the requirements of grain and other commodity purchasers mean that the standard of space required to store commodities is higher than was previously the case and, consequently, buildings are becoming more specialist in their nature. The all-year-round building that moved from machinery storage to potato storage to over-wintering livestock is increasingly a thing of the past, not least as the number of mixed holdings has reduced.

10.3.10 This trend towards specialist buildings makes grain storage important on arable farms. Good modern grain stores, sufficiently vermin and bird-proof to meet regulations, are an asset; fixed drying facilities may add value, particularly in specialist arable areas with multiple cropping. Elsewhere, on floor drying with mobile equipment may suffice. Cereal farmers may look to co-operative storage as a more cost effective, certainly less capital demanding, alternative with the potential loss in flexibility balanced by access to a range of marketing and agronomic services. In high value cropping areas, the availability of climate-controlled storage and, for those growing for the retail trade, processing and packaging space is an important element in maximising the value of the crop.

10.3.11 Traditional and obsolete buildings – If the farm buildings on any given holding are old, traditional buildings with no or limited modern buildings, this might, one would imagine, result in the capital value of the farm being a good deal less than a farm suitably equipped with modern buildings. However, again, the shortage of farms has reduced this impact. At the same time, traditional buildings can be converted into dwellings, holiday lets or as employment space providing useful supplemental income; indeed letting surplus buildings is the most common form of diversification on farms, although viewed through a forensic lens the return on capital on such investments may not be that attractive.

10.3.12 As with all buildings, location, particularly the proximity to and therefore interference with the farmhouse and the working farm buildings are an important consideration. The relaxation in planning requirements, notably the introduction of Class Q in England, is intended to make it easier to secure planning permission for residential use for redundant agricultural buildings, albeit with some strange architectural outcomes in some cases.

10.3.13 Many farms equipped with modern buildings have been badly designed, or more likely not designed at all with buildings added piecemeal and just filling the next available space. As farming developments continue, relatively modern buildings are now outdated. Size is often the problem – profits aside almost everything on farms has got bigger in recent years.

Buildings may now be too small, whether too low for access with larger machinery or with insufficient space to accommodate larger dairy cows, or simply too close together to enable access for larger articulated lorries delivering and collecting commodities. Systems may have gone out of fashion; slatted floor cattle accommodation, popular 30 years ago, is less so now and is expensive to convert to alternative uses. Slurry systems squeezed into the then available space may now be inadequate in the face of NVZ rules or other regulations and expansion may force relocation to another site.

10.3.14 In other circumstances, technological advances may simply have overtaken what was a perfectly adequate system. Thus, a dairy farmer moving to robot milking may find that rather more thought and investment is required, beyond the substantial cost of the new equipment, than simply putting the robot units where the parlour used to be.

10.3.15 This may mean that a purchaser, having acquired a new farm, will then look to invest in upgrading or replacing inadequate buildings. Whilst they may hope that some of this investment will be funded from a reduced purchase price due to the inadequacy of the buildings, they may be sorely disappointed as, again, the shortage of supply will probably outweigh the shortcomings of the farm in influencing the value of the holding.

10.3.16 Farm buildings and value – In practice, farm buildings are amongst the most problematic of rural assets to value. Valuers may be fortunate in finding suitable comparables for an entire farm they are valuing, but at some stage it is likely that they will wish to consider the value of the individual parts of the holding. Whilst they are likely to be able to find comparables for the farmhouse and an appropriate area of land and the balance of the land comprising the farm, there are very few comparables if any for the farm buildings for their existing use; any separate sales of farm buildings are more likely to be for alternative development.

10.3.17 Where such alternative development is relevant, then the potential for increased income may be used to calculate a latent value for traditional buildings, although it is perhaps worth considering whether buildings really would be developed just because current planning regimes will allow that change. In contemplating letting a building 80 metres from the house to a local car mechanic, income may outweigh inconvenience; if the building is only 8 metres away from the house, it probably will not.

10.3.18 In the face of this lack of evidence, some valuers will rely on their experience to apportion a value to the farm buildings, whilst others will adopt a more scientific approach, discounting the replacement cost of the buildings involved. Others, including the author, often take the rather broader-brush approach of simply reflecting the value of the buildings in the value per acre they apply to the land, although that will not satisfy those valuations which require a value to be attributed to the buildings, notably those required for loan security purposes. In the absence of any other information, perhaps the best evidence can be derived from comparing the value of a block of land sold with buildings and a similar block sold without. All methods have their pitfalls but also the distinct advantage that the apportionment of value is unlikely ever to be tested in the market.

10.4 Situation

10.4.1 The location of a farm has a significant bearing on values – whether the regional location, with farms with reasonable access to London or the major conurbations attracting a premium, or more immediate local concerns about accessibility. Potential purchasers will have different priorities, both between farming and non-farming buyers and within those categories, but many are likely to identify a preferred region first, although for some farming families looking to expand proximity to the current location will be critical. Once that selection has been made the following issues are likely to be of interest, although, again, given the paucity of supply, they may not have the significance which would have been attributed to them in previous editions.

10.4.2 **Position in relation to towns and infrastructure** – A holding convenient for towns and their facilities may attract a premium compared to a holding that is remote, although with improved accessibility to many areas, and the greater capacity for shopping and other commerce online, this is perhaps less of a consideration than it might once have been.

10.4.3 Similarly, proximity to livestock markets, which used to be a consideration, has, with the marked decline in the number of livestock markets, perhaps counterintuitively, become less of an issue.

10.4.4 Where higher value cropping is in mind, then processing infrastructure can matter. The closure of British Sugar's factories in the West Midlands and York substantially removed sugar beet as an option for most in those areas. The concentration of packing and processing facilities in areas such as the Fens, the Vale of Evesham, parts of Kent and Fife reinforce the farming economics that support them just as dairy processors and milk fields go together. There are similar concentrations of poultry farming in the West Midlands and Northern Ireland.

10.4.5 Farms right on the urban fringe are traditionally viewed as being more vulnerable to trespass, fly-tipping, theft and vandalism and livestock worrying. The generally increased number of footpaths on holdings near towns may exacerbate this problem whether through ignorance or carelessness on the parts of the users or malicious damage. Latterly, there have been some concerns over the risk to people from cattle grazing and the associated public liability issues for farmers.

10.4.6 That said, farms close to towns may offer far greater opportunities for diversification, whether through letting buildings or farm shops for example. They may also offer opportunities for more substantial and lucrative development, whether directly through urban growth or, looking to the near future, accommodating biodiversity offsetting or other conditions from developments elsewhere. Whilst the potential for direct development would feature in any analysis of local planning policies, these latter opportunities may be available to farms where larger scale development would be precluded, for example in metropolitan green belts.

10.4.7 Position in relation to roads – Good access to a reasonable road and in particular access for larger lorries delivering bulk purchases of feed and fertiliser is an advantage often not fully appreciated until one has to unload from an articulated lorry to a tractor and trailer to get under the railway bridge between the farm drive and the main road. Conversely, where a farm is close to a major road, fast-moving traffic is a major hazard to the movement of people, tractors, equipment and livestock. The growth in traffic means that relatively few dairy herds will graze milking cows on pastures separated from the buildings by a road unless that is the smallest of country lanes; demands on labour and the nervous system are too costly for most to attempt otherwise.

10.4.8 The same is true to a lesser extent moving other livestock and even with heavy machinery movements through silage making and harvest. Consequently, a farm in a ring fence, or at least with a substantial part within a ring fence, will be seen as more attractive.

10.4.9 A farm with the homestead close to the road will not be as quiet and may be more vulnerable to trespass or theft. It may be an advantage to be set back along a private drive, particularly where residential value is at a premium.

10.4.10 Where rural roads are very narrow, the network poor or travel limited by coasts and rivers, this can make for delays to accommodate passing traffic. In most cases this is not really an impact on the utility or value of a farm; however, anyone who has tried to farm, or indeed practice as an agricultural valuer, in a popular holiday area will know which roads to avoid during the summer holidays, and on occasion the weight of traffic can be a real challenge to farm journeys. That said, away from those areas a council road is an advantage providing access to fields at someone else's expense.

10.4.11 Farms with their own long drive to the council road have the cost of maintenance, although where the property has a high residential and lifestyle value this sense of scale and privacy may add to the attraction for purchasers. Indeed, the increasing replacement of five-bar gates with stone gate pillars and wrought iron fencing, redolent of a stately residence in a Wodehouse novel, are perhaps a subconscious illustration of the interest we take in the entrance to our properties.

10.4.12 In this age of diversified farm incomes good road access is important for farms with caravan, camping or glamping sites, fishing lakes, holiday cottages, farm business parks or the plethora of other farm-based businesses.

10.4.13 In another marked departure from previous editions of this book, nowadays virtual as well as physical access is important with the availability and reliability of mobile phone and internet services being important to farmers, visitors and farm-based businesses alike. That has been emphasised by the recent Covid-19 lockdown with good digital services enabling people to work effectively and communicate from home. However, even before that, connectivity was a significant issue, so that in areas with poor digital communications it is becoming increasingly difficult to let employment space.

10.4.14 In some areas, particularly lowland districts where direct access between adjoining fields may be interrupted by major drainage channels, the immediate location may have a marked impact on value. Thus, some sugar beet ground on the Fen accessible over a modern large highway bridge over the main local drain is likely to be more valuable than similar land where the only access is over a small eighteenth-century stone cart bridge unsuitable for large harvesters and regular heavily loaded trailers.

10.4.15 Farm approaches – The immediate approach to a farm can have an impact on value, both in terms of utility and aesthetics. Thus, in practical terms, a farm with a good level access which can be used in all but the very worst of weathers has an advantage over one at the top of a steep hill, or indeed one in an area vulnerable to flooding.

10.4.16 Given that we are considering farm property, aesthetics may not seem as obvious, but one of the most attractive coastal farms which the author visited was accessed through a very dilapidated and economically disadvantaged local authority housing estate. That effectively took that farm out of the lifestyle bracket and left it as a holding for commercial agriculture which, given its relatively small scale, was a significant bar on value. It is always worth considering the latter part of the journey to a farm, both as it is when inspected and, in an area where significant development is proposed, as it might be in the future as this can have a significant impact on attractiveness to a lifestyle buyer and hence value.

10.4.17 Farms on flood plains – The potential for flooding affects the utility of land as well as the access to the holding mentioned earlier. Growing and working seasons are shorter and that, in turn, limits the crops which can be grown. In a time of what appear to be more dramatic climatic events, even in the relatively benign climate of the UK, this has become more of a problem and even areas with heavily engineered drainage regimes managed by local drainage boards can be severely affected.

10.4.18 If a farmstead is itself on a flood plain, and is susceptible to flooding, this can be a serious problem and many, not just valuers, will recall the difficulties experienced by farmers during major floods in recent years.

10.4.19 The increased focus on water issues generally, whether the need to protect water quality, difficulties of water conservation and dwindling aquifers in some areas and the difficulties of flooding elsewhere are likely to focus interest on farms either in flood plains or upstream of vulnerable conurbations where there may be some scope to use farm land for flood attenuation. At present there are relatively few schemes encouraging water management and they are predominantly voluntary in nature. However, there is considerable prospect of them expanding, particularly as part of an emerging post-Brexit farm support regime. Whether voluntary or involving an element of coercion water management is likely to have an impact on the value of farms in the future.

10.4.20 Elevation and aspect – Whilst localised frost pockets, rain shadows and the like may create microclimates, farms at lower elevations are generally warmer, have deeper soil, can be accessed earlier, have longer growing seasons and are more productive than those higher up the hill. It may not be a scientific assessment, but living at roughly 150 metres above sea level in a narrow valley, plant germination, flowering season, and so forth seem to be between ten days and a fortnight behind land in the Vale of Taunton, five miles away.

10.4.21 Elevation is not the only issue however; the low lying moors of both Sedgemoor and East Anglia, with limited tree cover, particularly in the latter case, can be bitterly cold and exposed in the winter. In the west, the impact of the Gulf Stream warms temperatures, even as far north as the western highlands of Scotland, and that contributes to the capacity of land on the western coast, particularly in Cornwall and South Wales, to grow early potatoes in their frost-free climate while the more northerly climate of Aberdeenshire has made it a centre for quality seed potato production. Rainfall tends to be greatest in the hills of the west and lowest on the eastern coasts; while parts of the East Anglian coast can average below 12 inches a year, the Lake District, Snowdonia and the western Highlands, are among the wettest places in Europe, with that rainfall then a flood risk for downstream areas.

10.4.22 The impact of climate change has already seen some crops and activities extending their spread either higher up the hill or further north in the UK with the entry of premium French champagne houses into the farmland market in the south-east of England being a high-profile example and the northward movement in the limit on growing maize a more general one. Whatever the scientific arguments are most farmers will comment on the increasing lack of distinction between the seasons and more dramatic weather events, not least more concentrated storms and high volumes of rainfall, which are harder to manage at higher elevations and particularly on the sloping ground of the hillier counties.

10.4.23 The aspect of land is also important; land facing north and east is generally colder and thus later than land facing south and west which has more sun. That said, in very hot, dry summers the tables are turned with the colder land being far less likely to burn as badly. Land adjacent to woodlands can be shaded, limiting growth of both grass and arable crops. Indeed, it is rare for anything much to grow close to woodlands and under trees, except for stinging nettles. That said, wider field margins have reduced this problem a little as well as the difficulty of passing under overhanging branches. Dairy farming can be difficult in heavily wooded areas with more flies and the prospect of more cases of mastitis.

10.4.24 Steep or "banky" land generally is harder to farm, needing more powerful machinery where arable cultivation is involved or indeed for reseeding, fertilising or topping grass. Banks can be dangerous, particularly those hosting the unseen badger sett, which can upset a tractor. The steepest land is not suited to arable production and the more recent habit of growing maize or other arable crops on steep ground has led to problems of soil washing and erosion referred to earlier; this type of land is generally most suited for grazing purposes, or perhaps, looking ahead, tree planting.

10.4.25 Some valuers will operate in relatively tight geographical confines and become well versed in the merits of particular parishes in their patch and the peculiarities of different lowland moors or hill ranges – Dentdale having a "sunny side" and a "money side" on opposing banks of the Dee. Others may operate in more varied areas – West Somerset, for example, ranging from highly productive arable land growing potatoes and vegetables to hill farms of the top of Exmoor. Others still may operate over a wide area or indeed the whole of the country and encounter different farm types.

10.4.26 Whilst it applies to the whole of this chapter it is perhaps particularly relevant here to observe that the responsible valuer should be particularly careful to research an area before embarking on any professional work. Localised issues can have particular impacts on the marketability and value of a farm or land; some of that will be accessible from researching the internet, but some will be discovered through practical experience and careful inspection. Once again, the way one treats others will come to the fore; a local valuer, whether being asked for advice or in negotiation, is far more likely to respond positively to a courteous and respectful enquiry than to a brusque and hurried approach. That said, agricultural valuers have been rather more guarded with their comparable transactions than our urban counterparts, although that seems to be changing with greater publication of information through the internet.

10.4.27 Fencing, gates, ditches and drains – These elements of farm infrastructure are important, but again, in an under-supplied market with strong demand, their presence and quality whilst still important in practical management terms has come to have relatively little impact on value even on stock farms. Indeed, for most valuers these issues will be far more relevant in the context of tenant right than questions of capital value; however, questions of boundaries and drainage have a habit of escalating into disputes when they can have an impact on the cost of occupation, if not value.

10.4.28 Hedges provide a boundary to keep stock in and shelter from weather, both good and bad. That may lead to higher hedges on more exposed holdings, and the common assumption of cutting and laying hedges every seven or eight years may not apply to every holding. In some cases, occasional trees are left to mature to provide additional shade for stock, but they will also shade crops underneath. Unsurprisingly, hedges grow well on better soils and normally take approximately 10–12 years from planting to first laying. Traditions vary around different parts of the country with different arrangements of hedge, grass bank and stone bank. New hedges were traditionally of quickthorn, planted in two staggered rows, but there is an increasing tendency nowadays to include some other varieties to add to the biodiversity of the hedge and improve its function as a wildlife corridor.

10.4.29 On arable farms hedges are relatively ignored other than trimming. Whilst not needed as a stock barrier they can help with preventing windblow on very light soils.

10.4.30 Fences may replace hedges or guard poorer hedges on livestock farms, and generally the most suitable economic fence on a farm is medium/heavy gauge sheep netting with one or two strands of barbed wire above, on tanalised posts at 2-metre spacing to a height of 1.22 metres. A well erected and maintained fence should last for 15 to 20 years, although the effectiveness of tanalising seems to be by no means universal with some posts failing in very short timescales.

10.4.31 Similarly, gates are important for keeping stock in, and in some cases for keeping unwanted visitors out, although in the latter case something more robust than a five-bar gate may be required to maintain security and prevent fly-tipping or worse. Gateways, in common with everything else, need to be bigger to accommodate bigger machinery and be robust enough to withstand cattle. Steel gates are generally best unless you are near the coast where wooden gates will not rust from the salty atmosphere.

10.4.32 Where there are ditches and drains on a farm, they need to be maintained on an annual basis to keep them running and provide drainage and a wet fence. Open watercourses, natural or otherwise, should be guarded with barbed wire fencing to stop cattle treading the banks and pasture pumps can be used to facilitate drinking rather than the traditional drinking spot.

10.4.33 Farming standards – One would expect farming standards still to have an effect on farm values, but this is less and less the case. At the extremes, a very poorly farmed farm may see the fixed equipment neglected, the land in poor heart and weed-infested and boundaries neglected and overgrown. It will take more than one season to complete the physical work required and longer to restore soil health, both at considerable cost. However, it is unlikely that this farm will trade at a marked discount and at the other extreme even more unlikely in the current market for a good farm to trade at a premium.

10.4.34 The question of animal or crop health, particularly the impact of livestock disease, has routinely been discussed by valuers with no real evidence that a poor disease record has had much of an impact on value. Thus, at the height of the BSE problem, there seemed little evidence that land values, rents or grass keep prices were reduced by land being stocked by BSE cattle. Similarly, there is no real evidence of TB status, whether of individual farm or areas, impacting land values, not least perhaps because the disease tends to be localised, so neighbouring purchasers will be used to the problem and, as set out earlier, will only be looking to purchase in their home area.

10.4.35 Services – Given the rural location of most farms, it is unlikely that the full range of main services will be available. The absence of services, particularly mains electricity, may have an impact on value, particularly given the cost of connecting to a main supply. However, this is unlikely to exceed the cost of connection and in some cases will be less than that. Where there is mains electricity, which is the case now on almost all farms, a three-phase supply provides far greater flexibility in the range of grain driers and other equipment, which can be run cost-effectively on the farm.

10.4.36 Access to adequate supplies of potable water is important for dairy and intensive livestock systems in particular but also for dwellings on all farms. Ideally, water for stock will not be from the main as the costs of water to a dairy farm can be very substantial. Wherever the water supply originates, use should be made of recycled grey water for tasks other than watering stock where possible. On arable farms the ability to abstract water from ponds, reservoirs or rivers had considerable value in increasing the output of higher value crops. Increasing concerns over reduced water tables and flows in aquifers has brought abstraction from natural sources into sharper focus in recent years.

10.4.37 Historically, private water supplies have been cheaper than those provided by the privatised water companies; however, this is less likely to be the case as the owners of water supplies, faced with increasing costs of maintaining infrastructure and meeting regulations, begin to mirror water company charges.

10.4.38 Where a property relies on a third-party water supply, the valuer should ask for details of the agreement, if one exists, most particularly what responsibilities there are for the provider to maintain the supply. Very often these will be limited, with the supply capable of being disconnected at will. Given the costs of main connection, this might be expected to have an impact on value, but purchasers often take a philosophical view on the basis that "the supply has lasted this long so why should it be cut off". That is fine as it is not up to the professional advisors to take the risk; our collective job is to point it out.

10.4.39 Farm effluent is most commonly dealt with through private systems, and charges where effluent is discharged into main sewers from farms located on the outskirts of towns can be very high. The Environment Agency, in England, will charge for discharge into watercourses for those farms that have the licence to do so. Many no longer use that approach, but where the licence is available it may be worth maintaining it for flexibility for the future. The effluent will still need to be treated on the farm, and the capital costs of providing adequate storage capacity is a substantial burden when existing systems need to be replaced.

10.4.40 In some cases, access to substantial amounts of slurry may create the opportunity for viable development of an anaerobic digester (AD) to convert slurry and other feedstock into energy. However, the investment required is substantial, and for some the demands of managing a farm to keep a chemical process going as well as an intensive livestock enterprise with which they are often associated may be a step too far. Indeed, on occasion one is tempted to speculate whether the AD plant is being run to service the farm or vice versa given the investment and management commitment required.

10.4.41 As noted earlier, the additional service which has become increasingly important since the last edition of this book is access to broadband and mobile phone reception. Network coverage for the latter has increased markedly in recent years to the extent that reception in some rural areas is better than in some city centres. However, the rural broadband debate has become something of a cause célèbre with so much ordering and interaction with government agencies, particularly the government agencies handling the Basic Payment and other schemes, taking place online. It is difficult to say that this feeds through directly into farm values, but there is no doubt that it does when it comes to rural employment space. One of the first questions from a prospective tenant looking at office space to let on a farm business park is over the quality of the broadband, and there are plenty of prize-winning conversions lying vacant because of poor broadband connectivity.

10.4.42 Sporting – The valuation of sporting is explored further in Chapter 13. Arguably fewer farmers and lifestyle buyers are as interested in country sports as would have been the case a generation ago. However, many still are and the potential for good sporting on a farm will increase the interest from those keen on country pursuits whether shooting, fishing or horse riding.

10.4.43 Farm shoots tend to be relatively low key, but on some holdings the potential for shooting, whether within the holding or as part of a wider syndicate, will offer the potential either for some income or for the ability to invite guns onto the farm in the hope of reciprocal invitations in return. Good cover for game, whether from woodland or game crops, is important, but topography is critical for the best shoots.

10.4.44 Access to rivers or streams holding game fish in particular, or lakes which can be stocked for fishing, can be an added attraction and an additional source of income and sometimes value. This may work alongside other farm activities, with irrigation lagoons being stocked for coarse fishing for example. Farms which provide good opportunities for extensive off-road horse riding will also be attractive, offering both the prospect of personal leisure pursuits and in some cases additional income from livery activity or paid cross country courses.

10.4.45 Other sporting pursuits may also offer opportunities for farms to generate income and add value; the potential value of walkers and their spending power is well known in National Parks and many favour farm-based accommodation. Heretical though it may seem to some, mountain biking is a major economic activity particularly around such centres as those run in Natural Resources Wales forests. Neighbouring farms may benefit either from extending access or more likely from providing affordable holiday accommodation on camp sites or in camping barns.

10.4.46 Returning to traditional country sports, farms are sometimes sold with the shooting and other sporting rights reserved. This is a significant disadvantage at two levels: both where they are reserved to others but not used, thus preventing the landowner from using them, or where they are exercised by third parties who are bound to disturb the landowner's enjoyment whilst exercising those rights.

10.4.47 Development value and cottages – The promotion and valuation of development land is explored further in Chapter 15. Development potential is most likely to be found on land near to towns or villages, although this can be constrained by planning policies such as Green Belts. Where this potential is relatively remote it may be reflected in a degree of "hope value" in addition to the value of the property as a farm. The very substantial sums involved where there is the genuine possibility of development means that farms with tangible potential for development are seldom sold without some protection for the vendor, either preventing development without the vendor's consent or maintaining a share for them of any future uplift in value. There is an increasing habit, particularly amongst vendors in the public and charity sectors, to include such clauses in any event given the regulatory imperative on them to secure best value when disposing of property.

10.4.48 As mentioned earlier, the potential for conversion of redundant buildings, for residential, holiday or employment use, has added value to a number of holdings. Where such buildings have already been converted then the additional revenue would be expected to add to the value of the farm and that is often the case. It may not always be so, however, and perhaps not enough of those farms have been sold for proper analysis to take place; in this area, value is in the eye of the purchaser and some sublet cottages relatively close to the main house may be viewed as a welcome income stream by some but an unwanted intrusion by others. The planning regime, outside of protected areas, has become far more benign in respect of conversion of farm buildings to alternative use, and it will be interesting to watch the extent to which the notional investment value of the income stream is fully reflected in additional market value on the farm.

10.4.49 Additional cottages, particularly those not restricted to agricultural occupation, add value to a holding. That said, most vendors and agents considering disposal of a farm property with a number of cottages will seek to exploit this opportunity by selling them separately except where location close to the core of the farm precludes this happening. On a cautionary note, there are a number of residential developments of barns on farms which seemed a good idea at the time when planning permission was secured. However, having been sold off, they have since become a major nuisance to the operation of the farm, whether through inadequate thought given to access and other rights or a genuine nuisance caused by farmer vendors who struggling to remember that they have sold the barn and thus fail to change their farming operations to avoid conflict with the new owners.

10.4.50 Woodland – The valuation of woodland is explored together with sporting in Chapter 13. Woodland can be useful on a farm whether providing shade or shelter for stock, a windbreak to stop soil blowing soils or cover for game.

10.4.51 In the past, some farmers may have used homegrown timber for fencing posts and gate posts; however, without access to pressure treating plant it is more cost effective and convenient to purchase materials, although those with time, patience and skill may process hardwoods for posts. The increase in wood burning stoves in the last three decades has increased demand for firewood. Where woodlands are extensive, some thought may be given to wood pellet fired heating systems; homegrown timber, with the potential for reasonable continuity of supply, may be an advantage given that wood pellets are as costly as the equivalent oil supply in some areas. That said, whilst access to timber may prompt consideration of such systems, ultimately the decision is likely to turn on the relative generosity of government grants for the installation of renewable technologies.

10.4.52 In many cases, rather than seeking to restore unproductive areas to farm production it is easier and more beneficial to plant these areas, whether for biodiversity or firewood. Apart from the timber value, the prospect of which is remote when planting, and the shade and shelter provided, this will also be a conservation measure.

10.4.53 Easements, wayleaves, rights of way and other reservations – These various restrictions and reservations can have an impact on value where the rights protected are such that they impact markedly on the operation of the farm or the enjoyment of the dwellings.

10.4.54 The market is likely to be less concerned by easements and wayleaves in respect of pipelines or electricity lines, which are relatively common on farms, than by a right of way in favour of a third party. They have the potential to have significant impact where the right reserved passes close to the dwelling and interrupts privacy. Mobile phone masts, once considered an asset, may in future be regarded as a liability when rents are artificially capped under the Electronic Communications Code and mast operators' standard leases reserve entirely unfettered rights of access. The market might be more concerned where a property is subject to an unusual but possibly onerous liability with that for chancel repairs on some former glebe land property being one better known example (see, for example, *Parochial Church Council of the Parish of Aston Cantlow and Wilmcote with Billesley, Warwickshire (Appellants) v Wallbank*, which reached the House of Lords).

10.4.55 All these rights can prove a hindrance to further development, whether extending the farm buildings or converting traditional buildings, but so can restrictive covenants, clawback and overage clauses which may either prevent development of the property or reserve a charge out of any development proceeds in favour of a third party.

10.4.56 The effects of such restrictions are particularly marked on capital value; where equipment and rights of way are involved their impact should also be considered when assessing rental value.

10.4.57 General – Returning to the considerations set out by Gwyn Williams in previous editions, he identified "a few matters considered as a plus". Times and fashions change, but many will still be considered positive and important issues in the operation of the farm. However, at the risk of repetition, their presence or absence is seldom reflected in the value of the farm; some not mentioned earlier or worth repeating because of their significance include:

- a good hard internal spine road – good internal access is still important whatever the nature of the farm but particularly where regular stock or machinery movements are involved on heavier land
- a good layout with a well-located farmstead and fields of an economically viable size for the farming system
- the potential to lot the farm in various ways to best suit the prevailing market at the time that it is offered for sale
- the presence of any production quotas, now essentially sugar beet contracts. Milk and livestock quotas are now a thing of the past, but in some marginal dairy areas there may be a difference in value between those dairy farms capable of producing sufficient milk to make them of interest to milk processors and of generating sufficient profits to enable them to maintain the level of investment in new equipment required by most milk buyers.
- the scope for diversification, whether that involves letting buildings, adding value to produce or offering visitor services, location and the availability of surplus buildings will be key
- access to support schemes, although the current policy environment is opaque. It is unusual now to find land which is not registered for the Basic Payment, most likely applicable for future schemes, but clearly that is something to check when undertaking a valuation. Land is generally sold and hence valued without Basic Payment Scheme entitlements and there is a ready market for entitlements which can be acquired with none of the accompanying land occupation arrangements required in the early days of the Single Payment or milk quota. The impact of the progressive withdrawal of support through Basic Payment that is expected in at least England and Wales after Brexit remains to be seen.

10.4.58 Conclusion – Previous editions have included the following wholly reasonable comment:

> *The quality of the soil always, in the final analysis, determines the value of a farm.*

10.4.59 Logically, this ought to be the case, but increasingly over recent years the shortage of farms and agricultural land coming to the market has meant that distinctions in soil quality between the holdings, coupled with the increased technological capacity to deal with poorer soils, has reduced that effect.

10.4.60 Going forward, it may be the case that the increased interest in the health and conservation of soils will bring this back to the fore. However, other issues now have greater importance and, whilst the relative quality of the principal house, other dwellings, farm buildings and land will all be considerations, purchasers and thus valuers will often have increasing regard for location and proximity to favoured residential areas, privacy and the scope for generating diversified income.

10.4.61 The exit of the UK from the European Union and the impact on trade, access to labour and farm support has been an issue of debate since the 2016 Referendum. The nature of any long-term settlement on trade and labour is unclear at the time of writing, while the future levels, and perhaps more importantly, accessibility of future support regimes, is uncertain, to say the least. For some economists, including those with a relatively recently developed interest in agricultural economics, a more selective and less benign support regime will be just the change that the land market requires, increasing the supply of land onto the market in a way that would see values fall to economically sustainable levels. Again, that may be a logical conclusion, but it overlooks the extent to which the sale and purchase of land is a decision of the heart rather than the head; most farmers find it hard to part with land no matter how strong the logic is to sell and many Britons and others see the purchase of land as a tangible asset attractive at prices that also might not be assessed on a purely economic calculus. Ultimately, supply and demand will be the overriding influence on the level of the market and the value of rural property. The skill of the valuer is working out where within the range of values any particular property may sit.

10.5 Example valuation

10.5.1 Whilst the valuer will spend time considering the relative merits and challenges of the farm that they have to value, the valuation itself will be based very firmly on the current market as reflected by transactions for similar holdings. In the past, valuers developed a feel for value based on those comparables and simply reported the appropriate figure. Now it behoves valuers to demonstrate how they have reached that conclusion, whether explicitly providing an analysis of comparables for a funding valuation or having the evidence available to negotiate a settlement for capital taxation. That does not preclude the modern valuer from developing that market feel, but in addition now the proof must be demonstrably available.

10.5.2 To that end, it is important that valuers maintain their own databases of comparable transactions, not least as the number of transactions are far fewer nowadays than was the case for previous generations. Whether this is based on a simple hardcopy list or a more complex software program is academic, although the latter approach does lend for easier shared use, particularly in larger practices.

10.5.3 Individual valuers will adopt different approaches in analysing comparables and the subject holding. When starting out in the profession, some find themselves unduly exercised in striving to find the most appropriate comparable.

10.5.4 One useful if simplistic approach to overcoming this is simply to list the relevant comparables found in value order, based on price per acre, and then find where the subject property best fits in that list; it is an approximate and subjective method, but then so are most valuations no matter how much we might try to pretend otherwise. This approach gives an indication of potential value which valuers will normally then choose to test more vigorously by detailed analysis of comparables as explored later on.

10.5.5 In assessing comparables, a number of issues will affect their worth in the exercise; the nature and extent of the holding are obvious issues but so are proximity both in geography and time. Thus, the sale of the neighbouring farm dredged from records ten years ago may be helpful if the comparable record includes some analysis of the strength of demand for the property – a list of bids if best offers were invited, for example – but is hardly contemporary. The sale of a similar holding ten miles away and six months ago will be far more helpful. The more specialist the property, which might be code for the narrower the likely pool of purchasers, the more relevant nature and extent become compared to proximity of location or time.

10.5.6 In assessing the value of a farm, valuers are likely to consider both the value of the whole and the value of the constituent parts, not least as the latter approach enables easier adjustment of comparables to suit the subject property. The prospect of a comprehensive set of comparables is more remote than in the past, but there is nothing to stop the valuer making the relevant adjustments to match the subject property and indeed this is part of the skill.

10.5.7 Whether considering the whole farm or constituent parts some adjustments – the presence or absence of supplementary dwellings for example – may be simply achieved by adding or subtracting the market value of the relevant property. In other cases, such as quality of access or location, a valuer may ascribe a percentage addition or reduction in price to reflect differentials between the properties. Some may resort to spreadsheet adjustments, with the considerable merits of easy recovery for future comparison and repetitive use. Others may prefer a calculator and sharp pencil. Whatever the approach, ultimately the assiduous valuer will develop a feel for the market, which will both test and inform the arithmetic involved.

10.5.8 Turning to the value of the constituent elements of the property, the available evidence may vary considerably. The simplest area is likely to be the dwellings because there will be extensive information available on residential values. In contrast, perhaps the hardest area to compare are farm buildings. Practised valuers may happily ascribe a value to farm buildings based on their experience which can have little if any foundation in market transactions and is more likely than not simply an update of the figure suggested to them when they first came into practice.

10.5.9 The majority of valuers will complete their careers without ever seeing a set of farm buildings sold in isolation for use as farm buildings. In the absence of this market evidence, some will turn to a version of depreciated replacement cost, but that is no proxy for market value, not least when in an under-supplied market poor quality, poorly equipped farms may sell for as much as well-equipped units. In the absence of such information, a more rational approach, repeated in the example described later, may be to value the land and buildings together, not least given that occasionally land and buildings are sold and comparables can be easily analysed to extract the residential value. Whatever approach might be adopted, valuers are likely to find themselves having to apportion some arbitrary but uncannily consistent figure to farm buildings, not least as a number of banks seek such apportionments as part of valuations for secured lending.

10.5.10 The exception may be in respect of those buildings which are suitable for or already converted to alternative use. In the latter case, an investment approach may be adopted, and this is explored further in Chapter 15; in the former, where development has not yet been pursued, some may attempt a residual valuation, although a helpful comparable may well be a more reliable lifebelt for the valuer getting near to their depth in speculation. Whichever approach is adopted the most important variable, as explored previously, is the extent to which the market will value the opportunity, influenced, in turn, by the extent to which it can be pursued without compromising dwellings or the enjoyment of the rest of the farm.

10.5.11 Valuers find many different ways of recording their calculations, often reflecting the complexity and scale of the property involved. In the simplest valuations of bare land some simply prepare a list of comparables and select from them an appropriate value per acre for the subject; the more enthusiastic may derive a variety of mean and median calculations. In the more complex valuations of whole farm properties, careful consideration and adjustment to comparables, both for the whole and the parts, will be required and need to be recorded.

10.5.12 Faced with a complete farm to value, the typical approach is to value the constituent parts, in some cases geographically as if lotting the premises for sale and in other cases by quality or nature, as perhaps between arable and pasture land. This can be done simply by ascribing a single figure to constituent parts or considering the range of values which might apply depending on the strength of the market and the trade the property might attract. Thus, in the example calculation described later on,

the property has been split into constituent parts and a low
high value ascribed to each element, those in turn having bee
comparables. The resultant value for the whole might be dra
one range of values, the sum of the low for example or mix
on the attractiveness of the individual parts.

10.5.13 The example farm, known as West Farm, is used
considering investment valuations, but for these purposes:

> West Farm is a 100-hectare (250-acre) mixed farm in the south-west of England with a four-bedroom period farmhouse in fair condition, if a little outdated, adequate farm buildings mainly 25 years old and approximately 100 yards removed from the farmhouse, a mix of predominantly Grade III pasture and arable land all in a ring fence. The BPS entitlements are excluded from the valuation; there are no realistic opportunities for alternative development. The farm is being valued with vacant possession.

Example valuation – West Farm with vacant possession

In considering the property the valuer has decided to split it into the following elements:

- farmhouse and surrounding land extending to approximately 10 acres in total
- 170 acres arable land with the benefit of the buildings
- 70 acres pasture land with the benefit of the buildings

		Low	Medium	High
	£/acre			
Farmhouse & land 10 acres total		550,000	600,000	**700,000**
Arable land 170 acres Allowing for buildings	7,500 8,000 9,000	1,275,000	**1,360,000**	1,530,000
Pasture land 70 acres Allowing for buildings	6,000 6,500 7,500	420,000	**455,000**	525,000
Total		2,245,000	2,415,000	2,755,000
Overall value per acre		£8,980	£9,660	£11,020

10.5.14 The low, medium and high values may not be evenly distributed; in this case the valuer feels that there is likely to be a reasonable demand for the property, but given its location it is possible that it will attract interest from a London or south-eastern purchaser either retiring to the country or considering a weekly commute. In that case, the valuer feels that they would bid well beyond the market and, on reflection, considers that this interest is more likely to be reflected most in the house and concludes that the appropriate values are as follows:

Selected values	
House & 10 acres	700,000
Arable land	1,360,000
Pasture land	455,000
Total	2,515,000
Value per acre	£10,060

10.5.15 In practice, much to the dismay of those who favour arithmetical precision, that value is likely to be rounded to £2.5 million, equating to £10,000 per acre. Having established this figure from the assessment of the constituent parts the valuer can then compare this outcome with the comparables for the holding as a whole.

10.5.16 Elements of this approach may appear a little radical for some, and perhaps the more traditional approach to analysing values is as set out in the following:

Example valuation – West Farm with vacant possession – alternative approach

House & 10 acres	700,000
Farm buildings	100,000
Arable land	1,275,000
170 acres @ £8,000	
Pasture land	420,000
70 acres @ £6,000	
Total	2,495,000
Value per acre	£9,980

10.5.17 In practice, there is no default approach to calculating values or setting out those calculations. The test is that whatever approach the valuer adopts should be replicable across a range of properties, provide a thorough analysis of a robust evidence base and be intelligible to colleagues present and future who might one day find themselves obliged to pick up the file for whatever purpose. There should, however, be one element that is consistent to every valuer's approach; once all the mathematics are done step back and apply the sense check – would the market really pay this for that property?

11 Valuation of let property

11.1 Introduction

11.1.1 At the beginning of the twentieth century some 90% of farmland in the UK was let and a distinct minority of farms were owner-occupied. During that century, a wide combination of factors, including two world wars, developing agricultural holdings legislation and various iterations of capital taxation on death, saw a significant reduction in the tenanted sector such that by the early-1990s perhaps 30% of land was let. The introduction of Farm Business Tenancies in the Agricultural Tenancies Act 1995 saw an immediate reversal of that trend and the decline in the tenanted sector has been arrested, in England and Wales at least. The picture in Scotland is very different with annual CAAV/SAAVA surveys showing a continued decline from similar levels to England and Wales in the mid-1990s to approximately 20% of farmland, a trend associated by many with fears over the Land Reform agenda.

11.1.2 Whilst there has been a decline in the sector, many rural valuers will still find themselves called upon to value let farms and land whether for taxation or annual accounting purposes or for negotiations between the landlord and tenant over the future of the holding. Again, the picture is further complicated in Scotland by the introduction of the rights of relinquishment and assignation under the Land Reform Act 2016.

11.2 Basic principles

11.2.1 The basic principle with any investment valuation is to multiply the income received from that investment whether from rent, for a property, or dividend for an equity, by an appropriate yield; therefore:

> Capital value = income x yield

11.2.2 This basic theory of valuing a property let on a tenancy of any period will have been covered by most valuers in their training, typically in relation to commercial property.

11.2.3 The underlying principle applies equally to farms and land let on tenancies with the good news that the nature of the asset and rents and yields are generally rather more stable in agricultural investment markets than in commercial ones. Thus, the rural valuer is less likely to be involved in complex discounted cash flow calculations than his commercial investment cousin.

11.2.4 However, there is one fundamental distinction between rural and commercial markets. In the commercial market investors still hold sway, particularly in the prime markets. Consequently, in many cases the value of a let commercial property will exceed that of the same property with vacant possession. In contrast, the agricultural market, in common with the residential market, is dominated by owner-occupiers. Thus, the value of a let farm will, in almost all circumstances, be less than that of the same farm with vacant possession.

11.2.5 When the first edition of this book was published in 1985, valuers would often, as a rule of thumb, expect let values to be half of vacant possession values or sometimes even lower. However, that was a time of a single code for agricultural tenancies with security for the tenant and before the increased activity in rationalising tenancies which started in earnest in the 1990s. The intervening period has seen a significant increase firstly in values of the residential element of the holdings and latterly in the value of agricultural land and the significant fall in investment yields generally, particularly since central bank rates reached record low levels after the global recession. As a consequence, whilst the underlying approach may be based on the same principles the outcome now may be very different as developed further in the following sections.

11.3 Factors affecting investment value

11.3.1 Thus, in its simplest form, the skill of the valuer appraising a rural investment lies in assessing the potential rental income and the current yields in the market. These factors will, in turn, be influenced by the quality of the farm and many of the issues reviewed in Chapter 10 reflected in the rental value. However, the value will be affected by a number of other issues including the nature of the letting, the security of tenure and scope for possession, rent, the terms of the letting, the scope for alternative use and the reversionary value.

11.3.2 The nature of the letting – Since 1995, there have been two types of lettings of agricultural holdings in England and Wales: those under the Agricultural Holdings Act 1986, sometimes referred to as "traditional", "full" or "AHAs", and Farm Business Tenancies, those introduced under the Agricultural Tenancies Act 1995 (FBTs) and granted since 1995. In Scotland, the Agriculture Holdings (Scotland) 1991 acts in a way similar to the 1986 Act with more restricted forms of fixed term tenancy possible since

2003. Key features for valuers regarding agricultural tenancies in each part of the United Kingdom are reviewed in more detail in Part 4.

11.3.3 There are a variety of distinctions between the two codes, but the most important in the context of investment valuations are security of tenure and the basis of rent.

11.3.4 Tenancies under the Agricultural Holdings Act 1986 essentially offer the tenant security for life, except where he commits a major misdemeanour in terms of the tenancy or where the landlord is successful in securing planning permission for alternative development. In those circumstances where succession is possible two further such lettings may be enforceable against the landlord as successions (in Scotland a 1991 Act tenancy can continue as long as there are successors). Tenancies under the Agricultural Tenancies Act 1995 offer the tenant security of tenure for the term of the tenancy only, with the certainty for the landlord of being able to regain possession thereafter on service of the relevant notice.

11.3.5 Rents under the Agricultural Holdings Act are set in accordance with the statutory provisions set out in section 12 and Schedule 2 of the Act (see Chapter 26), and as observed in *Childers v Anker* there is no direct relationship between rents set under that formula and the open market. In contrast, the default position in FBTs under the Agricultural Tenancies Act is for rent to be assessed at market value (see Chapter 27), although some tenancies may include different alternatives including, since the Regulatory Reform Order in 2006 adopting the 1986 definition, particularly relevant where FBTs have been used in succession to AHA agreements.

11.3.6 In passing, there is now a generation of valuers for whom new tenancies have solely been those granted under the 1995 Act to whom the arcane prescriptiveness of the 1986 Act may come less easily than to those brought up under the 1986 Act or indeed its predecessor 1948 Act. Occasionally the two codes are compared, not least by those acting for some landlords, from whatever generation, with FBTs seen as modern tenancies, by definition, virtuous and appropriate to the twenty-first century and AHA agreements "old fashioned" tenancies no longer relevant to modern farming. This is a false and pointless comparison; both codes are entirely valid and the fact that the tenant under a 1986 tenancy may have greater security and hence as we shall see a greater stake in the value of the holding than his 1995 Act compatriot is just a fact of history and not, as some would have it, an abuse of the landlord's human rights. There are good and bad farmers on both forms of letting.

11.3.7 Security of tenure and the scope for possession – The prospect of possession will depend predominantly on the nature of the letting and whether it is under the 1986 or 1995 Acts. However, within those statutory frameworks there will be variations between individual lettings, particularly where 1986 Act tenancies are concerned.

11.3.8 The security of tenure for the 1995 Act tenant is limited to the end date of the tenancy, although that might be varied either by an earlier break clause, particularly in favour of the landlord, or in the rarer circumstances of a tenancy so long that there is the prospect of the tenant quitting either through death or retirement prior to the end of the tenancy agreement. Where there are break clauses involved, it is important to identify whether these are conditional on a specific event, for example the grant of planning permission in respect of part of the holding, or effective at a specific time subject to notice. In the latter case where undertaking a valuation the valuer should explore the likelihood of the break clause being operated and agree the appropriate special assumption with the client.

11.3.9 The situation may be slightly more complex where the letting is under the Agricultural Holdings Act 1986. In some circumstances, where the original tenancy was granted prior to 12 July 1984, there may be the prospect of one or two further successions, depending on whether there are any eligible successors, meaning that vacant possession might not be forthcoming for several decades in some circumstances. Conversely, where there is no prospect of succession the security of tenure will turn entirely on the anticipated time before the tenant quits the holding for whatever reason. In the clear majority of cases, this will turn on his health and choice, but there may be some occasions, where development is envisaged for example, where the landlord will be able to secure earlier possession.

11.3.10 Such modelling as might be attempted on the continuation of succession would suggest that given the initial tenancy must have been granted at least 36 years ago the majority of such tenancies will now be on their first succession and indeed many will be on their second. Projecting that forward, one might anticipate a sharp reduction in such tenancies from say 2050 with few actually lasting into the twenty-second century. That is the position in England; it is different in Scotland with 1991 Act tenancies, explored further in Chapter 24, where it is the tenancy which can endure rather than the tenant.

11.3.11 Whilst that may seem like good news for prospective future valuers, even in the twenty-second century they may not be entirely free of the 1986 Act. The vast majority of 1986 Act tenancies are let to one or more individuals and thus will eventually end on the death or retirement of the last surviving tenant or successor. However, where a 1986 Act tenancy is let to a company the tenancy will subsist as long as the company, so effectively creating a perpetual leasehold interest impacting significantly on the value of the freehold interest.

11.3.12 Thus, where the letting is under the 1986 Act and there is either no tenancy agreement or the tenancy agreement is silent on the question of assignment, the landlord is at risk of the tenant assigning the tenancy to a limited company and effectively isolating the landlord and their successors from any prospect of vacant possession in perpetuity. Whatever the nature of the instruction, simply to provide an investment valuation for financial accounting for example, the valuer noting these circumstances should immediately alert the landlord to the risks involved and recommend an immediate application under section 6 of the 1986 Act (in Scotland, s.4 of the 1991 Act).

11.3.13 Another, less rare, departure from the normal provisions in respect of security of tenure under the 1986 Act applies to those farms which are let by Statutory Smallholdings Authorities, often known as County Farms. For the relatively small minority of those holdings which satisfy the conditions, the holding may be capable of being determined by a "Retirement Notice" served under Case A which the Agriculture Act 2020 makes enforceable at the state pension age. That will be of interest to the relatively small number of valuers charged with undertaking valuations for local authority accounts, where ultimately it is interesting but academic, and those involved in negotiating a potential rationalisation settlement between landlord and tenant where it has rather more impact.

11.3.14 Returning to the more common position, this distinction between tenancies means that a valuer might be faced with a tenancy which has only one or two years left before vacant possession is available and another with 50 or 60 years before vacant possession might be anticipated. This has led to the emergence of two different approaches to the valuation of investments, valuation in perpetuity where the prospect of vacant possession is very remote and valuations on a term and reversion basis where vacant possession can reasonably be anticipated sooner. These are sometimes associated directly with the different tenancy types, so FBTs are valued on a term and reversion basis and AHA lettings in perpetuity, but that is rather too simplistic an approach and there will be many examples where term and reversion is appropriate for holdings let on AHA tenancies as well.

11.3.15 Rent – The assessment of rent differs between the two types of tenancy, but in terms of investment value the question is the same: is the current rent appropriate or should a different rent be adopted in valuing the tenancy over the long term? In many cases with AHA lettings which are, by definition, mature, the difference between the current rent and the rent an arbitrator would fix under the statutory provisions is very small and it may be entirely appropriate to adopt the current rent in the valuation. In other cases, there may be an argument for assuming a different rent and occasionally reflecting the difference between the current rent and the more appropriate rent which might be forthcoming at the next review.

11.3.16 Where some variation is required with an AHA tenancy, it is quite likely that this is because the rent may be historic and there is room for an increase. However, there may be circumstances where the rent is too high, not least perhaps where this includes interest on a recently completed landlord's building and, in such a case, it may be necessary to adopt a lower figure. Whilst finesse is required in assessing these figures, it is easy to get too carried away with very fine adjustments which have relatively little impact over time. That said, low yields exacerbate the value of such differences: for example, a £5 per acre difference in rent, between say £70 and £75 per acre, on a 100 acre block of land valued at 3% in perpetuity will make a 6% difference in the capital value.

11.3.17 The distinction may be more marked with a Farm Business Tenancy where the difference between a tendered rent, which is essentially the basis for market value, and the rent currently being paid by a long-established tenant may be substantial. In certain circumstances, this might have a very significant impact on value and, where passing rents are significantly below market value, it would be appropriate to explore the rationale and the background with the landlord. It may be that these have little or no impact on the approach you adopt to your valuation, but it is important that your client understands this, particularly if you are poised to make a heroic assumption that the rent of the farm used to assess capital value should be double the current rent.

11.3.18 The terms of the letting – Again, there are differences in terms between the two codes, but beyond that the terms of the letting may vary from tenancy to tenancy within codes as well. Some terms may have little or no impact on the value of the holding, but perhaps the most relevant are the terms in respect of repairs and maintenance. A holding let on full repairing and insuring terms is significantly more attractive to an investor than one let on the statutory default Model Clauses where there may be significant liabilities for the landlord, particularly where the holding includes a farmhouse. In some cases, it may help to consider net as well as gross rent when calculating value, an issue addressed further later on.

11.3.19 To the extent that the terms are reflected in the rent, this will have an impact on value, but the reduced risk may also have an impact on the yield used when capitalising the rent.

11.3.20 Other clauses, including restrictive user clauses or requirements to farm in accordance with a strict set of environmental or other provisions, may also have an impact on the rent and through that the capital value.

11.3.21 One area which could have a fundamental impact on value are provisions referring to alienation, particularly any bar on assignment. Where the tenancy is an FBT then the absence of clauses barring assignment, whilst unattractive to some landlords, is not fatal to the reversion, the landlord will still be able to determine the tenancy at the end of the term. However, where there is no bar on assignment in a letting, under the 1986 Act the landlord is at significant risk as set out in paragraph 11.3.12 described earlier.

11.3.22 The prospect of alternative use – In many instances, a let holding will have limited potential for any use other than as an agricultural holding, although with complete holdings, including dwellings and buildings, there will often be prospects of significant additional income or capital receipts for the landlord if they can secure possession either of the whole or those constituent parts. Where the prospects of possession are good as, for example, where the tenancy includes provisions for the landlord to recover possession for development, these opportunities should be reflected in investment value. However, as always considering the practical issues and personalities recovering a surplus cottage, for example, will be far easier where the tenant is amenable, often as part of a wider negotiation.

11.3.23 More significant prospects for alternative use, particularly development either of land or buildings, may have a very significant impact on investment values for two reasons. Firstly, the prospect of development may make it easier for the landlord to secure vacant possession, subject to the terms of the tenancy. Secondly, the reversionary value may be very different from that which would be the case were the reversion solely to agricultural use. Where significant development opportunities are involved the valuation may change in nature from the assessment of an agricultural investment with added value at reversion to a deferred development valuation where a very different approach may be required.

11.3.24 Reversionary value – Where possession is available in a timescale which makes the term and reversion approach appropriate, then plainly the reversionary value will have a significant impact on the value of the let farm as an investment. In some instances, particularly for valuations for financial reporting in a charity or public sector context, it may be appropriate for the reversion to be assessed at the rack rental value, assuming the holding was let on an FBT. However, in the majority of cases the reversion is normally assessed at Market Value for the freehold assuming vacant possession.

11.3.25 Mindful that the reversion may still be many years off, some valuers may wonder whether it is appropriate to forecast the vacant possession reversionary value at the end date of the tenancy. However, there are two difficulties with this approach: projecting future values with any accuracy and then having to discount for inflation over the period. Given those difficulties, many valuers adopt the current value as their reversionary value, on the not unreasonable assumption that over time general inflation and increases in property value will be broadly in balance.

11.3.26 One temptation for valuers considering the reversionary value is simply to assess the holding as it is let at the time of the valuation. However, that is coincidental. The valuer should consider how the property would be offered for sale on the open market were there vacant possession and to value it accordingly. In some circumstances, where the valuation is of a substantial portfolio for financial reporting, for example, it may be difficult for the valuer to justify that amount of consideration when the fees involved may be limited by competitive pressures. In such circumstances, it will be important for the valuer to explain to the client the limitations involved in any such valuation.

11.4 Example valuations

11.4.1 For ease of comparison the following valuations all assume the same holding to be valued, West Farm, which featured in Chapter 10 on Freehold Valuations; by way of background,

West Farm is a 100-hectare (250-acre) mixed farm in the south-west of England with a farmhouse, adequate farm buildings and a mix of predominantly Grade III pasture and arable land all in a ring fence. The BPS entitlements are held by the tenant and excluded from the valuation, and both landlord and tenant have complied with their various obligations under the lease; there are no realistic opportunities for alternative development.

The current value of the farm assuming vacant possession is £2.5 million.

11.4.2 All the valuations have been approached on a gross rent basis other than in Example 2 seen later. The relationship between the various values is explored further later in this chapter.

11.4.3 Valuation subject to a tenancy under the Agricultural Holdings Act 1986 – As explored in section 11.3.1 the approach to the valuation of a farm let under an AHA tenancy is likely to turn on the security of tenure of the tenant, and potentially his successors, and the likely delay until vacant possession is achieved.

11.4.4 In this case it is assumed

the farm is let under a 1986 Act tenancy which commenced in 1982 on Model Clauses terms. The tenant is aged 63 and is assisted by his son, aged 26, who works on the holding and is considered likely to qualify as a statutory successor. The passing rent is £17,500 (£175/ha ~ £70/acre) and is considered to be the appropriate rent for the holding.

11.4.5 In these circumstances with the potential successor it might be reasonable to approach the valuation by capitalising the rent in perpetuity. Returning to the previous section, the two questions for the valuer are "what rent should be adopted?" and "what yield should be used?". The first has been answered so the outstanding question is one of yield. For many years, the accepted range of yields used for valuing farms let on and AHA tenancy was between 3% and 5%; latterly some AHA investments with little prospect of early possession have been traded at yields below 2%. The yield selected will depend not only on the nature and strength of the investment but also on comparative yields for other investments. In the present market of record low central bank base rates, the pressure if anything is for the yield to be below the traditional level, although there are circumstances, as we shall see later in this chapter, where higher rates have traditionally been used. Turning to the investment itself, with let farms the amount of fixed equipment and the consequent liability for the landlord will be a key influence on the yield used, but the wider market will remain the dominant factor.

11.4.6 The yield reflects the rent divided by the capital value, the inverse of the yield; the multiplier used to capitalise the rent is referred as the Years Purchase. Years Purchase may be adopted for a term of years or in perpetuity, for single or dual rates (the latter providing a sinking fund for wasting assets such as leasehold interests) and with or without taxation. In most cases valuers will refer either to their calculator or, even more simply, to *Parry's Valuation Tables*, available either in hard copy or online, to find the relevant Years Purchase. However, where the Years Purchase in Perpetuity (YPP) is required, this can be derived by the simple sum of dividing 100 by the yield selected: so at 5% the YPP would be 20; at 4%, 25; at 3%, 33.33; and at 2%, 50.

Example 1 – Valuation using gross yields in perpetuity

11.4.7 As set out earlier, yields have reduced in recent years from the traditionally accepted "average" figure of 3% to nearer 2% and in the current record low interest markets closer to 1% and occasionally below even that. However, to adopt a more recognisable figure in perhaps vain anticipation of this circumstance returning nearer to "normal" in the future this yield adopted is 2% and the calculation at its simplest would be as follows:

Rent	17,500
Years Purchase in Perpetuity at 2%	50.00
Capital Value	875,000

11.4.8 This calculation is based on the gross rent, making no allowance for the landlord's costs. Many would argue that a more appropriate approach is to make an allowance for these costs so that it more properly reflects the net income received by the landlord. For completeness, this is addressed in the following examples. However, it should be noted that there are very few agricultural investments sold so there is relatively little comparable evidence, but such limited evidence as is available is normally quoted on a gross yield basis or the information provided is such that it is only possible to derive a gross yield accurately. In the circumstances it would be inappropriate to apply gross yields to net rents and some adjustment should be made. Rather than making adjustments for assumed landlord's expenditure liabilities and then further adjustments to the yields there is considerable attraction in simply adopting the gross yield approach.

11.4.9 That said, adjustments might be made for a number of costs including management, maintenance and insurance, and, where considering the purchase of an investment farm, purchase costs. Whatever approach is adopted it is important to clarify which costs have been reflected, particularly in terms of purchase costs which may be included for some valuations and not in others. This apparent difference can cause confusion amongst some clients and valuers and again it is not uncommon for valuers simply to argue that they have reflected the approach the market would take including purchase costs in the yields chosen, whether gross or net.

Example 2 – valuation using net yields in perpetuity

11.4.10 In this case the rent has been adjusted to reflect the various costs referred to earlier. For ease of comparison and to illustrate the potential anomaly, the yield has been retained at 3%, but as set out in the next table some adjustment should be made.

Rent		17,500
Less costs		
Management, say 8%	1,400	
Landlord's repairs and insurance, say 20%	3,500	
Deductions		4,900
Net rent		12,600
Years Purchase in Perpetuity at 2%		50.00
Capital Value		630,000
Less Purchasers costs, say (11.4.14)		33,300
Capital Value allowing for purchaser's costs		596,700

11.4.11 The outcome of these adjustments is to reduce the value by 28% by allowing for landlord's costs and 32% in total allowing for purchaser's costs as well. On that basis, plainly some adjustment is required to the gross yields which might be deduced from reported sales. The valuer will use his or her judgement of the particular circumstances and the landlord's liability to adjust the yield; and indeed, in some cases, where major structural repairs are required to the farmhouse, for example, it may be appropriate to make a specific capital deduction for that risk. These adjustments may be substantial; in passing the net yield would need to be adjusted to 1.44% to achieve the value of £875,000 from Example 1.

11.4.12 It is worth reflecting briefly on the cost adjustments made. Valuers have typically made an adjustment of between 5% and 10% for management costs and the level of charge is likely to vary depending on the nature of the holding and the liability on the landlord which will be reflected in additional work for the managing agent. However, these costs are likely to understate the costs of management because in most cases there will be work outside of the management contract, whether revenue work, undertaking rent reviews for example, or capital work, supervising building investment. Analysing a range of management contracts with which the author has been involved, the total charge for both capital and revenue work was nearer 13% to 15% of the rent roll and this might be significantly higher for smaller, complex estates: for example council smallholdings estates.

11.4.13 Turning to repairs and insurance costs these will again turn on the extent of fixed equipment with a farm of the type used in the example carrying rather more costs as a percentage of rent than a larger arable farm in East Anglia. Those valuers involved in estate management can use analysis from their own managed estates to assess the proportionate costs which in the example given are likely to be between 20% and 25%, but it is important to consider the extent to which an individual farm will have a different cost profile to a portfolio of a number of farms.

11.4.14 Purchaser's costs in this instance have been based on Stamp Duty Land Tax, calculated on the value of £630,000 and 2% for agents' and legal costs.

11.4.15 In both previous examples, the passing rent has been assumed to be the appropriate rent; however, there may be circumstances where the passing rent is either lower or higher than an appropriate figure. As indicated previously the position may be particularly extreme where the letting is a Farm Business Tenancy, although the term involved will be shorter and hence the impact on value potentially less. Where the valuer feels an alternative rent may be appropriate then some are assiduous in reflecting the different periods before and after the rent review. Whilst there is much to be said for being thorough unless very significant rents are involved there is relatively little difference between splitting the valuation and simply adopting a different rental figure.

Example 3 – valuation using gross yields in perpetuity reflecting different rental values

11.4.16 In this case, the same assumptions are made over the letting as set out in the following example, but two alternative approaches are adopted to the valuation:

> The farm is let under a 1986 Act tenancy which commenced in 1982 on Model Clauses terms. The tenant is aged 63 and is assisted by his son, aged 26, who works on the holding and is considered likely to qualify as a statutory successor. The passing rent is £17,500 (£175/ha ~ £70/acre), but the valuer believes the rent, which could next be reviewed in two years' time, should be £20,000.
>
> (£200/ha ~ £80/acre)

Example 3a – splitting the valuation to allow for the rent review

Current Rent	17,500	
Years Purchase, 2 years at 2%	1.94	
Capital Value		33,950
Rent in 2 years	20,000	
Years Purchase in Perpetuity, deferred 2 years at 2%	48.06	
Capital Value		961,200
Total Capital Value		995,150

Example 3b – adopting the alternative rent from the outset

Adopted Rent	20,000	
Years Purchase in Perpetuity at 2%	50.00	
Capital Value		1,000,000

Example 4 – valuation on a term and reversion basis – AHA letting

11.4.17 In the previous examples, the common basis of the valuation has been to capitalise the rent in perpetuity, based predominantly on the prospect of succession. However, were there no prospect of succession it would be more appropriate to value the holding on a term and reversion basis.

11.4.18 In this case it is assumed

> the farm is let under a 1986 Act tenancy which commenced in 1986 on Model Clauses terms. The tenant is aged 63 and is assisted by his son, aged 26, who works on the holding, but the tenancy does not carry succession rights. The passing rent is £17,500 (£175/ha ~ £70/acre) and is considered to be the appropriate rent for the holding.

11.4.19 The decision for the valuer is how long the tenant might remain on the holding until his retirement, in this instance, given the nature of the holding, the conclusion is that this might be another 12 years. Given the shorter term and the uncertainty over the possession date, a higher yield has been used, arguably in some cases the margin over the rate previously used is too high, but it serves to illustrate the risk and the need for a margin while also reflecting the fact that the anticipated capital growth, implicit, if hidden, in the bullish rate used for the valuation in perpetuity, is valued separately here in the reversion so that that the valuation of the term is simply that with no allowance for a future capital receipt.

Example 4a – single rate term and reversion

Value of Term		
Current Rent	17,500	
Years Purchase, 12 years at 4%	9.39	
Capital Value of Term		164,325
Value of Reversion		
Vacant Possession Value	2,500,000	
Present value of £1 in 12 years at 4%	0.6246	
Capital Value of reversion		1,561,500
Total Capital Value		1,725,825

11.4.20 In some instances where the length of term and thus the period until the reversion is longer the valuer may decide that it is appropriate to use a higher discount rate in assessing the present value of the reversion than the yield. Thus, were the term assumed to be 20 years rather than 12 the valuer might adjust the rates as follows:

Example 4b – differential rate term and reversion

Value of Term		
Current Rent	17,500	
Years Purchase, 20 years at 4%	13.59	
Capital Value of Term		237,825
Value of Reversion		
Vacant Possession Value	2,500,000	
Present value of £1 in 20 years at 4.5%	0.4146	
Capital Value of reversion		1,036,500
Total Capital Value		1,274,325

11.4.21 Valuation Subject to a Tenancy under the Agricultural Tenancies Act 1995 – One of the distinctions of a Farm Business Tenancy letting under the 1995 Act is the certainty of possession for the landlord subject to the agreed term of the tenancy and service of the relevant notice. On that basis, the valuer can be certain of an enforceable end date of the tenancy and a Term and Reversion approach is far more appropriate, even in one of those very few cases where there is a very long-term FBT.

11.4.22 On that basis the calculation is essentially the same as in Example 4 from earlier; assuming in this case

> the farm is let under a 1995 Act tenancy which commenced in 2005 for a 20-year term on Model Clauses terms. The tenant is aged 63 and is assisted by his son, aged 26, who works on the holding, but the tenancy does not carry succession rights. The passing rent is £30,000 (£300/ha ~ £120/acre) and is considered to be the appropriate rent for the holding.

Example 5 – valuation on a term and reversion basis – FBT letting

Value of Term		
Current Rent	30,000	
Years Purchase, 8 years at 4%	6.73	
Capital Value of Term		201,900
Value of Reversion		
Vacant Possession Value	2,500,000	
Present value of £1 in 8 years at 4%	0.7307	

Value of Term Capital Value of reversion	1,826,750
Total Capital Value	2,028,650

11.4.23 Where the remaining term is relatively short, perhaps less than three years, it may be that valuers will take a very simple approach by simply making a discount from vacant possession value. Thus, a farm with, say, three years until vacant possession might be valued at say 95% of vacant possession value. The key test in this approach is how willing a purchaser will be to accept the delay before securing possession. In passing, were one to apply the Term and Reversion approach to West Farm as in the example from earlier but with a three-year term the resultant value would be approximately £2.3 million, or an 8% reduction.

11.5 Comparative values

11.5.1 The previous examples all provide answers to the same question: "what is the value of West Farm?". However, the answers differ depending on the circumstances of the letting and the approach adopted by the valuer. The latter is particularly relevant, and valuers raised in the era of valuing agricultural holdings as a letting in perpetuity may find some difficulty with the Term and Reversion approach.

11.5.2 There are circumstances where this can create difficulties particularly where valuations are provided for financial reporting purposes in the local authority and charity sectors. Thus, common practice with smallholdings in particular has been to assume that any farm falling vacant would be re-let and thus holdings let on Farm Business Tenancies have been valued as perpetual investments. Whilst this approach may reflect the value of the estate if it is retained this is not always the case in practice with farms being sold or merged when they fall vacant. Even where farms are retained in exactly the same form such an approach will still fail to properly reflect value unless it allows for the uplift in rent arising from changes in tenancies from AHA to FBT on reletting.

11.5.3 This is not the only difficulty in this area. Elsewhere, some institutions may be driven by portfolio value even where there is no prospect of sale as where, for example, the farm is to be retained for its potential for development in a couple of decades time. In those cases, the potential reduction in value when a farm is let on a longer tenancy is seen as a problem by accountants and investment managers, which can manifest itself in successive short-term lettings rather than a longer tenancy with a break clause. That can appear to be an artefact of how the assessment has been done, arguably to the detriment of landlord and rural economy alike.

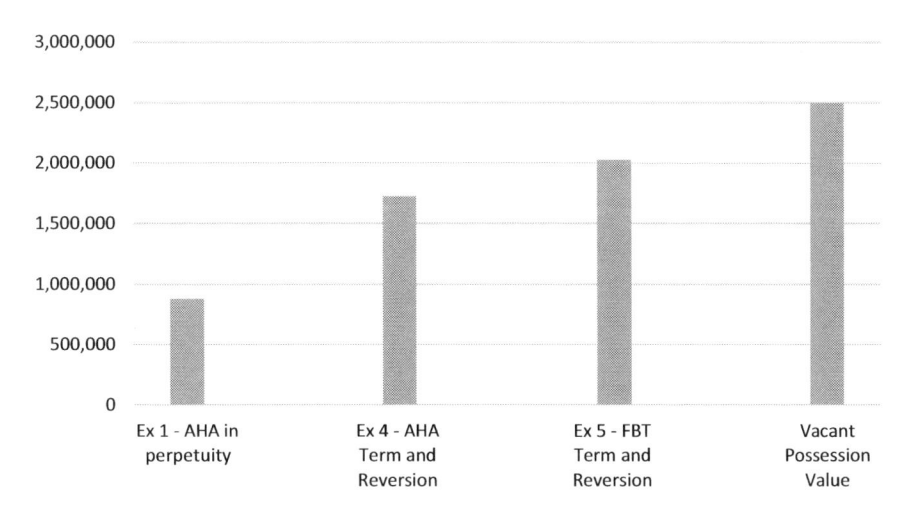

Figure 11.1 Comparative Values Investment Valuations – West Farm Example

11.5.4 The comparative valuations derived from Examples 1, 4, 5 and the vacant possession value of West Farm are set out in Figure 11.1.

11.6 Marriage value and negotiations between parties

11.6.1 In times past landlords and tenants tended to leave agricultural tenancies to run their course, not least given the security of tenure afforded by the Agricultural Holdings Act 1986. However, the combination of significant increases in property values, enhancing the vacant possession premium and pressures for economies of scale in farming and the flexibility provided by Farm Business Tenancies has prompted parties to explore the opportunities to release the Marriage Value inherent in any letting. To an extent this is simply a development of sales to sitting tenants, but pressure on landlords to improve financial performance, amongst local authority and traditional institutions in particular, has seen an increased focus on the prospects to negotiate sales and surrenders.

11.6.2 Whilst any third party might purchase a let farm only, the landlord and the tenant can release the Marriage Value which exists between the two parties' interests and the vacant possession value and investment values.

11.6.3 Returning to the example of West Farm used throughout this chapter, the Marriage Value will differ depending on the nature and term of the tenancy. The two extremes amongst the examples are represented by Example 1 – the AHA letting with the anticipation of succession – and Example 5 – the FBT letting. The Marriage Value (in this case equivalent to the vacant possession premium) in each case is calculated as follows:

Example 1 – AHA Letting		**Example 5 – FBT Letting**	
Vacant Possession Value	2,500,000	Vacant Possession Value	2,500,000
Investment Value	875,000	Investment Value	2,028,650
Marriage Value	1,625,000	Marriage Value	471,350

11.6.4 In practice, it is unlikely that the Example 1 calculation would be used, although the tenant's agent may argue that it is the correct approach particularly where a succession is possible. In practice, the mere fact that the parties are talking, whatever the motive, brings the prospect of a Term and Reversion approach into play, although if succession was in prospect, and thus vacant possession was deferred 50 years, adopting 3% as both a yield and discount rate the value would only increase to approximately £1.07 million. In this case the value from Example 1 has been retained to illustrate the extreme position.

11.6.5 The extent to which the parties might share this value, whether by the tenant buying the holding at a discount from vacant possession value, or the landlord paying the tenant to quit, is a matter of negotiation between the parties. This will turn on the security for the tenant, the funds available and the anxiety of the parties to achieve the transaction. The starting point is often, unsurprisingly, to split the Marriage Value 50/50 between the parties, but that is by no means a given. In most circumstances there will be greater incentive for one party than the other. As many tenants will only contemplate this if they have no successor, the tenant's share might often be less than this. However, adopting that textbook 50/50 approach, transactions might be as follows:

Example 1 (AHA letting)

The parties have agreed that the Tenant will purchase the farm from the landlord subject to his tenancy, on the basis that Marriage Value is split 50/50 the purchase price would be calculated as the investment value, which any third party should pay, plus the share of Marriage Value as follows:

Investment Value		875,000
Marriage Value	1,625,000	
50% share		812,500
Purchase price (subject to tenancy)		1,687,500

11.6.6 In practice, the Marriage Value is so significant here that the Tenant may seek to bid a rather lower share of Marriage Value to secure the farm. However, the Landlord, equally conscious of the discount from vacant possession value, is likely to push for the Tenant to pay a higher share. Again, the price paid will turn on the aspirations and capacity of the parties.

11.6.7 Price may not be the only issue between the parties. Landlords seeking to protect their position may seek to impose restrictions on future development and a clawback on any "profit" if the Tenant resells the farm relatively quickly thereafter. Tenants and their agents are likely to resist such restrictions, commonly arguing that this will impact on the willingness of banks to support a purchase. That may often be the case, and banks may resist such arrangements when first considering a loan proposal. However, that reaction is commonly founded on an, at best, partial appreciation of the issues and values involved by the banker and it is very unlikely that a bank's interest will be constrained by such clauses to the extent that their loan-to-value criteria are breached. Whatever arrangement is ultimately reached will have an impact on value, with an unconstrained sale clearly more valuable than one which reserves, say, 50% of the profit on any sale to the Landlord for the next 20 years.

Example 5 (FBT letting)

The parties have agreed that the landlord will pay the tenant to quit enabling him to sell the holding with vacant possession; the parties have agreed to a 50/50 split of Marriage Value as follows:

Marriage Value	471,350	
50% share		235,675
Compensation to tenant to surrender tenancy		235,675

11.6.8 In this case, the opposite argument to the earlier example may apply; the tenant has relatively limited interest in the holding and once the term gets below five years the landlord may be much less inclined to make a payment to the tenant. On that basis the landlord may try to negotiate a lower settlement.

11.6.9 The ultimate value will turn on the aspirations, intentions and freedom to act at the time. That may vary, along with the funds available to reach a settlement, depending on the nature of the transaction. A landlord *in extremis*, seeking to raise funds for essential repairs, may have rather less freedom for manoeuvre, but a rather more pressing need, than one seeking to secure possession of farm property to facilitate where, for whatever reason, he cannot secure possession without resort to negotiation. Equally, a tenant fearing prospective ill health or under financial pressure may not be in a position to hold out for a full half share of the vacant possession premium, even if the landlord is unaware of his predicament. The evidence given by the District Valuer in the *Baird* case, reviewed in Chapter 32, was of tenants being paid between 18.5% and 50% of the vacant possession premium for the surrender of their 1991 Act tenancies.

11.6.10 The calculation is interesting, but the prerequisite of any such transaction is for the parties to recognise the respective rights and liabilities of their opposite number and approach the transaction with an open mind. This is where a good agricultural valuer can prove his or her worth by explaining the issues, opportunities and likely values realistically to clients which in some cases may mean overturning strong prejudices which clients may have to paying or giving value to the other party.

11.7 The valuation of agricultural tenancies

11.7.1 The values discussed so far in this chapter concern the effect of the tenancy on the capital value of the property, whether that is then expressed as a discount from the vacant possession value or as a surrender payment, in the context of a specific transaction.

11.7.2 However, there may be other occasions when an agricultural valuer has to assess the value of a tenancy, whether in a statutory environment, specifically for taxation or in compulsory purchase proceedings, or where a tenancy figures as an asset in a partnership or family settlement, whether on death or divorce. Most recently the new provisions for 1991 Act tenants in Scotland for rights of pre-emption and relinquishment and assignation will require an assessment of the value of the tenancy. This is explored in greater detail in Chapter 32.

12 Valuations for insurance

12.1 Introduction

12.1.1 Agricultural valuers may often be asked to produce "valuations" for insurance, whether as part of a valuation for finance security purposes or to assess the cover required as part of insurance renewal. Whilst typically referred to as valuations, these are far removed from the normal process of valuation, having no direct relationship with any value in the marketplace, instead being an assessment of building cost. In practice, these exercises should more properly be called "rebuilding cost assessments" or "reinstatement cost assessments".

12.1.2 They should also be treated with a degree of caution; there is considerable difference between a detailed cost assessment which might be undertaken for insurance purposes and those produced as an adjunct to a valuation for lending. These latter assessments are not and should explicitly be identified as not being comprehensive cost assessments fit for insurance negotiation purposes. The fee basis and time available to surveyors undertaking finance valuation work is such that they cannot produce a detailed cost assessment and thus these should simply be seen at best as a check on the present level of insurance cover provided for the security property.

12.2 The process

12.2.1 Whilst the insurance market has changed considerably, in some cases with impacts on the client or valuer as explored further later, the basis of assessment is very much as it has been for many years, involving:
(a) establishing the nature and extent of the buildings involved,
(b) identifying the method of construction and materials, in particular where either are unusual,
(c) assessing to what extent the existing buildings, if wholly or partially destroyed, would be replaced or alternatively what modern equivalent may be provided in their place,
(d) checking any specific requirements, whether for buildings (associated with a current trade being conducted at the property for example) or as a consequence of listed building status or planning requirements,

(e) assessing an appropriate building cost either using a component approach or, particularly with assessments for security valuations, adopting a unit cost per square metre or square foot derived from relevant estimates or published figures, including the new building cost estimates published by the CAAV and
(f) making the necessary additional allowances for demolition, particularly where the existing buildings contain noxious material such as asbestos, VAT and fees.

12.2.2 Where it is anticipated that buildings of traditional construction, for example, a range of nineteenth-century brick buildings, would be replaced with a modern equivalent, the unit cost is likely to be substantially reduced. However, modern buildings are not always a cheaper option, not least as in many cases the space required has increased very substantially, the eaves height required to accommodate modern farm equipment being an obvious example.

12.2.3 In assessing the cost, valuers should have particular regard to issues which may impact on the cost of construction, including, but not confined to

- any listed or otherwise protected status of any building or location: being in a National Park for example. This may be particularly relevant where listing is Grade 2 star or above which may require not only rebuilding the existing structure but also the use of specific materials, in some cases, sourced from specific limited sources; in one instance a client was only allowed to use oak from an identified estate.
- any difficulties with access, in particular those which will involve the use of smaller than normal vehicles for transporting materials or, in some cases even transport other than road vehicles (helicoptering materials to awkward or remote locations as an extreme example) and the double handling of materials which may be involved.
- the potential need for full planning permission, whether due to the scale of the replacement buildings or previous development on the property using up the Permitted Development Rights.
- the relationship between neighbouring buildings and particularly with a long-standing farmstead which has been developed over time, whether individual buildings can be replaced independently from the others.

12.2.4 Other factors to be considered include:
- the extent to which cover should be provided for any consequential costs and losses (hiring milking equipment for example if a parlour were damaged by fire) or loss of rent with a let farm
- the impact of a partial loss, including the prospect of averaging, reducing the amount which can be claimed (should overall cover not be sufficient) and ensuring that again there is sufficient cover for any

consequential loss which in some cases will be just as severe as would have been the case with a total loss
- the potential conflict between building works and normal farm activities (for example, access for cows for milking on a dairy farm) or harvesting on an arable unit, which might add to costs
- in a similar vein, particularly on livestock units, the need for contractors to abide by biosecurity measures, whether statutory or imposed by supermarkets or other buyers.

12.3 Developments in the insurance industry

12.3.1 Developments in the insurance industry may also create difficulties for those involved in assessing the rebuilding cost of fixed equipment. New contractual arrangements between insurance companies and retained loss adjustors or the use of preferred contractors for certain specialist works may have an impact on the cost of work, not least due to the compliance requirements of insurance companies' retained panels which are likely to limit preferred contractors to larger firms with significantly different cost bases to smaller local contractors more likely to be retained by farmers or estate owners.

12.3.2 This may lead to the risk that the amount insured, which will be perfectly adequate using local building contractors, will not be sufficient, sometimes by a significant margin, to cover the cost of the work being undertaken by a retained preferred provider.

12.3.3 Consequently, the commentary on any building cost assessment should refer specifically to the assumptions made and should state that it discounts any impact on either the direct rebuilding cost or the associated professional fees arising from any contractual or procurement arrangements specified by the insurance company.

12.3.4 Where the rebuilding cost assessment is prepared as part of arranging insurance, rather than a security valuation, the valuer should draw the client's attention to these potential issues and the need for them to seek clarification from the insurance company or broker. These may, in turn, require some further adjustment to the cost assessment on the part of the valuer. Furthermore, in these circumstances the agricultural valuer should give very careful consideration to whether he or she has sufficient relevant experience and expertise to provide a comprehensive and accurate building cost assessment, particularly where there are any historic, listed or specialist buildings involved.

Further reading

Building Reinstatement Cost Assessment for Farms and Estates (CAAV, 2018)

13 Woodland and sporting

13.1 Introduction

13.1.1 Whilst most valuers in general rural practice will principally focus on agricultural land and property (and nowadays diversified rural properties), they will sometimes be called upon to comment on woodland or sporting interests, particularly where these are ancillary to other rural property.

13.1.2 These are both specialist areas and, mindful of the duty of care to clients, valuers should not attempt to provide formal valuations of these properties unless they have the necessary skills and experience to undertake such a valuation. Even for those with the relevant skills and experience it is worth reflecting whether that experience is sufficiently current and relevant to the property in question to enable them to make a professional assessment of the value of the asset involved.

13.1.3 Bearing that caution in mind, the following chapter is intended merely to provide an overview of the approach to valuing these specialist assets. As in every other case it remains for the valuer themselves to conclude whether it is appropriate for them to accept an instruction.

13.2 Woodland

13.2.1 Introduction – Two clear markets for woodland have developed in the last two decades:
- the market in commercial blocks of timber, whether lowland hardwood or mixed plantations or very large blocks of upland coniferous forestry, the latter often in more remote locations in Scotland and Wales; and
- a "lifestyle" market in smaller blocks of amenity woodland typically purchased as much for leisure and quiet enjoyment as timber production – where individual property agency businesses have developed focused on that market.

13.2.2 The two markets are generally populated by very different vendors and purchasers, with the former typically the province of investment trusts and larger commercial timber purchasers whilst the latter is very much a market for private individuals. In turn, most rural valuers will feel that they have the necessary skills to value amenity woodland, drawing as that process does on the established valuers' skills of the collection and analysis of comparables.

13.2.3 Farm woodlands – Farm woodlands are unlikely to be valued as commercial woods; rather they are more likely to be of a scale where they would be of greater interest to the lifestyle market. Indeed, in some cases, where they are located close to the roadside with independent access, farm woodlands may have a premium value in the lifestyle market. Often, however, farm woodlands will be so integral to the farm that they cannot be marketed separately in which case the value may be limited, although where there is potential for shooting on a farm well-positioned woodland will be an asset.

13.2.4 One unfortunate characteristic of recent years is the general lack of management of farm woodlands which, absent sporting, were long seen as of little interest. The various woodland planting schemes have focused grants on planting rather than management. Standards of management may again be better where there is some sporting interest or where biomass is now a factor, but otherwise much potential timber output has been lost from farm woodlands.

13.2.5 Amenity woodlands – Amenity woodlands are generally valued on a comparable basis and the value will derive as much from the amenity, location and ease of access as any potential timber value. Whilst many agents may occasionally have the odd block of woodland for sale, there are also a relatively small number of specialist woodland agents who tend to dominate the market. Their websites will indicate the asking prices for blocks of woodland which in some cases may rival or exceed the value of agricultural land in the same area.

13.2.6 As with other properties, the smaller lot sizes generally attract higher prices per acre on the simple premise that it is the absolute cost of the purchase which is critical to most purchasers. They will have a certain sum available to invest and, whilst hoping to get the best value for their investment, will be less concerned about the individual price per acre. In common with small areas of agricultural land this is very much a discretionary purchase and consequently the prices achieved may sometimes defy logical analysis.

13.2.7 Additional features (the presence of a pond or stream for example) may enhance the value, but relative ease of access combined with a degree of privacy appear to be the principal influences on value. As with other property, proximity to towns can be an advantage, not for any particular utility in woodland management but simply because of being accessible to a larger population of potential purchasers with disposable income. Some purchasers will be interested in the potential to harvest firewood for the ubiquitous wood burning stove, but for many the attraction is no more or less than some private space in which to enjoy their leisure

13.2.8 Commercial woodlands – The approach to the valuation of larger blocks of commercial woodland may involve a consideration of comparables, but it is also likely to include an assessment of the productive capacity of the woodland. This will involve a detailed analysis of the potential timber output of the property projected over the term of the established crop and in turn will require an in-depth knowledge of timber markets, silvicultural practice and the growth attributes of the particular woodland and crop.

13.2.9 Whilst few valuers outside of the specialist practices would wish to become embroiled in the valuation of many hundreds or thousands of acres of uneven aged upland forestry, there is an awkward size of plantations, beyond the amenity market, where commercial timber value is important, but the market may still have a relatively local nature. In those circumstances, a mix of skills may be required with comparables providing a stronger guide. Armed with the relevant evidence local valuers may feel more comfortable to attempt a valuation of such property, not least where this forms part of a larger portfolio: for example as part of the valuation of a rural estate for Inheritance Tax purposes.

13.2.10 In the case of larger blocks where the valuation will go beyond a simple comparable value per acre, the starting point may be to split the value between land value and the value of the growing timber. The land is then valued on a comparable basis and the first challenge may be to find sufficient comparables for the land (or "prairie") value. Again, scale is likely to be an issue as is the extent or otherwise of infrastructure available on the property and the comparables. The prairie value will also be required where valuations are being produced for Inheritance Tax and Woodland Relief is being claimed.

13.2.11 The additional value of the standing timber may then be assessed by reference to the costs of establishment for a young plantation or the volume and market price of the timber for a standing crop.

13.2.12 In the latter case, evidence from standing timber sales, particularly auctions, whether live or more recently electronic, is a helpful source of comparable information. Those involved in the timber market will naturally have access to more detailed and comprehensive price information.

13.2.13 The appraisal will involve forecasting income and expenditure in thinning, felling, replanting and the maintenance of forest rides as well as other infrastructure, but at the same time there will be significant decisions affecting value around the discount rates to be used in constructing the relevant discounted cash flows. The forestry valuer needs an appreciation of financial forecasting, the market for forest and alternative investments and the Internal Rate of Return being sought by investors in the market at the date of the valuation.

13.2.14 In the larger, uneven age plantations the valuation may involve a mix of methods, produced on a compartment or even sub-compartment basis. Valuers active in this area will use relevant valuation software to assist them with the assessment. Woodland owners and management companies may well hold much of the necessary information on forestry management software, some also employing precision management and mapping techniques. Ultimately, however, establishing market value is more than simply the function of various arithmetic exercises, and the woodland valuer, in common with all others, must know their market and the likely demand for the asset they are appraising.

13.2.15 Woodland leisure – There have always been ancillary uses of woodland – stalking being the most obvious – but latterly there has been a significant diversification in leisure activities in woods led by the Forestry Commission and Natural Resources Wales but increasingly also embraced by private and third sector owners.

13.2.16 These leisure activities can be broadly split into two groups: informal activities which generate little of any income for the owners (general public access using public rights of way or permissive paths being the obvious example) and more formalised activity whether individual access or events which can generate income and hence capital value.

13.2.17 This latter group may again be divided between activities for which there is a direct charge and others which give rise to charging opportunities. The former may often involve the owner granting licences to activities such as tree-top trails involving high wire routes or Segway trails whilst the latter, creating mountain biking trails, for example, may involve lettings to a bike hire outlet or café which are ancillary but essential services. Thus, in the latter case the Forestry Commission in Wales (now NRW) and Scotland have created mountain bike centres investing in trails of various difficulties which cyclists can use without a direct charge nonetheless generating income directly for the owners through car parking charges and indirectly through lettings to cafés and bike shops. This is a model which has now been embraced by private sector owners, albeit with the constraint of viewing any such development from a solely commercial viewpoint whereas the Forestry Commission could justify their first investments in Wales and the 7stanes project in Scotland on non-financial as well as financial grounds.

13.2.18 The majority of these enterprises are in their early stages and, other than perhaps in the course of preparing annual accounts, there has been little need for valuations. However, it will not be too long before valuations are required whether for sale, finance or taxation purposes. Where the services involved are delivered by third parties, hosted by licences or lettings, the basis of valuation may be very similar to other investment valuations. However, both in those cases and when considering income from car parking, for example, it will be important to reflect the operating costs of the activity which attracts visitors to the site in the first instance. That may involve maintaining mountain bike trails or the scientific and silvicultural work required to maintain an arboretum but, in common with other activities considered in this chapter, valuers must be assiduous in ensuring that they consider the costs required to generate income and value.

13.3 Sporting property

13.3.1 Introduction – In many cases sporting rights will be a supplemental but minor element of farm value and most rural valuers will be able to assess the additional value involved. However, where the very best sporting is involved, whether stalking, driven shooting or fishing, the trading and valuation of these rights is practised by specialist valuers and sporting agents.

13.3.2 In common with other rural assets, valuations are based on the assessment of comparable transactions, but in the case of sporting rights this is often assessed, and indeed expressed, on the basis of the bag, thus per brace of grouse shot or per stag or hind or per salmon. This makes records of previous bags an important element in the assessment of value, and where these records are missing it may take a considerable time for values to be re-established to appropriate comparable levels.

13.3.3 One of the complications in valuing sporting rights is the variety of different lettings; thus shooting may be sold for a day or let for a season and so may carry different costs. The position is even more complicated for fishing with beats being let on a traditional basis, per rod per season or on a timeshare equivalent with rods being sold for particular weeks in a season for a term of years. Whilst there may be some novel ways of marketing, sporting rights country sports are steeped in tradition and the best shoots and river beats still carry a kudos which can translate into a significant premium in value over apparently comparable properties.

13.3.4 Information on rents may be available, either by direct comparison or interpretation of other letting arrangements; however, evidence on freehold sales is more limited. Whilst a number of river beats are offered for sale, principally by specialist sporting agents or larger management firms, there are very few sales of shooting rights and consequently it may be difficult to establish an appropriate yield to be adopted when seeking to establish a capital value for shoots. At the same time, it is important to ensure that the (often substantial) costs of exercising the rights are fully accounted for. In a long-term lease the tenant is likely to be responsible for costs which are thus reflected in the rent; however, where rents from shorter term lets are used as comparables, it is important that the full costs of the landlord are reflected, particularly employment of a keeper or ghillie and associated accommodation costs.

13.3.5 In common with other rural property, proximity to conurbations and ease of access can add value. That said, by their very nature some of the best sporting estates are in remote locations and consequently accessibility is a relative term and for Highland estates and sporting rights in particular, proximity to airports may be the most telling measure.

13.3.6 Sporting rights can exist separately from land, but they can only be exercised by access to that land and, on occasions, that access may be a constraint on the use of the land. Intensive lowland shoots, for example, are likely to limit flexibility of cropping on the land over which they are exercised, and grouse moors need to be kept free of forestry planting. To that extent, there can be a risk of an element of double counting of value if the farmland is valued as if unencumbered by the sporting and then full value is ascribed to the sporting rights as well.

13.3.7 Sporting may be an ancillary use in some areas but in others, most notably in the Scottish Highlands, sporting may be the principal or only value. In these areas, the decision of the Scottish government to remove the exemption for shooting rights from non-domestic rates, which has existed since 1994 (and earlier in the rest of the UK), has been seen in some areas as a potential threat both to asset values and the economics of sporting. At the time of writing Assessors, who undertake rating valuations in Scotland, are still undertaking rating assessments, but it appears that the impact may not be as dramatic as some apocryphal reporting first suggested.

13.3.8 Sporting is a leisure activity and the social element (the shoot lunch for example) has always been an important element in the enjoyment of a day's sport. This has become more relevant in recent years as sporting activities have increasingly been offered as part of a leisure experience or package as closely associated with the holiday industry as with rural estate management. There will be some purchasers for whom sport will be the key but there are others who will be more influenced by the quality of the entire experience.

13.3.9 To this extent, the income generating capability of a sporting asset may turn in part on the quality of hospitality and accommodation, which will very often be offered by third party partners of the sporting owner. Whilst the underlying sporting asset will always be key to the value, it is important for the valuer faced with a commercial operation offering those rights through sporting packages to consider how much of the value is inherent in the asset and how much is in the ancillary service which may lie outside of the control of the owner of the rights. To that extent, the sporting valuer is facing the same challenge as a business valuer – how much value is inherent and transferable and how much is in goodwill? – which is as notoriously fickle and unstable in the sporting sector as in many others.

13.3.10 Lowland shooting – In lowland areas driven shooting is for pheasant and partridge with, in simplistic terms, the former needing a mix of open land and woodland and undulating terrain and the latter dry, clear arable ground.

13.3.11 Shooting estates need to be able to provide a reasonable number of days shooting per season, generally for between eight and ten guns. That requires a relatively large area whether in a single ownership or by neighbouring farms and estates acting together. In the former case, there may be some additional value in the sporting rights; in the latter, where much of the value turns on the activity of the owners or a syndicate to which the rights are let, the value for individual properties may be very limited.

13.3.12 Experienced valuers, particularly those who also have an interest in shooting, will be able to identify the respective characteristics of property which make for good driven sport; the mix of woodland and open ground (particularly for pheasant shoots), topography and elevation. This assessment can be augmented by good records, particularly of the number of birds put down, the frequency of shooting, the number of guns and the number of birds shot.

13.3.13 Where sporting is let, care needs to be taken in assessing the rent and its constituent elements. It is common practice for syndicates to be asked to provide an owner's day, or days, and the precise terms of this arrangement, the number of days, the number of guns and the anticipated number of birds shot will all have an influence on value. Similarly, there are a variety of hybrid arrangements for dealing with the costs of keepering and accommodation for keepers, particularly where a syndicate has taken over from the owner in managing the shoot.

13.3.14 Upland shooting – In upland and moorland areas driven shooting of grouse is still regarded by many as the epitome of country sports, reflected in the value of the best grouse moors. As with other assets the value of the moor will turn on its suitability for use, thus how good the habitat it provides for grouse. This will be a combination of management, topography and the type of shooting, whether driven or walked-up; again, recorded bag numbers will be important.

13.3.15 In contrast with lowland shoots, the prospects for alternative use for the land will be limited – either sheep grazing, which can be an ancillary use if tightly controlled, forestry planting or wind farm development. However, the latter two both have the potential to conflict with grouse moors. Consequently, the problem of potential double counting of value, referred to in the previous section, is far less likely to be an issue here, and the values of moors are typically assessed solely on the sporting potential.

13.3.16 Whilst there are substantial areas of land involved, given that the quality of the shoot rather than the extent of the area is the critical issue values are often expressed as a value per brace shot. Moors are seldom traded so comparable evidence can be rather historic and further complicated as often they may be included as part of a larger estate sale with principal house, other dwellings, farm or forestry land and other sporting interests. Distilling the cumulative proceeds of sale into a value for the constituent elements can be a difficult process not least given the impact of marriage value, and the motivations of those relatively few individuals who can afford to purchase such properties can be hard to distinguish.

13.3.17 Deer stalking – Again, there is a distinction between upland stalking for red deer, and lowland stalking for roe, fallow or in some areas muntjac deer, in woodland settings.

13.3.18 Whilst red deer are found in the south-west of England, in Norfolk and some limited other areas, the high value stalking areas are again in the Highlands of Scotland. Deer forests, as they are known, are distinctly short of trees, being open hill and mountain and usually higher up the hill than the grouse moors. In the majority of cases there is little alternative commercial use for the land, although many forests are in relatively high mountain areas where hill walking and mountaineering are also popular but generate little direct income other than for estates with holiday cottages or other accommodation. As with grouse moors values are often expressed in bag terms, values per stag shot. Stalking is routinely let by the day and any assessment of value must have regard to the costs of running the forest, predominantly employment and associated costs. Again, relatively few deer forests are offered for sale and typically there will be other property involved requiring some analysis to identify the value of the stalking itself.

13.3.19 Lowland stalking has limited value, being perhaps a useful additional income for some woodland owners, but as important for population management, particularly where there is conflict between growing deer populations and neighbouring farms. In the vast majority of cases therefore there is very little if any additional value created by the existence of this stalking.

13.3.20 Game fishing – The right to fish for game fish, that is salmon, sea trout and trout (with grayling sometimes included), can vary enormously in value from the best Scottish salmon and English chalk stream beats which are beyond the budget of all but the fortunate few to the smaller upland streams of the north and south-west, where a day's fishing may cost less than a cinema ticket.

13.3.21 Once again value turns very much on the existence of comprehensive records, not least as, for salmon fishing at least, capital values are assessed and quoted on the number of fish caught on average over previous seasons. As mentioned earlier, fishing rights are now let in a variety of different packages, from periodic tickets from a day to a season, exclusive lets or time-share arrangements. It can sometimes be difficult to resolve these different approaches into a realistic set of comparables, but again an important element of any valuations is ensuring that the costs, in this case predominantly those of the ghillie, are fully recognised.

13.3.22 Recent years have seen the development of a number of fishing passport schemes, for example those on the Wye and Usk rivers and in the south-west of England, which enable relatively low-cost access to some premier rivers albeit on some of the lesser known beats. It might be thought that this would have an impact on the value of letting or selling more exclusive fishings, but one suspects as with other sporting properties it is that exclusivity which creates value, whether in the sales or lettings market.

13.3.23 In contrast to salmon rivers, the value of trout fishing is assessed on the length of the river bank over which the rights are available and whether that is for both banks (double bank) or only one (single bank).

Whilst values may not be so overtly reported in terms of past catches, these are still the key contributor to value alongside location and ease of access.

13.3.24 Historic catch records reveal some remarkable seasons and indeed days' fishing from earlier in the twentieth century. However, in most cases a combination of factors, whether climate-related, the impact of sein netting or fish farms, have seen a significant reduction in catch numbers. While this might be anticipated to have an impact on values, its relatively universal effect means that the best rivers from the past seem to remain the most valuable rivers today. Whether that remains, particularly for some southern chalk streams where demands for water may have dramatic impacts on aquifers and thus water levels, remains to be seen.

13.3.25 Coarse fishing – There are many canals and rivers where coarse fishing can be enjoyed for little if any cost beyond the Environment Agency rod licence. It will be unusual for a rural valuer to have to assess these rights. However, there may be instances where coarse fishing ponds or lakes have been developed on farms, sometimes as an ancillary use for irrigation reservoirs. Where these are relatively small scale and low key, perhaps let to the local angling club to manage, the additional value is likely to be limited. However, more commercial facilities may attract a greater value, to the extent that a separate valuation may be required as explored in the next section.

13.3.26 Fisheries – There has been a significant growth in the number of fishing lakes from relatively low-key sites to highly commercialised and professionally managed fisheries often targeted at enthusiastic anglers pursuing specimen fish.

13.3.27 Some of these are farm ponds, whether long established or new diversifications, others, including some of the largest, restored gravel pits, but whatever their origin larger fisheries, particularly those capable of being sold separately, may attract considerable value. That value will turn on both the extent of the property involved, whether there is an owner's residence or holiday accommodation, a tackle shop and catering as well as the lakes, and the commercial success of the venture demonstrated by the income generated and consequent profit. The valuation will be more akin to that of a trading business than a rural property, although the underlying quality of the site, location and proximity to centres of population and the infrastructure will contribute to value. Some fisheries concentrate on either coarse or trout fishing whilst others may have a mix of ponds extending the season and the attraction.

13.3.28 Occasionally local agents may be called upon to sell such properties, but their valuation is again a specialist venture with very few firms routinely involved in this work. In the more commercial lakes, offering trophy species such as carp to the enthusiastic angler the fish stock itself may be of considerable value (its ownership following a sale of the land by a receiver was an issue in *Borwick*), an issue easily lost on those of us tempted to have a go at a valuation of a local fishery just because it is in our patch: something which might not subsequently be appreciated by our insurers and so good reason to refer to the specialist valuers.

14 Diversification

14.1 Introduction

14.1.1 The trend for farmers to seek to diversify their income, first examined in detail by John McInerney and others in the 1980s, has grown markedly in recent years to the extent that by most estimates more than half of farm businesses now have some form of diversified income though often not much more than storage. This may be very close to traditional farming activity, contracting or sporting activity for example, or rather newer activities such as woodland burials or paintballing.

14.1.2 As Peter Prag points out in his excellent book on the subject, *Rural Diversification (Estates Gazette 2002)*, there may be a number of reasons for diversification; including:
- checking the decline in farm income
- supplementing inadequate income
- exploiting an opportunity or ability
- planning future expansion
- arranging joint ventures; or
- facing changing circumstances.

14.1.3 The rationale for farm diversification has been the focus of much academic study, which occasionally seems to strain rather too hard to force what appears to be a simple business decision into a complex socio-economic theory. That is not to say that the reasons for diversification are of no interest to the valuer, however, as this will have an impact on the nature of diversification, the mechanics of how the diversified business is owned and managed and the relationship between the farming activity and the diversified business. This latter issue may offer some particularly interesting work for valuers. The dynamics of that relationship will change as the diversified activity grows in terms of financial performance, demands on the physical resources of the farm and the time input of the family members principally involved.

14.1.4 That shift may well cause tensions within the family and between the generations where, typically, the older generation is farming and the younger generation is managing the diversified business. The patriarch of the farming family may not take kindly to being told by his daughter that he needs to keep the lane clear of mud to avoid upsetting guests arriving at the wedding barn; he may be even more put out when she points out that the profit from her business is greater than that from the farm. Helping to manage the business and other relationships to enable continuity and succession in the overall family business may test even the most finely tuned mediation skills.

14.1.5 Whilst simple financial contribution will not be the only issue, and the simple and probably incalculable value of the farm environment as host to a wide range of diversified enterprises should not be overlooked, business consultancy and assisting with occupational arrangements is likely to be an increasing part of the rural valuer's workload.

14.1.6 Peter Prag's book addresses these and a range of other issues in far greater depth and with greater expertise than is available here. Whilst the publication is nearly 20 years old, the underlying principles remain the same and the challenge and need for professional advice is even greater. In the context of this book, focusing on valuations, the pertinent questions are "to what extent do diversified enterprises add value?" and if they do, or even might, "how does the valuer attempt to answer that question?".

14.2 Capital valuations

14.2.1 A valuer may be required to value a farm with diversified assets for all the usual range of reasons, from matrimonial dispute to security for a loan, but for those not specialising in diversification the most common occasion may be where it forms part of a deceased's estate for Inheritance Tax purposes.

14.2.2 Whatever the purpose, many of the normal tenets of valuation practice will apply, but it is worth observing that to date many farm diversifications, which may generate reasonable revenue, have not necessarily fed through to an increase in the capital value of the farm.

14.2.3 It seems self-evident that an activity which generates revenue has the potential to have a capital value and indeed much commercial property valuation is based on that premise; the annual revenue stream, a rent, produces a value with the variables turning on the quality of the property, the strength of the tenant and the perceived attractiveness of that asset class. However, in those cases the revenue is a consequence of the space available provided by the property itself. In the case of diversified activity, whilst the diversification may be hosted by or contingent upon the farm and the farming business that may not create any added value.

14.2.4 Diversification may take many forms but will commonly fall into one of a number of broad categories (though this is not exhaustive):
- off-farm working – contracting or direct employment elsewhere (the latter not strictly diversification of assets)

- adding value – whether through growing alternative crops or processing commodities for sale on the farm or elsewhere
- tourism – whether traditional bed and breakfast, letting holiday cottages, camping, glamping, caravanning or log cabins
- sporting – traditional country sports, equestrian activities, off-roading or paintballing
- events – weddings, parties and conferences
- lifestyle – spas, mindfulness retreats or alternative therapies
- letting space – from burial grounds to building conversions
- renewable energy.

14.2.5 In many cases, the success or otherwise of the enterprise and its ability to generate revenue and potentially therefore value, is more contingent on the expertise and entrepreneurial flair of the operator than the property. The business, particularly one relying heavily on the consumer's experience and thus the proprietor's approach, may have limited transfer value to a purchaser of the farm. Indeed, the operator may be able to relocate the business to new premises were they to leave the farm. That suggests the prospect for additional capital value is very limited. In other cases, holiday lettings and particularly letting buildings for alternative uses, the revenue is being generated directly from the existence of assets on the farm and there is far more prospect of value.

14.2.6 That said, this is not always a given; a vendor's successful holiday cottage business may be an unwelcome chore and an intrusion of privacy for a purchaser otherwise interested in buying the farm. Location will be a critical issue, and there is a far greater prospect of value where buildings converted for letting are removed from the farmhouse and remaining agricultural buildings and particularly where they are capable of being sold separately without significantly impacting on the value of the remainder of the property.

14.2.7 The issue of off-setting, the use of one part of the holding for one activity, having an adverse impact on the enjoyment and hence the value of another part of the holding is a critical consideration for the valuer. The letting of a barn conversion as office space may cause a marked reduction in the value of the farmhouse if the buildings are in close proximity. If that impacts on the optimum lotting for a future sale, the net impact of the barn conversion on the value of the entire property is either negligible or, worse still, negative.

14.2.8 Where the diversification involves "traditional" commercial lettings of space for employment use then the agricultural valuer can turn to the normal methods of investment valuation learnt in the first valuation lectures of their university degree, or for those of a different vintage their correspondence course. These are explored in depth in a number of publications referred to later but by way of a simple example:

East Farm is a 200-hectare (500-acre) mixed farm in the West Midlands with a five bedroom Grade II listed farmhouse, a range of traditional buildings now converted to office use some 100 yards from the farmhouse with a separate access, adequate modern farm buildings approximately 80 yards removed from the farmhouse, a mix of predominantly Grade III pasture and arable land all in a ring fence. The traditional buildings extend to some 300 square metres net internal (approximately 3,230 square feet). They are let on a single lease with 5 years unexpired on Internal Repairing Terms for a rent of £22,000 per annum (£6.80 per square foot) with no further rent reviews. The tenant is a well-established firm of local accountants with a predominantly rural client base. Mobile phone and broadband signals are good. Comparable lettings in the area have recently been agreed at £7.25 per square foot and there is reasonable demand for units of this size.

Example valuation – building lettings

		£		
Term – 5 years				
Gross Rent			22,000	
Deductions	Landlord's Repairs	2,500		
	Insurance	1,200		
	Management	1,100		
Total			3,800	
Net Rent				18,200
YP for 5 years @ 7%				4.1002
Value of Term				**74,623**
Reversion after 5 years				
Gross Rent			23,400	
Deductions	Landlord's Repairs	2,500		
	Insurance	1,200		
	Management	1,170		
Total			3,870	
Net Rent				19,530
YP in perpetuity deferred 5 years @ 7%				10.1855
Value of Reversion				**198,922**
Total				**273,545**
Say				£275,000

14.2.9 As with all valuations this is a combination of collecting data, evaluating comparables, in this case of lettings rather than sales, evaluating risk and applying an appropriate method. The key issues in assessing the valuation are

- is the rent sustainable during the current lease? In this case, there are no further reviews.
- what are the prospects of reletting at the end of the current term? And at what rent? There is evidence of lettings at a higher rent per square foot and good demand. That suggests that the buildings could be re-let at a slightly higher rent whether or not the current tenant wanted to continue. That will not always be the case with rural employment space; there are some very good barn conversions which have foundered after the first round of leases have come to an end with tenants either returning to urban locations, finding the romance of a rural office did not quite outweigh the lack of facilities, simply unable to manage the increasing data traffic through their businesses on slow rural broadband or just retiring.
- what are the landlord's liabilities? In this case, the landlord is responsible for the main fabric of the building and although he may not spend £2,500 every year he will want to put some funds aside for long-term repairs (e.g. refurbishing a roof).
- what is the overall risk and what is the appropriate yield? In earlier times, faced with some incidental income from lettings on a farm, valuers may well have used a simple rule of thumb of 10 YP (years purchase) on the rent, which in this case would have given a value of £220,000. However, in this long-running, ultra-low interest rate environment a lower rate, more akin to commercial rates, is more appropriate.

14.2.10 The mechanics of the valuation are well rehearsed in other publications, but it is perhaps worth pausing to consider another issue, understandably not addressed elsewhere: what does this mean in the context of the whole farm?

14.2.11 East Farm is twice the size of the example used in Chapter 10, West Farm, and has a better house and the benefit of this range of buildings which will be ignored for the time being. Allowing for a certain discount for scale (there are simply more purchasers who can afford a 250-acre farm than a 500-acre farm), one might adopt the same value per acre as West Farm (say £10,000 per acre). That would give a value for the farm of £5 million, meaning that this let building is worth approximately 5% of the value of the whole. Whilst an income of £18,000 per annum before tax is not to be lightly ignored, being the current equivalent of and far more resilient than 200 acres of BPS payment, the impact on the capital value of the property is not that significant.

14.2.12 A purchaser less enthusiastic about sharing their space with a firm of accountants, or indeed any other profession, might not see the value that the valuer would naturally derive from an assessment of the income stream. If they do not want to let the premises in the future, then beyond any potential personal use there is no additional value. Where that income is less resilient due to changes in the marketplace that value will be diminished further. Conversely where there is the prospect of improved demand – and many would argue that post-Covid-19 demand for rural offices would be stronger not weaker – values should increase, but not if the prospect of lettings to third parties is unattractive to potential purchasers.

14.2.13 Returning to an earlier theme, location is critical. In this case, we know that the buildings are 100 yards from the farmhouse and, indeed, further away than the farm buildings with a separate access. The impact on the house is marginal and it may be that the buildings could be sold without impacting on the rest of the farm. In those circumstances there is no real argument to offset any of the value of the house to reflect the presence of the buildings; but what if the offices were 20 yards from the house and followed the farm drive until 10 yards before the front of the house? What would the impact on the value of the house be then? And would there be a net increase in value as a consequence of the office letting?

14.2.14 Where a small number of cottages and barn conversions are let for holiday use, then it may be possible to assess the value from sales comparables. These will be particularly relevant in popular tourist areas where there is a regular turnover of sales. However, with larger holiday complexes, particularly where other facilities including swimming pools, clubhouses and licensed restaurants are involved the value will turn very much on the success or otherwise of the business and the valuer will be entering the realms of the business valuation and the complexities of the profits method of valuation.

14.2.15 Given the growth in on-farm diversification, more agricultural valuers are becoming involved in this area of work. Business valuations are a distinct sphere of valuation activity and the RICS is particularly strict in terms of the supervision, qualification, experience and technical competence of those holding themselves out to be "Business Valuers". Traditionally focused on licensed and leisure property, business valuation now extends to a variety of other properties where the physical assets form part of the service including nurseries, whether for plants or children and nursing and retirement homes.

14.2.16 Once again, the profits method is beyond the scope of this publication, but valuers becoming engaged in such work will find detailed explanations in various texts, some of which are identified later on. The critical issue is that the valuer should understand the industry involved; most rural valuers can appreciate that as the first thing many look for in an assistant is an understanding of agriculture. Familiarity is no substitute for in-depth knowledge – an affinity with public houses does not automatically make a good licensed premises valuer – and where significant diversified businesses are involved it may be appropriate for the rural valuer to retain expert advice. This may be both prudent in terms of best advice to the client and create an opportunity for the future; at some stage a business valuer may be asked to value a farm-based diversification but be challenged if there is a substantial residual agricultural property. In such cases they may well reciprocate by seeking your advice.

14.3 Rental valuations

14.3.1 Whilst not unknown, it is still relatively unusual for farmers to let land for diversified activities. Letting buildings is more common, as indeed illustrated in the example in the previous section, but very often that will be for a "traditional" business use, offices, workshops or the like and most other business uses of buildings will fit somewhere within the very broad range of storage/workshop at one end and high-tech office at the other. To that extent, assessing the rental value of building space follows the same principles as in the wider market for employment space. It is good practice for the rural valuer to maintain an up-to-date list of comparable lettings of converted buildings, and indeed lettings in the local town which will often set the tone of the market, in the same way as they would farm sales or 1986 Act rents.

14.3.2 Assessing the potential rent of part of a farm to host a field archery instructor or quad bike course is rather more challenging with comparables being few and far between. However, one can still draw on the basic principles of valuation to assess the parameters within which one might set a valuation or frame a negotiation:

- what is the benefit to the potential tenant (or licensee)?
- what is the loss of utility or interruption of enjoyment, if any, to the landlord (or licensor)?
- how many alternative hosts are there in the market?
- what demand is there from other providers?
- is any investment or infrastructure provision required from landlord or tenant? If so, how much and how will it be funded? Will it have any residual value or utility if the use does not last?

14.3.3 Very often initial answers to the questions may be rather slight. However, in the internet age research is much easier than was once the case and the agent may also have a good source of information in the potential tenant. Most prospective tenants seeking to establish their business are willing to be open about their proposals if they see that will be reciprocated by the agent and owner and if they are not, or indeed if they do not find that reciprocal approach from an agent, all parties should pause for thought on whether or not this is a relationship that will work.

14.3.4 In the absence of many or in some cases any robust comparables, then the parties must either look for a proxy rent for a similar use or adopt a basis which best reflects the value to the occupier, most commonly based on business turnover. That again demands that the agent has or can rapidly develop an appreciation of the key attributes and risks of the business sufficient to test the prospect of success and the merits of the opportunity for the client. Where prospects seem good there may still be a significant gestation time for the business, both through investment prior to trading and in the initial trading period. In those circumstances, there may be merit in having an escalating rent approach so that the tenant can retain a greater share of any surplus through the initial high cost stages and create a more robust business capable of paying a greater rent as performance improves. This is a worthwhile approach provided there is a real prospect of success and it is not just used to reinforce a less-than-robust initial business plan.

14.4 Business rates

14.4.1 Business rates are covered in greater detail in Chapter 20, but they are worthy of mention here because many diversified activities will be subject to business rates and they can be a disproportionate influence on a decision to diversify. Farmers, enjoying the exemption from business rates for agricultural land and buildings, have traditionally had a morbid fear of anything which might threaten that exemption and involve the payment of business rates. Consequently, rural valuers have had limited exposure to such issues and so it is not a topic which needed to be covered in any detail in previous editions.

14.4.2 However, a number of cases in the horticultural and agricultural sectors coupled with increased diversification activity, means that this will be more of an issue for current and future generations of valuers. This can lead to some interesting areas – the discussion of pop festivals (and the rating thereof) by the tweed-clad members of the CAAV's Valuation Compensation and Taxation Committee remains one of the author's fondest memories – but more often the issue turns more on routine valuation principles than rock and roll.

14.4.3 The significant amount of work which was generated by successive rating valuations of the first two decades of this century passed many, but not all, rural valuers by. The impact of more routine updating (though the timetable has extended more recently) and stringent efforts by the government to reduce the scope for appeals means that workflow is unlikely to be repeated to the same scale, although any substantive amendment to the rating system will almost inevitably create anomalies in the rural and urban sectors which valuers will be called upon to address. However, the more likely involvement for rural valuers is when diversified activities are first rated or, based on experience from the 2017 revaluation when new industries such as renewables, come properly under the rating spotlight.

Further Reading

An Agricultural Valuer's Guide to Business Tenancies (CAAV, 2012)
Property Valuation, Peter Wyatt, Second Edition (Wiley Blackwell, 2013)
Reviewing a Business: An Introduction for Agricultural Valuers (CAAV, 2019)
Rural Diversification, Peter Prag, Second Edition (Estates Gazette, 2002)
European Business Valuation Standards (TEGoVA, 2020)

15 Development

15.1 Introduction

15.1.1 The pursuit of development and diversification opportunities for rural land is a core area of work for many valuers and property consultants, both those who would naturally identify themselves as rural valuers and our commercial cousins.

15.1.2 The development of rural land for residential or commercial use has long been a feature of changing land use, particularly in the urban fringe or near key transport hubs but also where new towns have been designated, common practice in post-war Britain and coming to the fore again as part of much needed expansion of the housing supply. Similarly, mineral exploitation has been an important revenue opportunity for many landowners. However, during the currency of the five editions of this book, the range of such work has expanded to include the development of traditional, and more recently modern, farm buildings for residential purposes, alternative uses, diversification into other land-based activities and, latterly, renewable energy development.

15.1.3 There has also been an expansion in the range of services involved with development and, consequently, of the breadth of skills required of the valuer wishing to work in this area. Alongside the old faithful skills of valuation and transactional work, which both require different nuances in these areas, there is much work to be had in the promotion of sites for development and in advising on the fiscal, physical and family impacts of farm diversification.

15.1.4 This chapter explores some of these issues, starting with an introduction to some of these other skills and turning to the approaches to valuing various forms of alternative land use. Whilst many of the core skills are similar to those used in other areas of rural practice, there are significant differences in development work, particularly when it comes to large greenfield developments. As with every other facet of our work, valuers can only advise effectively, and thus responsibly, if they have an appropriate level of expertise and experience.

154 *Valuations*

15.1.5 As an added tinge of excitement, headline capital values are likely to be substantially greater where development is involved. However, considerable caution is required here; the gross development values, which are the values that are often quoted are only part, and indeed a misleading part, of the story. Wide variations around the country according to local housing market and other conditions anyway mean that development land values reflect many more variables than agricultural land. However, that gross development value will not be the value that the landowner receives as, instead, the net value to the owner will be substantially reduced by a variety of factors, including:

- professional costs in securing permission,
- development charges paid to the local authority, whether Community Infrastructure Levy, Tariff or otherwise,
- obligations under planning agreements, for drainage and other matters;
- land set aside for affordable housing provision, community facilities, public open space and employment use with nil or negligible land values,
- depending on the scale of development, major contributions to school provision and community facilities,
- additional construction costs involved in meeting unforeseen ground conditions, flood mitigation; and
- with the Bill's proposal Environment, the need to produce a biodiversity gain, whether on site or off (see Chapter 17).

15.1.6 As a result, the net development land value (and so the return to the landowner) for a large site, with significant infrastructure costs, might be nearer 10 to 15 times agricultural value than the 50 to 100 times which might more often be quoted in the press and public discussion as the value of the land with residential permission, an essentially academic figure as far as the landowner is concerned. Nonetheless, the potential values involved, and particularly the difference between gross and net, serve as a timely reminder to check one's level of Professional Indemnity cover.

15.2 Planning and development

15.2.1 In most cases, construction works can only be undertaken and land and property can only be put to alternative use where planning permission has been secured for that use. Whilst there are a few exceptions, particularly for some temporary uses such as events, whether sporting (such as motor cycle scrambling) or entertainment (such as music festivals), these do not of themselves tend to change the value of the property; rather it is a demonstrable trading history that will give some value to those uses. In contrast and with the restricted opportunities for development, securing planning permission for alternative use will add value to land in many cases, with the highest premiums having been for out-of-town retail and residential use.

15.2.2 Successive governments have sought to improve and particularly speed up the planning system since the modern regime was introduced in 1947 and that trend seems set to continue, not least given the current obsession with infrastructure development and the problems with housing supply. A detailed description of the planning system is beyond the scope of this publication and, given this propensity of governments to fiddle with the regime, is likely to be out of date before publication. However, in summary:

- the central government, whether the UK government for England or the devolved administrations, set out broad strategies for development in National Planning policies; these will include policies on rural development and recent iterations have included specific policies for the development of farm buildings for alternative use
- local planning authorities (LPAs), which may be metropolitan, district or unitary authorities or national park authorities, should set a development plan for their neighbourhood, although most authorities are normally lagging behind both government targets and their own timetables
- in so doing they will normally put out a call for specific sites to be proposed for residential, employment and other uses with the aim of meeting the required five-year or other development targets set by central policy
- that offers landowners and developers the opportunity, firstly, to put forward sites for development and, secondly, to challenge the emerging development plan when, as is often the case, the LPA fails to demonstrate an adequate supply of sites
- the development plan should both identify specific sites and set out policies to deal with smaller developments, whether these include the construction of buildings or simply change of use, with these latter policies of particular relevance to the development of smaller rural sites, whether small parcels of land or building conversions
- as set out earlier, development for alternative use will normally require planning permission, although for some rural development, generally for farm buildings, it may be possible simply to use the notification system relying on permitted development rights (which vary around the United Kingdom)
- planning permission may be granted initially in outline, with various matters reserved for detailed permission where some matters may still be reserved and will almost always be subject to conditions
- where larger developments are involved, the LPA will also require the developer to enter into a section 106 agreement (s.75 in Scotland) which, in the context of residential development, for example, may

include the provision of affordable housing, and, subject to the LPA's policy Community Infrastructure Levy (CIL) in England or Wales, or tariff regime, financial contribution to education and other costs and infrastructure investment
- alongside the s.106 agreement, many local authorities in England and some in Wales have a charging regime for Community Infrastructure Levy (Scotland has legislated for an Infrastructure Levy). This is a charge intended to provide funds for local and regional infrastructure costs. Whilst intended to preclude the need for individual site negotiations (an aspiration only really held by those who knew nothing about the development process in practice), CIL has failed in this respect with most residential developments still involving a site-specific negotiation between LPA and developer. The common result is that a fixed CIL tariff and site-specific obligations prove to squeeze down the negotiable contribution to affordable housing if the development is to remain viable (a fertile area of work for valuers able to undertake viability assessments);
- under the 2020 Environment Bill, developments in England will also have to deliver biodiversity net gain, with a target of 10% gain, that will add further to costs and complexity of development, add a further pressure to the viability argument and ultimately, most likely, reduce the receipt for the landlord; the potential for this to offer value opportunities for other landowners is explored in Chapter 17;
- development will only be possible where a full planning permission has been secured including settling the s.106 agreement, paying CIL where due and agreeing conditions or reserved matters which tends to make for a fairly lengthy process and in the authors' experience the longest planning discussion with a local authority has been over 20 years for a residential-led urban extension
- once granted, a planning permission has a life so that ordinarily, if work on the development has not commenced within three years in England and Scotland (five in Wales and Northern Ireland) it will lapse.

15.2.3 It is important that any valuer advising on the potential for development, or the value of a development site, has a full understanding of the planning policies which affect any site with which they are involved. The timing and nuances of the planning process will differ with each local planning authority, and the policy background is likely to be complicated by the extent to which the LPA has saved policies from earlier plans. That means that a valuer may have to consider a suite of different documents, with various different statuses from draft to adopted, to gain the full understanding required.

15.3 Site promotion

15.3.1 The planning regime enables landowners and developers to promote sites as part of an emerging development plan or outside of that process, whether on the back of plan policies or on the basis that the LPA's plan is inadequate. This offers opportunities for the rural valuer either working with planning consultants, architects or others to promote sites directly or to negotiate strategic arrangements with others to promote land on behalf of clients.

15.3.2 Whether an individual landowner would be willing to promote their own site and their valuer feels comfortable in advising them will turn, in part, on the scale of the site and the nature of the development. Whatever the approach, the valuer will need to manage the client's expectations in two specific areas, timing and value. Planning work is seldom straightforward and the local planning authority which, typically, will be under-resourced and under no real time penalties, is likely to take considerably longer to deal with matters than the client might expect. This can be particularly frustrating where the Planning Committee has resolved to grant planning permission and the only outstanding matter is the s.106 agreement which may still take months and sometimes years to complete.

15.3.3 Where the owner is likely to retain the property after development, so perhaps a change of use for diversification or the conversion of traditional buildings, then the owner will pursue planning in their own names, whether directly or through an agent. In that case, the valuer may be involved in the design of the scheme and in preparing the supporting arguments. Where the diversification involves a separate business, there will also be a role for the valuer in settling occupational arrangements, as explored further later on.

15.3.4 Similarly, where relatively small-scale development is involved, a barn conversion scheme for three of four residential or commercial units for example, then the owners are still likely to pursue planning themselves, even where the property will be sold once permission is forthcoming. The issue becomes more complicated where a larger scale development is involved. In these circumstances, owners may well feel that unless the prospect of securing planning is almost certain they would rather not risk the considerable costs involved in promoting a site and would rather seek an alliance with a third party who will both be willing to take that risk and have greater experience and entrepreneurial flair in securing planning permission. The larger the site, the more likely it is that third parties will be involved, with only the very largest private or institutional landowners willing to pursue major development prospects on their own account. Indeed, the threshold at which third party developers or promoters are likely to become involved seems to have been lowered in recent years, reflecting perhaps the cost and complexity of securing planning permission.

15.3.5 There are a variety of different approaches to arrangements with third parties, but the three main structures are Promotion Agreements, Option Agreements, and Conditional Contracts. The mechanism used tends to depend on how speculative the site is thought to be with Promotion Agreements used for the most speculative sites and Options and Conditional Contracts more relevant as the likelihood of planning permission becomes more certain. A conditional contract is most commonly used after the site has been exposed to the market for sale and might be particularly relevant where a site is sold either zoned for development or with Outline Planning Permission and the purchaser then secures Detailed Planning Permission under the contract.

15.3.6 The Promotion Agreement is simply an agreement between parties to secure planning permission whereas the Option Agreement and the Conditional Contract both include terms for a sale. The main characteristics of the different arrangements are as follows:

- **Promotion agreements** – involve the landowner entering into an exclusive agreement with a promoter simply to seek planning permission for development. The land is then offered for sale in the open market if planning permission is secured. The promoter will bear the speculative costs of seeking planning permission and will recoup those costs plus a fee, normally calculated as a percentage of the sale price, from the sale proceeds.
- **Option agreements** – involve the landowner entering into an exclusive agreement to sell the property to a developer if the latter is successful in securing planning permission. Again, the developer will bear the planning costs and recover these as a discount from the sale price if the property is ultimately sold; they will also take an entrepreneurial share, usually reflected by a discount from market value in the sale price. The majority of option agreements offer the developer the option to purchase but do not commit them to doing so.
- **Conditional contracts** – involve the landowner and developer entering into a contract for sale subject to certain conditions, normally the grant of satisfactory planning permission for development. Contracts are exchanged and move to completion on the conditions being satisfied.

15.3.7 As might be imagined there are a range of hybrids between these different options, most typically Promotion Agreements which have the scope to move to an Option or Conditional Contract on a first refusal basis.

15.3.8 Negotiating these different agreements can be a time-consuming exercise, not least as there are so many potential variables involved. However, there are a number of key considerations which apply to a greater or lesser degree across all three mechanisms, as follows:

- *term of the agreement* – Promotion and Option Agreements tend to be for longer periods than Conditional Contracts given that more planning work will be involved; agreements should be specific about the end date and any extension period, most typically where an application for planning permission has been submitted but not decided.

- *critical dates* – where the agreement involves submitting planning applications there should be at least a timetable and preferably set dates by which various key milestones will be met.
- *initial fees* – option agreements include a relatively small fee for the landowner; this is seldom significant in terms of overall value but may carry more weight with the landowner than logic would suggest should be the case. It is important to be clear whether this payment is deductible from any ultimate sale price.
- *Promoter's Fee/Developers' Discount* – Promotion Agreements will include a share of the sale proceeds as a fee for the Promoter for securing planning permission, similarly Option Agreements will enable the Developer to purchase the site at a discount from Market Value. The shares will vary with the terms of the agreement and the nature of the site, but the Promoter's share will generally be higher at perhaps 15% to 20% of sale price compared to the Developer's Discount in an Option Agreement of perhaps 5% to 10%.
- *cost recovery* – both Option and Promotion Agreements will enable the promoter/developer to recover their speculative planning costs; it is important that these costs are carefully defined and do not include any inappropriate costs as, for example, a Promotion Agreement which enables the Promoter to claim in-house staff time as well as third-party costs might be seen as double counting given the share of sale price which they will receive.
- *obligations of the parties* – whilst it may seem superfluous, the parties should be obliged to act in mutual support in seeking planning permission; thus a landowner should not object to the scheme but more particularly there should be a positive obligation on the Promoter/Developer to use their best endeavours to pursue planning. This avoids the problem which occurred occasionally in the past, where developers took options on a number of sites in order to gain more control over submissions to the Local Planning Authority and then promoted the site which offered them the best advantage whilst simply mothballing the rest and denying those sites to others.
- *minimum price* – perhaps the trickiest part of an Option Agreement or Conditional Contract, not least as the landowner will expect this to be a minimum whilst the purchaser will have this as a target. It is important that the price should be expressed in simple terms ideally as an absolute price or a price per acre. Where price per acre is used, it is important to be clear whether that is on gross or net acreage and gross or net price.
- *contract price* – a Conditional Contract will include a contract price but may also include provisions for adjustment subject to the outcome of any conditions; an Option Agreement will include valuation clauses where the sale price cannot be reached by agreement between the

parties. In both cases the price may be adjusted to account for extraordinary costs, for example, dealing with unstable soil conditions, and this can be a particular area of conflict between parties, not least as such problems are not always anticipated in the heady days when the prospect of planning permission becomes stronger.

- *valuation clauses* – the valuation clause should have as few variables as possible and be relevant to the property itself. As with other valuation clauses this may mean that an unexpurgated Market Value definition is not the best for the valuation clause and the landowner may, for example, want the expert to take account of a Special Purchaser's interest. Valuation clauses should provide for reference to a third-party expert and, in passing, one of the difficulties for those dealing with large development sites is the paucity of experts able to deal with the complex viability assessments required for a major scheme.
- *tax and CIL* – landowners may wish to include provisions to enable them to withdraw from a sale should there be an adverse change in capital tax provisions (typically Capital Gains Tax). Purchasers will resist such clauses, seeking at the least to recover their abortive costs. Similarly, agreements should address how such issues such as Community Infrastructure Levy and s.106 agreements are reflected in the minimum price and contract price.
- *overage and legacy provisions* – in many cases, landowners will want to protect their position should the developer seek to improve the development after the sale, whether under an option agreement or conditional contract. Such a clause may refer to the number of units or, in more sophisticated arrangements, by reference to the Gross Development Value (GDV), that latter approach having the benefit of covering both an increase in the scale of development and, in residential developments, any reduction in the share of affordable housing or changes in the tenure which may have a significant positive impact on the GDV.

15.3.9 The arrangements will involve valuer and solicitor working closely together to achieve a comprehensive but practical agreement under whatever arrangement. This can be a struggle, particularly where the developer starts with their "standard" agreement, commonly accompanied with the assertion that this is inviolable. One of the difficulties is that while the landowner and development partners' interests are closely aligned in securing planning permission, they are thereafter in direct conflict over the price to be paid, with the landowner's share vulnerable to additional planning or development costs none of which will impact directly on developer's profits.

15.4 Site assembly

15.4.1 Arrangements for site promotion become more complex where there is more than one landowner involved, particularly the case with larger schemes. There are two critical issues: creating an environment within which the different landowners can act together and, essential to achieving the former, reconciling the different clients' financial interests.

15.4.2 The former is generally achieved through a consortium agreement, which sets out a commonality of purpose between the various owners, creates a mechanism for decision making, generally through voting rights and appoints an independent chairman, often a valuer with experience in property development. The latter is managed through a land equalisation agreement which, in its purest form, recognises that the proposed development only works as a single entity; thus, the vendor whose land is solely allocated for residential development will not see any benefit of the value in that unless another landowner is willing to commit their land for a public use with limited or no value, such as a school site or dedicated green space.

15.4.3 Complex site assemblies require goodwill and a willingness to recognise the merit of joint working from the various landowners. In the authors' experience there will normally be some more reluctant partners, and this is where the consortium Chairman earns their fee, encouraging recalcitrant owners to work with others. Indeed, in the most complex and contentious land assembly schemes, the majority of vendors will find more commonality of purpose with the developer than with their reluctant owner partner.

15.4.4 One particular issue with land assembly is the tax treatment of the disposal, particularly in the context of a large site involving phased development, which will be the case with the vast majority of consortium sites. It is too easy in the excitement of a development coming to fruition after many years of planning and preparatory work to overlook the incidence of Capital Gains Tax in particular. It is important that consortiums consider this at the outset, when planning is still possible rather than at the point of sale when it may be too late to alter the structures that will dictate how and when the tax will be due.

15.4.5 Site assembly for strategic developments may be a long and complex process, even where matters seem relatively straightforward. Again, personal experience may colour one of the author's judgement, but one strategic residential scheme the author – full of youthful enthusiasm – introduced in the first month of his joining a practice was eventually sold 22 years later.

15.5 Development valuations

15.5.1 Whilst much of a rural valuer's development work will involve guiding clients through the maze of planning considerations and marshalling the team of consultants required, the ability to value property with development opportunities is an important skill for rural valuers.

15.5.2 This is an important and complex area of work, and the capital values involved, particularly with larger development opportunities, can be very significant, well beyond the value of most farm properties.

15.5.3 Rural valuers are most likely to encounter development opportunities involving the conversion of rural buildings or development of sites on the fringes of towns or villages, although those in mixed local practices may also be asked to comment on town centre developments in smaller settlements. Similarly, the proposed development is most likely to be residential, although some local workspace may also be involved. In some cases, the core skills of valuation, market understanding and the ability to source and properly interpret available comparables will be sufficient. However, the larger the development opportunity, and the ultimate capital value, the more likely more sophisticated market knowledge and technical skills will be required, particularly where developments are phased over a long period.

15.5.4 Traditionally rural valuers have struggled to say no to a job and there is a natural antipathy to the idea of turning down any instruction, particularly where the property is in the valuer's local patch and belongs to an established client. However, the values involved and the complexity, which may not be apparent to a valuer who does not regularly undertake development valuations, means that the prudent valuer will pause for thought before accepting an instruction for development work. During that pause, it is important to check one's professional indemnity insurance in terms of the skills required, the level of cover for any single claim and the level of excess. In some circumstances, where the valuer is confident of the task, but it is outside of the normal scale of business undertaken, it may be appropriate to take out specific insurance to cover the project in hand. That may seem unduly cautious, but the feeling of having committed unnecessary expenditure to securing additional professional indemnity insurance is likely to be more comfortable than wishing in hindsight one had an extra £5 million cover.

15.5.5 Whatever the scale of valuation, valuing a site with the benefit of planning permission at least gives the valuer some certainty over the prospect of future development. Whether or not that development is viable, and thus valuable, is a different consideration. However, there will be occasions when a valuer is instructed to value a site without the benefit of a current planning permission or of permitted development rights. In those circumstances, the valuer is likely to be asked to take a view on the prospect of development, which can involve some complex and highly speculative assumptions about the scale and nature of development, the associated planning conditions and costs, likely timing and fundamentally the likelihood of success.

15.5.6 Any such assumptions, if intelligible and palatable, will be Special Assumptions under the provisions of the Red Book. However, valuers will have to think very carefully whether there is a sufficiently realistic prospect of such assumed events materialising in practice for them to qualify as Special Assumptions. Site-owning clients can be particularly unreliable accomplices in such considerations. Unsurprisingly they may have a rather optimistic view of the potential for development and even the most phlegmatic farmer may behave more like a bullish estate agent when contemplating the development prospects for a long-collapsed barn on their farm.

15.5.7 Small-scale development – The majority of rural valuers will feel that valuing a relatively small-scale rural development site is well within their compass, whether this involves some traditional buildings for conversion to residential or employment use or a paddock with scope for half a dozen houses.

15.5.8 The preliminaries of the valuation will be very similar to other capital valuations, and in some cases so will the mechanics of the valuation itself. There will hopefully be some comparables available and many valuers will adopt a comparable approach. However, there will also be occasions where a residual valuation is a more appropriate method, as explored further later on.

15.5.9 Whichever approach is adopted, there are a number of key issues beyond the physical factors and location of the site which could have a significant impact on the valuation. These are particularly focused on planning and include

- *planning status* – does the property have the benefit of a full planning permission, including the relevant s.106 agreement where appropriate, or deemed consent under Permitted Development Regulations? Or is that not yet secured?
- *planning permission* – where permission has been secured, what precisely is provided for in the development? What s.106 or other requirements or charges are involved? Is a purchaser/developer able to meet these requirements from within the site or are they contingent on other land or infrastructure developments?
- *practical development* – can the site be developed in isolation or does it require access to other land, whether to reach or locate services or for access to and from the highway (see 15.5.27 later on)? If not, have agreements been secured over other land and at what cost and if not, what might be required? Have agreements been secured with all the relevant service providers?
- *location* – how close is the development to other property? This is particularly relevant in the case of a traditional building conversion on a farm. Will that compromise either the continued working of the farm or the enjoyment of the retained farmhouse or the newly developed property? In one of the most awkward cases in the author's experience, complaints from the occupiers of a barn conversion close to a set of farm buildings led to the abandonment of dairy farming, an outcome which, coming to the case with the benefit of hindsight, was entirely foreseeable but which was not considered by the vendor when planning permission was sought, then taking the view that "any purchaser will know what they are buying next to a farm".

- *planning applications/promotions* – in the absence of an extant planning permission, in addition to all the previously mentioned considerations what is the relevant progress of any application? How long might it take to secure a decision? What would the proposed scheme comprise? Even more significantly, what is the likelihood of success?

15.5.10 The variety of issues in play, including the general inclination of owners to retain land with such potential, will make the prospect of identifying reliable comparables more difficult than when valuing, say, a block of arable land. This is where the residual method of valuation will come into its own, deriving value from an assessment of the anticipated outputs from the scheme rather than sales only remotely comparable to the undeveloped site.

15.5.11 The principle is that the valuer establishes the current value of the site by considering the ultimate value once developed (Gross Development Value, or GDV) subtracting the various costs of making that development, including building costs, fees, finance costs and developer's profit, and allowing for the risks on the way with the resultant figure being the value of the site.

15.5.12 In the past, the broad rule of thumb quoted by experienced valuers was that the site value would be approximately a third of the Gross Development Value with that GDV split roughly one third build costs, one third finance and profit and the final third site value. However, that rule no longer applies with increased building costs and, in particular, increased planning and other compliance costs eroding the site value. The potential for variation, particularly where planning conditions and associated costs are concerned means that it is unwise to rely with any certainty on a rule of thumb. That said, at the time of writing if a residual valuation of a small residential development opportunity produces a value more than 25% of the GDV it is worth rechecking the mathematics involved.

Example valuation – residual valuation part West Farm

The owner of West Farm is promoting a small paddock extending to approximately one-third of an acre for residential development. The proposed scheme, which has been favourably received by the Local Planning Authority, is for two pairs of semi-detached houses. A valuation is required on the Special Assumption that planning permission will be forthcoming on reasonable terms.

The valuer has undertaken some preliminary analysis of the scheme and the local residential market and has established that:
- based on the plans, the houses will be quite large semi-detached houses extending to roughly 110 square metres (1,200 square feet)
- the likely sale value of the houses is £325,000, equating to £2,955 per square metre (£270 per square foot)

- planning permission, if granted, will have no requirement to provide affordable housing, but there will be a total CIL cost of £7,000 per unit).

At a relatively simple level a Residual Valuation might work as follows:

Consideration			
Gross Development Value 4 semi-detached houses @ £325,000	1,300,000		
Less costs of sale @ 2.25% (agents/legal/marketing)	29,250		
Net consideration			1,270,750
Costs			
Construction			
4 x 110 sq. m @ £1,325 per sq. m	583,000		
Site/infrastructure costs @ £10,000 per unit	40,000		
Architects/surveyors fees @ 8% (on construction costs)	49,840		
		672,840	
Planning			
Planning costs	20,000		
CIL/Tariff	28,000		
		48,000	
Finance & Profit			
Finance costs (interest only @ 6%)	18,530		
On say 50% of construction cost & estimated land cost for 9 months			
Developer's Profit @ 20% of GDV	260,000		
		278,530	
Total Costs			999,370
Residual Site Value (incl. costs)			270,380 (20.8% of GDV)

15.5.13 Having completed a residual valuation the first step should be to compare the outcome against similar transactions to see how realistic that seems to be. That comparable analysis can be more complex and produce much wider apparent disparities than when dealing with a block of land for example. There are far more variables here: the nature of the site, the number of units, the constraints of the planning permission and the attractiveness of the permitted development. If anything, scale and the ultimate house product are the key issues here which will have a greater impact on the merits of the comparable than location or time. There are significant differences between a development of select detached houses and smaller terraced units (even if the agent describes the latter as mews cottages). Similarly, there are differences in costs, timing, scale of operation required and thus finance in developing 4 and 14 units.

15.5.14 The analysis of comparables may give rise to some significant apparent disparities and there are a number of variables which can have a significant impact on value, beyond the initial question of what the houses might be worth once developed, not least:
- what are the likely levels of construction cost? Here there are economies of scale to be had by the smaller builder developer using much of their own labour which may mean that building cost estimate books or websites will overestimate the actual cost.
- are there any particular difficulties with the site which will require remediation? That is assumed not to be the case here, but it is a risk with farm and other building conversion and particularly brownfield sites.
- are there any planning difficulties which may add to costs?
- how long might the development take and what impact will that have on finance costs? As noted later on, this is often treated in a more sophisticated fashion, with different tranches of finance for construction costs and land purchase. A degree of experience is required to estimate the likely development period. The well tried and tested approach adopted in this example, to estimate how long the development might have to fund 50% of the cost, given those funds will be drawn down in asymmetric tranches through the development period, might not always be appropriate. The larger the project the more important this element becomes.
- what level of developer's profit is required? In negotiation over options, for example, developers will talk about required levels of profit in the order of 25%, very often expressed in terms of a target that the land buyer involved would very much like to relax to assist the landowner but which is fixed "by the board". In practice, levels can range from 15% to 25% and sometimes beyond turning on the strength of the market and the nature of the risk.
- the residual value in this example is strictly the cost of land, so value plus acquisition costs for the developer, namely fees and SDLT (LBTT in Scotland, LTT in Wales). The view may be taken, however, that these costs form part of the speculative activity of the purchaser for which he charges his profit and hence this is effectively the value.

15.5.15 A further real case that develops the example described earlier was the subject of the 2019 Tribunal decision in *Foster*. This was an Inheritance Tax case over a parcel of pasture with development potential but no permission and reliant on another owner for access. By chance, the death was in a brief period when the English LPA did not meet its requirement for a five-year supply of housing land and so was exposed to speculative applications. The Tribunal, as in other such cases, keenly wanted direct market evidence from comparables, but was not satisfied with what was submitted and concluded that it was driven to use a residual approach. It assessed the number of plots that might be practical, deducted costs such as those mentioned earlier and then considered the risks in achieving the planning permission (coloured by the timing needed for a bat survey) and also what might have to be paid to have the necessary access to the site. The Tribunal deducted 60% from the net figure to allow for the planning risks and then 50% of the remainder for the access issue, on a ransom assessment, leaving just 20% of the net figure as taxable.

15.5.16 The next step, having tested the outcome against comparables, and particularly at the early stages of any development project, is simply to test the outcome against existing use value. There is a commonly held assumption that any planning permission will enhance the value of a property beyond existing use. Whilst that is very often the case in the rural market, it is not always so. For those valuers operating in market town economies who might occasionally stray into a more urban development project in secondary and tertiary locations in particular, the existing tired property let on a low rent but with a reasonable tenant may be far more attractive than a new vacant development where the project costs mean that the rent required to achieve viability is far above that for alternative properties on offer. Changing requirements, such as for flood protection, may outweigh the extra value a residential development might seem to offer over an employment one.

15.5.17 This is the simplest approach to a residual valuation, other than perhaps taking a figure between 20% and 25% of GDV and hoping for the best. For those valuers who become regularly involved in this sort of work, or aspire so to do, more detailed study of the explanations of residual valuation in core valuations textbooks is strongly recommended. Here valuers will discover a more sophisticated approach to finance costs in particular, not least where the development is for properties to be retained for investment as is more commonly the case in urban situations.

15.5.18 Strategic development – Whilst there are issues of uncertainty and interpretation in valuing relatively small-scale developments, the challenges increase exponentially when the site concerned is a major development site, whether an urban extension or a new settlement. The variables, particularly in terms of planning conditions, the cost of the s.106 agreement and Community Infrastructure Levy (CIL) are far more significant and the range of potential values significantly greater.

15.5.19 Looking back to the twentieth century there were still some local valuers, whether in general or rural practice, sufficiently confident, or foolhardy, to attempt valuations of more substantial residential development sites. At that time planning was generally more homogenous, with far fewer variables on housing mix and much smaller demands on what was then termed "planning gain", particularly contributions to infrastructure and affordable housing. However, marked slumps in the development market in the 1990s and then again with the credit crunch coupled with a rash of professional indemnity claims, including some highly speculative claims led by consultants reviewing loan books for banks on an incentivised basis, caused many valuers to reconsider the wisdom of undertaking such work.

15.5.20 Similarly, market variations have been far more extreme in the twenty-first century than has previously been the case. Depressions in the housing market in the latter part of the twentieth century saw significant fluctuations in the market for residential development land but, whilst demand may have been subdued, land would generally sell, albeit at a depressed price. The impact of the financial crisis in 2007 was of an entirely different magnitude. Developers did not simply slow down development, as had been the case in slumps in the 1980s and 1990s, for example; rather, major housebuilders put their entire land-banks up for sale, generally off market, and in some extreme cases land values fell into negative territory. As always, the effects varied around the country.

15.5.21 This created shockwaves for landowners and stakeholders across the development spectrum. As usual, there was something of a lag in parts of the development community recognising the trend and the author had the unnerving experience of advising one local authority that the unsavoury combination of the collapsing market and their rather ambitious infrastructure and planning brief meant that the combined site value of the two major urban extensions in the district was **minus** £900 million.

15.5.22 Whilst a comparable approach will still be helpful as a test in some cases, the value of strategic sites involving large scale and long-term developments can vary hugely on location, infrastructure and planning costs and scale. Where detailed valuations are involved, whether for funding, tax planning, price setting for option agreements or many other reasons, specialist development valuers are likely to turn to more sophisticated discounted cash flow (DCF) methods to establish the present day value of a site which may take 20 years to develop out in full.

15.5.23 In the vast majority of cases, such DCF valuations will be undertaken using bespoke software to undertake what are effectively a succession of discounted residual valuations. Those programs ease the mathematical effort required, although they are by no means infallible. They are less well equipped to deal with negative values as in the case cited earlier for example and, more particularly, they are governed by the same universal acronym as all other IT software, "GIGO" – garbage in, garbage out. They are only as effective as the valuer identifying the necessary variables and, subject to the age of the valuer involved, their more recently qualified or specialist assistant who understands the nuances of such software.

15.5.24 A detailed consideration of the valuation of large-scale strategic development sites is beyond the scope of this publication and the author. However, for those tempted to indulge in such arcane practices – and the difficulty for many rural valuers may be in deciding whether to engage with those cases on the margin between small scale short-term sites and large-scale long-term sites – key considerations for a residential led development include

- *proposed development* – including the mix between residential and other uses, whether market uses, local retail centre or workspace, or social uses, public open space or education or other facilities and the residential mix, between market and affordable housing and the type of affordable housing whether to rent or to buy, which can have a marked impact on value;
- *phasing and timescales* – the phasing of market and affordable housing and the timescales for infrastructure and other expenditure, in the local retail centre, for example. These issues have a significant impact on cash flow and funding requirements, and developers will be anxious to generate receipts from house sales before becoming embroiled in too much other expenditure.
- *market capacity and demand* – where significant housing numbers are involved, assumptions will have to be made on the capacity of the market and the number of units likely to be sold, generally estimated as monthly or quarterly sales. This requires something more sophisticated than a damp finger in the wind. Experienced development valuers will seek advice from local estate agents over the capacity of the market to absorb new houses and the house types most likely to attract keen demand, which does not always coincide with the local planning authority's view of house types required.

15.5.25 The challenge for the valuer, armed with that information, together with all the cost estimates and sales value by house type, is to assemble a rational flow of income and expenditure, apply an appropriate range of discount rates and derive a realistic value. Whilst the concept of valuations seems automatically to imply a single outcome, sensitivity analysis on pricing and sales volumes, varying economic forecasts impacting on borrowing costs and discount rates and different building cost inflation estimates are all likely to lead to a range of value estimates for the early phases of strategic sites, even before one embarks on some heroic assumptions as to the future price of the individual house types in the face of local competition.

15.5.26 In common with all other valuations, the process becomes more logical and manageable with practice but, by comparison with most valuation exercises in the rural world, there are many more variables and much more information to manage. This demands a rigorous approach to the mechanics of the valuation process, an affinity with numbers and an awareness of wider economic circumstances. As with all valuation work, initial sorties are best undertaken with the reinforcement of an experienced practitioner.

15.5.27 Development and ransom value – As mentioned previously, one of the complications in developing a site will be access, whether to services or to and from the highway. In some cases, the site will have the benefit of direct access to both. However, this is often not the case, and some agreement will be required with adjoining owners to enable development. In the larger strategic developments, this can often be resolved by consortia and land equalisation agreements. Where smaller sites are involved, a simpler agreement is more likely, between site owner and neighbouring landowner who controls the access, although the service provider's input is essential where services are involved.

15.5.28 Where one site controls the access needed to another, any positive resolution will usually be by commercial negotiation when the respective strength of the parties' positions will be fundamental. However, valuers also have the advantage of legal precedents, most notably the compulsory purchase case *Stokes v Cambridge Corporation* (see also Chapter 21). In that case, the Corporation controlled the access without which Mr Stokes' land could not be developed. The Corporation's land could also be developed as part of the scheme and, in assessing the compensation overall, the Lands Tribunal concluded that the value of the access in that case was 30% of the uplift in the value of the land as a result of the development that was enabled by the access. That case, and the decided percentage, has been handed down through generations of valuers involved in negotiating access agreements for development, not least because it has the great benefit of being applicable to a wide range of situations, not just access.

15.5.29 The *Stokes v Cambridge* principle is commonly applied to road access, but it can also apply with services and the author has used it in assessing and dividing the additional development value where two neighbours, one with a more dominant position, joined together to increase the scale of development. The actual percentage of the uplift that is used will depend on the circumstances. The wider and more practical the options are for the subject site, the less anyone else can hold it to ransom and the smaller will be the share of the uplift that accrues to them.

15.5.30 The approach in *Stokes v Cambridge* has been applied in other, perhaps less well known but equally relevant, cases from *Ozanne*, where the share of uplift to the access providers, who between them owned the only access, was 50% to *Kaufman* where two access areas received 15% and 10% respectively.

15.5.31 Looking to the future, the principle that the partner providing whatever is required to facilitate or enhance development should share in that enhanced value, rather than simply being paid the market value of their land might equally apply, for example, to the provision of off-site biodiversity net gain that will enable development elsewhere. However, in such a case the share of that uplift might be less because, whilst there may be only one of two practical options for physical access to a site, there might be a larger number of landowners able to offer biodiversity net gain, though possibly fewer who would be willing.

15.6 Renewable energy schemes

15.6.1 There have been a number of changes in the rural economy over the 35 years since the first edition of this book was published in 1985. Amongst the most evident of these is the development of renewable energy, which has been most marked in the dozen years since the fourth edition was published in 2008.

15.6.2 The various renewable technologies tend to be bundled together, but there have been three distinct stages of development: large scale hydro-electricity, the development of onshore wind farms, where the first farm at Delabole in Cornwall was launched at the same time as the second edition in 1991 and the subsequent development of domestic and field scale solar, anaerobic digestion (AD), hydro and also offshore wind. New technologies are developing all the time and the latest generation includes different varieties of energy from waste and 'battery' and other storage mediums which contribute to the balancing of supply and demand on the grid on a millisecond basis.

15.6.3 The growth in renewable energy means that it is now second behind gas powered plants as the main source of electricity for the UK (28.1% from renewables through 2018). That is likely to continue to grow with the pressure to meet climate change targets, new developments and the repowering of first-generation wind farms with more efficient modern turbines.

15.6.4 Whilst the technology may be new, many of the demands on the valuer are very traditional, negotiating the terms for an option, lease or very occasionally sale, securing planning permission and valuing the asset as well as occasionally trading it, although much of that seems to be taking place between businesses rather than necessarily involving a valuer or agent.

15.6.5 The variety of different assets and technologies involved demand different levels of involvement from valuers. Offshore wind fields developed by operators under licence from the Crown Estate will have little impact on the rural valuer until that electricity is brought onshore when traditional easement and compensation negotiations will arise for the cables. Field scale solar, onshore wind and storage sites will demand considerable input in representing clients who have the opportunity to host these activities.

15.6.6 Site finding – The majority of renewable energy involves natural resources, principally sun, wind or water. That suggests that the first thing to look for is an abundance of each but, while that is certainly the case, it does not mean that only the sunniest or windiest sites will be suitable. Elsewhere the energy comes from industrial processes, AD and other forms of energy from waste, where proximity to feedstock is the important variable. In either case, the availability of a good grid connection and capacity in the local and regional grid is essential if the project is to rely on exporting power to the grid rather than be used on or by the site. This is a frustration for landowners in some parts of the country where what would have otherwise been a very suitable site is discounted through lack of grid capacity or an available connection.

15.6.7 Some valuers have found employment with various renewable development companies engaged in site finding, firstly for onshore wind, then solar sites and, more recently, for power storage. One early feature of the renewable sector was the extent to which such site-finding businesses did just that, gathering a portfolio of option agreements, with varying potential for successful development, with a view to selling those to renewable developers. That seems less prevalent than was previously the case, but it remains a sector where much development will often be undertaken using the business structure of a Special Purpose Vehicle (SPV) created specifically for each individual development. The SPV may sometimes include the landowner as a participant in it where they want to take a position in the development.

15.6.8 There are some rural valuers, who have specialised in this area, who will understand the technical issues which make for a good wind farm or solar site, the two technologies most commonly encountered by valuers to date. Headline data on windspeed and sunshine is available from various internet sites, but otherwise specialist consultancies will be able to provide site assessments both from historic data and through on-site monitoring. This may not seem essential as most developers will undertake their own detailed surveys during the predevelopment period to support a planning application. However, some owners, wishing to develop sites themselves – farmers looking to single turbines or roof mounted solar for example – will need to gather such information.

15.6.9 Beyond the natural resource required, key attributes, extending to both natural and industrial renewables, include:
- an adequate, accessible and cost-effective grid connection (normally electricity other than for the gas to grid sites) – a connection can often be surprisingly expensive
- a telecoms connection for telemetry, less challenging now with modern communications than previously
- wayleaves for power supply and communication
- reasonable access for development and maintenance
- the ability to make the site secure
- a site which avoids, depending on the technology:
  - flight paths, for both aircraft and migrating birds
  - electromagnetic and microwave interference

- a site which mitigates as far as possible, again depending on the technology:
  - visual impact
  - noise impact (although turbine gearboxes are much quieter now than was the case
  - visual flicker – particularly close to dwellings
  - impact from traffic movements where regular feedstock deliveries are involved

15.6.10 For natural resource renewables, that tends to suggest sites which are remote from dwellings in particular. Industrial sites are more likely to be either on farm, that being the case for the vast majority of AD plants or on industrial estates for some waste from energy sites.

15.6.11 Leases – As with minerals, much of this activity takes place under lease with landowners retaining the reversionary interest in the site and negotiating option and lease agreements has been a fruitful area activity for rural valuers.

15.6.12 The nature of those leases and the financial rewards will vary with the different renewable types, but as with all work there are important nuances requiring some adaptation to accommodate different issues and, in particular, to protect the client for the future. In an area where change is happening at considerable pace, technologies still on the virtual drawing board may be available for commercial exploitation during the term of the lease being negotiated. Precision is required, not least with user clauses, to circumscribe what is and is not permitted now in the tightest terms to preserve the best position for the landowning client for the future.

15.6.13 It is unlikely that developer and landowner will move straight to a lease; the developer is likely to want the security of an option to protect its position whilst undertaking the complex and often lengthy planning and feasibility work, detailed negotiations for grid connections and the like.

15.6.14 Options for renewable developments will have many terms which are consistent with those for residential or commercial property development, with the same characteristics of a relatively small initial payment for the landowner often based on the scale of development based on the projected scale of generation.

15.6.15 Again many of the lease terms will be recognisable from traditional property leases. The duration of most leases will also be recognisable from previous generations of "institutional" leases ranging from 20 to 25 years generally reflecting the anticipated life of the equipment. Larger scale solar farms with the benefit of support payments may want longer terms. That said, decisions about the future may come rather earlier than might be anticipated. Early wind farms were let on institutional length leases, but the technology, improving efficiency and outputs from new turbines has advanced so that it is worth replacing first generation turbines well before they are worn out. That, in turn, is likely to involve revisions of the terms of the lease for matters such as the permitted development, length of term and the rent package.

15.6.16 There will also be some distinct elements reflecting particularly the particular renewable technology involved; these might include, but are not limited to:
- user clause – as set out earlier, sufficiently detailed to enable the landlord to protect their position so that the use cannot be extended beyond the agreed use without consent from the landlord
- a rent provision based on
  - rent per acre, or more likely and appropriate
  - a royalty on electricity generated, subject to a minimum payment
  - or a hybrid of the two – with the rent per acre acting as the minimum payment supplemented when an agreed percentage of output exceeds that figure, that percentage now tending to rise over time for longer leases as the developer recovers its outlay in the earlier years and, where necessary
  - arrangements for rent sharing where the site covers more than one landowner
- a confidentiality clause
- provisions for site development
- reservations for access service wayleaves over land outside the lease
- provision for assignment – perhaps to a nominated company in a group
- arrangements for management of land within the lease but still in agricultural use
- provisions for repowering
- an indemnity for the landowner
- options for extension
- detailed provisions for decommissioning the site
- provisions for a bond to protect the landlord in the event that decommissioning is not completed effectively

15.6.17 Two particular areas are perhaps worthy of mention. At the risk of repetition, the user clause should be tightly defined to avoid granting rights for a wide range of future technologies unforeseen and thus not provided for financially in the negotiation. Thus, a generic user clause for a battery barn that, for example, refers to *"the use of the site for the storage of electricity"* enables any future technology which will store electricity many times more efficiently or at much greater capacity than that provided for in the rent to be used on the site.

15.6.18 The second area is decommissioning, yet to be seen for most renewable leases but potentially an area of substantial risk for the landowner. Wind turbines are relatively discrete, if very large, structures; solar farms are the opposite. The former will need some heavy engineering equipment to remove them, capable of causing substantial damage to the surrounding land and, if contractors lack care, properties on and around the access route. The concrete base might be considered separately. Solar farms, consisting of miles of steel framing and cable and acres of glazed panels, may not require the same scale of equipment, but they will need considerable and very assiduous effort to clear all the debris from the site.

15.6.19 In some cases, this may not be a problem, with, for example, one generation of panels or turbines following the next. But what if the site does have to be cleared, whether at the landowner or the tenant's behest? Arguably anything capable of being built is capable of being effectively dismantled; however, it is difficult to imagine the site clearance contractor, with profits depending on speed of operation and the scrap value of the material and no long term interest in the future of the site, being quite as careful as the installers, who were building a scheme with a design life of 20 or 25 years.

15.6.20 The practical risk is with the landlord. While there is recourse through the Courts there is normally not much future in suing a shell Special Purchase Vehicle tenant with limited assets at the end of the lease or indeed may already have been wound up – or, in default of the tenant, the contractor. Consequently, as well as ensuring that the lease makes the strongest possible provision for decommissioning with no room for misinterpretation in the future, the landlord should insist on an insurance bond to cover decommissioning and future legacy costs. Advice should be taken on the likely costs of decommissioning the site; whatever the figure now, it is likely to be eye-watering when projected forward 25 years.

15.6.21 Some will argue that that shows an excess of caution and that the reversionary value of the site with its "developed" use will far outweigh decommissioning costs. That seems optimistic, although, interestingly, approximately two thirds of respondents to the Grant Thornton Renewable Energy Survey, a good source of industry data, expected the salvage value of equipment to match or exceed decommissioning costs, obviating the need for a decommissioning reserve. That may prove to be the case, but any problems that arise are likely to be for the landlord who would then suffer most if there is no decommissioning bond. Even if there is one, it might still not guarantee a full clearance of the site as there is likely to be a point where clearing the last elements of the installation is not cost effective. That residue, however small, and particularly in the case of underground debris, can have an adverse impact on the utility and value of the site.

15.6.22 Renewables and agricultural tenancies – Given the long-term nature of most renewable development, this may be seen as an enterprise for freehold owners rather than the tenanted sector. Where a let farm does offer potential for, say, a wind or solar farm, then it is highly likely that the landlord will wish to regain possession of that land, with all the attendant complications of securing possession for any development. However, in the case of smaller scale single turbines or roof mounted solar, it may be that the farm tenant, who will see a direct benefit from generated electricity perhaps for on-farm use, will wish to pursue the scheme. On some holdings with intensive livestock, AD may even be a prospect, although the levels of investment involved means that this will be a challenging investment decision for landlord and tenant alike.

15.6.23 Whilst the change in support regime has dramatically reduced the number of new commercial roof mounted solar installations, the rural valuer may well encounter this as a tenant's improvement, the parties hopefully having considered matters together at the time the work was done, at the end of the tenancy – especially where there is a Feed-in Tariff agreement, it may be less apt for it be a tenant's fixture. Aside from domestic panels, many poultry buildings have their power needs supplied by solar panels on their roofs, while controlled environment onion stores will take energy from solar panels just when it may most be needed. Each of those options brings the final benefit of using self-generated electricity to substitute for buying it in, with the larger saving in that over exporting power to the grid.

15.6.24 There is nothing different about the basis of valuation; this is the same for the solar panels as the grain store on which they might sit. However, it might be appropriate to value the revenue stream on a Dual Yield approach, otherwise normally a foreign concept for rural valuers. The suggested distinction here is that the panels are essentially wasting assets and, once they degrade sufficiently to stop working, there is no value and potentially a liability for disposal, whereas the building on which they sit is likely to have a residual value for much longer.

15.6.25 There are further complications where the panels are sitting on a landlord's building, although in most cases landlords, tenants and their agents may be able to have mature conversations and reach pragmatic arrangements to enable the development to take place. Who knows how many such developments were left on the "too difficult" pile, however, simply on the basis that the tenant thought it was too novel to be acceptable? That perception is changing with new generations of agents and clients, albeit perhaps not fast enough.

15.6.26 Valuation – As with all assets, valuations may be required for a variety of purposes and at various different stages in the lifecycle of the project. The basic preliminaries are the same as other property valuations with the exercise covered by the Red Book, by regulation for those valuers who are chartered surveyors and as a commonly accepted reference point for others, although with the TEGoVA Blue Book taking its place for European valuers.

15.6.27 Thus, the same principles over securing instructions, inspections, assumptions, special assumptions and reporting will apply, although with particular application to the challenge of renewables. Guidance has been provided by both the RICS and the CAAV with the most as set out in Further Reading later on.

15.6.28 The requirements for valuation may vary with the different stages of the project:
- once trading, valuations might be required for disposal or acquisition, whether of the site or the leasehold interest, or as part of a refinancing package
- before and during trading, valuation might be required for funding purposes
- at any stage, valuations may be required for Inheritance Tax, matrimonial or other family or business disputes or financial reporting

15.6.29 Whilst the underlying asset, or combination of assets, land and equipment, may stay the same, the interest to be valued may vary particularly in a leasehold situation between the landlord and the tenant's interest, in some on-farm situations complicated where landlord and tenant are linked; for example, a family farm business operating the wind turbine on a site rented from the family landowning members. That in turn will decide the basis of value, whether that is Market Value, Fair Value or Worth.

15.6.30 A further complication may be the benefit to be valued. RICS guidance, and common practice, directs the valuer to consider income streams, whether for the operator or the landlord. However, there will be some circumstances where there is significant additional value, whether that is simply from the cost saving on self-generated electricity or, less tangibly, the operational value that comes from a secure electricity supply. Thus, one Irish dairy, for example, entered into a solar farm arrangement to provide 20% of the plant's electricity requirement, partly on a competitive cost argument but more particularly to enable further expansion of the plant in an area with very limited electricity capacity. That utility, or security value as it might be called, is more than simply the sum of the electricity supply bills saved over the next 20 years and arguably would be best valued based on the future profits of the business.

15.6.31 As mentioned earlier the timing of the valuation will be important. Broadly this might be thought to fall into three main phases: conceptual, development, completed and then, in due course, obsolete or legacy. However, the RICS Guidance Note of 2018 identifies six stages:
- pre-consent
- pre-construction
- mid-construction
- post-construction
- mature trading and end of life

15.6.32 In common with all valuations, a good deal of research and data gathering is required, although in this case rather more is likely to be learnt from inspecting documents than the site. Site inspection is, though, still essential, not least to ensure that the turbine, AD plant or other asset is really there. That may sound slightly far-fetched, but one of the authors, commissioned by a commercial enterprise to value a variety of interests which had been acquired "off plan", was obliged to report that, whilst there were some assets, they were not necessarily where they were supposed to be – with a substantial impact on value.

15.6.33 Having satisfied oneself of the physical existence of the site, or where developed, the plant which sustains the interest to be valued, there is a long list of other data to be collected, including, beyond such traditional questions as ownership and title, and depending on the stage of the project and the nature of the asset being valued:
- option and lease terms
- development access
- service and supply wayleaves
- planning consents and conditions
- community fund requirements
- the grid connection offer, availability and cost
- power supply contracts
- incentive agreements and subsidy payments
- development costs and timescales
- technology warranties – including any warranty mismatch; for example, between the life cycle of the equipment and the warranty
- generating capacity forecasts
- generating records
- management agreements
- business rates
- capital and revenue expenditure forecasts
- feedstock supply contracts (for AD, with potential problem of having a long-term obligation met by short-term feedstock contracts)
- digestate disposal agreements
- duration forecasts
- decommissioning bonds

15.6.34 Alongside that data, the valuer may need to make an informed judgement, again, subject to the interest and the nature and stage of development on matters including:
- the covenant of the Special Purpose Vehicle tenant and the operator
- advances in the chosen technology
- the degradation and effective economic life of the technology in place
- future electricity pricing for generated and exported power
- repowering opportunities
- end of life/legacy issues

15.6.35 That is likely to lead to a valuation report that is liberally distributed with assumptions and special assumptions, particularly where the valuation is at the conceptual or development stage. It is even more important than usual to reinforce the extent to which reliance has been placed on the accuracy and veracity of information provided by the client.

15.6.36 As with almost all things in life, except perhaps rural valuers' attire, there are fashions in renewable energy and particularly in funding. In an earlier life, one of the author's more intriguing tasks was to sign off appraisals for wind turbines, typically single or two or three-turbine developments on leased sites on farms, funded through crowd funding. At a time when returns from other investments were already falling to what at the time seemed remarkably low levels, funding targets were met in between two days and a week.

15.6.37 Having collected and established this wide range of data, the valuer is challenged with which is the best valuation method to adopt. The RICS guidance note manages to explore the entire range of methods normally available; comparable, investment, profits, residual, Depreciated Replacement Cost and Discounted Cash Flow, observing that they may all have some application before concluding, not unreasonably, that in most cases the Discounted Cash Flow approach is likely to be most appropriate, albeit requiring a reliance on assumptions including:
- factual – contracted revenue and costs
- forecast – energy yield/expenditure
- market influences – electricity prices
- return – investor sentiment and appetite
- support – subsidy regime

15.6.38 As always with DCF valuations, one of the key considerations will be the discount and inflation rates used to forecast the net present value. Here, as in all areas of valuation, an appreciation of the industry and the ambitions of the parties involved is important and specialist knowledge is important. Again, industry surveys provide some indication of the target yields.

15.6.39 Whilst DCF is likely to be most appropriate for valuing the operator's interest, whether that is as a tenant or owner-occupier, there are complications, most notably:
- interpreting complex rent formulae and other rent terms
- modelling generating capacity
- potential risks including grid frailty and changes to incentives
- the stage of development.

15.6.40 Here the requirement for detailed modelling of future performance may be slightly less demanding when valuing the landlord's interest, but only marginally so. Considerable conjecture will still be required, and it may be that a DCF approach is still the most appropriate.

15.6.41 A further nuance in many cases, particularly with wind turbines, will be the potential for repowering with the potential to significantly improve both the output and efficiency of wind farms. Technological development can be fast paced; in the six years between 1999 and 2005 similar turbines improved output by approximately 20%. Similarly, of 13 wind farms repowered the original output of 51.9 megawatts increased to a repowered 151 megawatts, generated from 80 turbines as opposed to 122 turbines in the first generation. That adds a further complication to the assessment where repowering may be a possibility.

15.7 Minerals and waste

15.7.1 Mineral working has been a characteristic of the rural economy for generations, whether for building or aggregate, often sand and gravel, or more specialist uses ranging from high grade sands for glass manufacture to specialist minerals for mobile phones and other modern essentials. Once exploited, the resulting voids may also offer value in a variety of ways, from waste disposal to leisure.

15.7.2 In the vast majority of cases, these minerals are exploited by quarrying companies either occupying land which they own or, very often, land which they occupy on specialist mineral leases. In a limited number of cases, the quarry may be occupied by the farm and estate and exploited for aggregate or building materials for home use. Whatever the arrangement, value will turn on two issues, the demand for the mineral involved and the ability to secure a useful planning permission for the working or its extension.

15.7.3 In common with other topics in this chapter, this is a specialist area but one significantly further removed from the remainder of the property market than, for example, the residential development market. To that end, any rural valuer contemplating the valuation of minerals for whatever purpose should consult with one of the firms of specialist mineral surveyors, many of which have now grown into substantial engineering and environmental consultancies.

15.7.4 In the same way that the rural valuer will expect to understand the demand for agricultural commodities, the mineral surveyor will appreciate the relative quality and demand for minerals. In the current market, with the government looking to infrastructure development as the engine for economic growth even before the Covid-19 pandemic, one would imagine there would be an almost insatiable demand for minerals, good for quarrying companies and landowners alike. However, this is a volatile sector and the recent rapid development and equally rapid, but hopefully temporary, demise of the Drakelands tungsten mine on the edge of Dartmoor shows how rapidly things can change.

15.7.5 On that basis, one may wonder whether there is anything further to be said. However, there will be some rural surveyors for whom an appreciation of the sector will be important, whether because they act for estates who have the benefit of valuable mineral rights or they operate in areas such as the Thames Valley or the Mendips where the mineral industry, sand and gravel and hardstone quarrying respectively, are important and accepted parts of the economy.

15.7.6 Many mineral consultancy firms will also deal with the procurement, management and assessment of waste sites, whether these are exhausted mineral workings or more sophisticated modern waste treatment plant. Again, rural surveyors may become involved in their management role or occasionally in greenfield development where some urban extension sites in particular may border, or include, former waste sites. To that end, the remainder of this section considers some of the headline issues which may impact on the assessment of such sites.

15.7.7 As in other specialist areas the temptation for the general practitioner comes where a mineral working or waste site figures as part of a larger instruction. Faced with the valuation of a small estate for Inheritance Tax purposes, one might be tempted to question what harm there can be in having a quick guess at the 30-acre quarry site when you have all the skills to value the rest of the property. Arguably none, if the guess happens to be correct, but what if it is not? Worse still, you will not know whether it is right or not until you have to defend it. As elsewhere, there is much to be said in referring the matter to a specialist; after all at some stage a mineral surveyor will want a hand with a farm valuation so this is both a prudent and professional way of proceeding, not least securing the right answer for the client, an investment in networking for the future.

15.7.8 Mineral rights – Mineral rights may run with the surface ownership, or those rights may be reserved and owned by a third party, including amongst others the Church and various long-established academic institutions as well as where it is part of manorial waste or held by the owner of a common. In other cases, mineral rights will be reserved to the state, notably rights to coal, oil and gas, or they may be Mines Royal with the right to win gold and silver reserved to the Crown and now managed by the Crown Estate.

15.7.9 Where rights are reserved to a third party separately from the surface, the impact on the surface land value will turn on the terms of any such reservation and, more particularly, whether the minerals are ever likely to be worked, which will reflect in turn the extent of the deposits and the prospect of planning permission. The reservation of mineral rights does not, of itself, suggest that there are any winnable minerals; it is more likely that this is a consequence of a universal approach to property disposals, by the historic landowner, sometimes centuries ago.

15.7.10 In some cases, rights may be reserved subject to the payment of compensation to the surface owner when this may be some recompense if the rights are exploited. Whatever the situation it is important for the valuer, undertaking the valuation of a farm or land for any purpose, to confirm the ownership of rights. Where they are reserved, then even if the prospect of minerals being worked is infinitesimally small there may still be some marginal impact on value, particularly for those properties in the lifestyle market. The absence of complete control, rather than the presence of any real threat, will be a distinct disadvantage to some potential bidders.

15.7.11 Where the rights are owned then the prospect of their adding additional value is fairly remote in most cases, unless there is a real prospect of minerals being won and won in such a way that the activity will not significantly impact on the value of the retained holding. This is the stage where the valuer, rather than reaching for a silage valuation to extrapolate the volume of potential mineral deposits, should call in an expert.

15.7.12 Deposits – Meaningful deposits are likely to be identified in the Minerals Local Plan whether simply identified as potential deposits or strategic reserves. Once the preserve of county councils, Minerals Plans in England are now set at county council, unitary authority or national park level. These plans sit below the national policies which are set separately at national rather than UK level and are part of the authority's wider Mineral and Waste Development Framework.

15.7.13 Further information on deposits is held in the data compiled in the British Geological Survey which maps the location of extraction sites, the extent of mineral deposits, planning permissions and licences and statutory designations. The inclusion of a site in the Minerals Local Plan would suggest at least the potential for commercially viable exploitation. Planning policies in some districts will support mineral extraction in some areas but bar it from protected landscapes, although upland national parks tend by definition to have supported mineral working in the past.

15.7.14 The existence of historic working and the potential for associated grandfather rights can be particularly relevant in the minerals sector. The Drakelands Mine, lying within the Dartmoor National Park, owed its redevelopment at least in part to the previous mineral operation.

15.7.15 Whatever the planning status, one of the most significant hurdles to clear will be local antipathy to the development. Quarry workings, whilst operated to much higher regulatory standards now than used to be the case, do not make the best neighbours, generating lorry movements and seen as likely to cloak the neighbouring land, properties and passing livestock in a coat of dust. Even the best managed can be noisy, unavoidably so given the scale of equipment used, and, at the extreme boundary of effects, one of the authors used to manage an estate where the local drinkers were suddenly disturbed by a boulder being blown through the pub roof from some "controlled" blasting in the neighbouring quarry.

15.7.16 Aggregates Levy, introduced in 2002, was in part designed to offset these and some of the environmental and other impacts of quarrying. However, the fixed charge on sand, gravel and crushed stone (relieved in Northern Ireland) does not go directly to the affected parties and consequently has done little to reduce animosity to mineral development.

Whilst also intended to encourage the recycling of minerals, china clay and slate waste, for example, some would argue that it has simply exported environmental impact and economic activity which would otherwise have taken place offshore to those administrations which do not have similar taxes.

15.7.17 Managing environmental impact will be an important element of securing planning permission, whether for new sites or extensions to existing sites, which might be lateral, taking in more land, or vertical, taking out new benches within the confines of the existing quarry. A further particular consideration will be the future reinstatement for the site. In some cases, this may be relatively straightforward, as evidenced by the shallow sand and gravel workings along the M4 allowed to convert to lakes, sometimes with leisure uses. Elsewhere, traditionally, reinstatement has been to agricultural use, although, often poorly done, this seldom if ever produces much beyond some fairly sour grazing. In the current climate, ultimate reinstatement to forestry or conservation sites may be both more beneficial and more cost-effective.

15.7.18 In many cases, there may be a final phase before that ultimate reinstatement, with part of the workings being used for associated industrial working, an asphalt plant in a roadstone quarry, for example, or for waste storage or recycling. Given the cost of managing demolition material and excavated soil, that latter use may offer good returns, although the move against landfill has reduced some of the demand.

15.7.19 Mineral leases – In many cases, a rural valuer's engagement in minerals will be through managing leased mineral sites, which may often extend to little more than being sent weighbridge tickets against which to raise a royalty demand. However, as always, it will be important for valuers to check and understand the lease, not least when it comes to a rent review or lease renewal.

15.7.20 As with any other form of lease, there are a wide variety of arrangements for mineral workings, although most, depending on their age, will involve:

- a fixed term with the potential for extension for another fixed term subject either to:
  - identified deposits not being fully worked
  - planning permission being secured for extension – although this may also be covered by an option to extend on negotiated terms
- a term beyond the end date for mineral working for the completion of reinstatement and future management of the landscape
- a relatively detailed user clause, often incorporating elements of the operator's plan put to the Local Planning Authority and covering issues including hours of working, timing and regularity of blasting (where used), provisions for access and egress routes,
- provisions for the development of processing facilities on site, sometimes with separate provisions as to rent

- a rent package combining:
  - a basic site rent, paid for the full area whether worked or restored or awaiting working (sometimes at distinct rates) and
  - a royalty payment for every tonne of mineral extracted
- provisions to deal with a downturn in demand, based either on a minimum extraction target or a minimum royalty payment
- provisions to review elements of the rent package
- an obligation on the operator to provide full details of the output to enable calculation of the royalty
- full indemnity for the landlord, including in modern leases against contamination
- provisions for determining the tenancy
- the restoration plan, including arrangements for the removal of the processing plant
- perhaps a share of fees for waste received
- provision for a bond to be paid by the operator in favour of the landlord, essentially to protect the landlord's position should the operator not complete the restoration

15.7.21 Again, when it comes to negotiating a lease, often done on an option basis to start with, the valuer should turn to an expert more experienced in the minerals market, not least for the best appreciation of the strength of covenant of the operator. However, the rural surveyor will have skills to bring to the negotiation:
- what will be the optimum restoration plan? To what extent might that fit into a wider estate plan for example?
- what will be the impact on adjoining agricultural and sporting activities and any residential property? Is that sufficiently compensated for by the likely returns?
- how will this impact any diversified activity whether operated by the landlord or another tenant? Arguably the attraction of a Montessori Nursery in a farmhouse might be a little diminished by the new quarry next door and that will that impact have to be addressed in any transaction?
- is the land required accessible? In physical terms, does it have road access? In legal terms, what is the status of the tenancy? If it is a 1986 Act tenancy will a Case B notice work or, as so often the case, is it contingent on a part resumption clause in a tenancy agreement or an otherwise costly negotiation?

15.7.22 In very general terms the same issues and the same terms for leases may apply with a waste site, with the notable distinction that material is being brought in rather than taken out. Those sites tend to fall into four main categories, sometimes with more than one on the same site:
- waste collection sites; waste transfer or civic amenity sites
- waste treatment sites, particularly recycling and green waste composting facilities

- energy from waste plants
- waste disposal sites; landfill and underground storage sites

15.7.23 For extant landfill sites, there will be particular concerns about the nature of material being imported and the long-term legacy risk of contamination. Increasingly, waste sites will be industrial processes, producing a product capable of being recycled, with no long-term dumping on site. In those cases, valuers versed in the market will have an appreciation of comparable rents which will turn partly on issues which any property valuer would recognise, such as location, access and convenience of the site. However, they will also be vulnerable to volatility in the marketplace, with technological advances and changes in industrial processing impacting on the demand for waste management and the profitability of the business.

15.7.24 Planning can be equally as challenging for waste operators as quarry companies. That factor, together with a number of companies operating in both sectors, means that quarries towards the end of their economic life as a mineral site may be proposed for waste processing. Where this is the case, the landlord's agent should clearly consider the impact of the activity, which is likely to prolong the occupation of the site and weigh the financial benefit against the potential impacts.

15.7.25 The proposed waste use is sometimes represented as marginal financial activity by the operator, with the quarry being chosen on the basis of the long-term relationship rather than anything else. That may be the case. A quarry site, which is there because of the mineral deposits, may be rather less convenient for a waste site, but the supply of sites for such activities is limited, and that should not be overlooked in considering any proposal.

15.7.26 In some cases, a landfill waste site will be no more than a steep-sided valley (with a variety of dialect names, sometimes "goyle" in the West Country or "clough" in the north) on a farm, with the owner striking a deal with an operator to import inert waste, often from building sites. This may be a very effective and lucrative way of creating level productive agricultural land, but the licensing rules involved, particularly in terms of volumes of material which can be deposited, are strict and the penalties very significant. Again, specialist advice is required, particularly where there is any prospect of over-exuberance on the part of the operator, or the landowner, leading to the limits being breached.

15.7.27 In a few cases, the quarry or waste operator may wish to acquire the freehold of the site, possibly where significant investment is required. On the very rare occasion that this suits both parties, and a lease is by far the more common way of working, then this is more likely done by an option rather than an outright sale with uplift on the grant of planning.

15.7.28 Valuation – Where the mineral deposit is being worked under the provisions of a lease, an investment approach will be appropriate, capitalising the anticipated income from the lease in the same way as one would

with an agricultural investment. However, there are marked distinctions including, but not confined to, the significant challenges that
- income is uncertain and potentially volatile given that the demand for the mineral may vary (as with the economic cycle or the fortunes of the area within a convenient radius)
- the reversionary value of the land involved is likely to be very substantially reduced from its value before the mineral activity commenced
- there may be an opportunity for alternative use at the end of the workings – one quarry in south Wales for example has recently been sold to a diving school – but that may be highly speculative
- in some cases, there may be the long-term risk of subsidence or contamination

15.7.29 In practice, therefore, this is more likely to involve a Discounted Cash Flow approach to the valuation to reflect the vagaries of the income stream. That will be particularly the case where the deposit is to be exploited in phases over a long period. That, in turn, needs an appreciation of current and future demand. Whilst that will be easier where a quarry serves a predominantly local and construction market, for example, it is still likely to be the province of a mineral expert.

15.7.30 Some specialist valuers will be involved in the valuation of sites for the quarry companies, whether these are freehold or leasehold interests. The detail of such valuations is outside the scope of this publication, but for those interested the concept of wasting assets comes into play as explored in the RICS Guidance referred to later on. That note draws the valuer's attention to the distinction between valuing a quarry, void or waste site in isolation or as part of an operating business. A similar distinction might be drawn between valuing a let quarry in isolation and as part of the landlord's wider interest.

Further reading

The On Farm Generation of Renewable Energy (CAAV, 2011)
Property Valuation, Peter Wyatt, Second Edition (Wiley Blackwell, 2013)
Valuation of Assets in the Commercial Renewables Sector (RICS Guidance Note 2018)
Valuation of Mineral Bearing Land and Waste Management Sites (RICS Guidance Note 2016)

16 Livestock, machinery, growing crops and produce

16.1 Introduction

16.1.1 A variety of assets that are not part of the land may also have to be valued, whether on death, as part of a farm sale (if sold "lock, stock and barrel"), between outgoing and ingoing tenants, in stocktaking, a partnership dissolution, divorce or for other purposes.

16.1.2 Animals are a significant asset and separate stock of value on livestock farms. Modern agriculture requires machinery, and some machines can cost hundreds of thousands of pounds. Farm produce may be held against sale or, like silage, made and kept for own use.

16.1.3 The chapter offers a basic introduction to what have tended to become more specialist areas of work but ones for which the agricultural valuer should have an understanding as these are fundamental parts of farm business. It is very useful for those training to watch auctions and dispersal sales carefully to see what animals and machinery are being sold for what prices and what may have driven those prices. The farming press is a good source of current information of prices of arable produce and animals sold through markets for slaughter. Breeding livestock may more often be seen at more specialist ram sales, bull sales or dispersal sales. Dairy cow prices are best seen at dispersal sales. Pigs and poultry will now rarely be on public sale in any volume.

16.1.4 For livestock, it is increasingly the case that livestock auctioneers do little other work and may specialise within that, in particular species or classes of animals, while high value animals (such as high genetic value bulls or dairy cows, not only pedigree ones) are the province of a few specialists.

16.1.5 The agricultural valuer will ordinarily meet only machinery that has been purchased, usually used, some well used, second hand or very second hand. Many farms have a miscellaneous collection of older implements in various clumps of nettles. While new machinery may have listed prices, buyers can often achieve significant discounts and may also be trading in the equipment being replaced as part of the purchase. There are practices that specialise in machinery auctions, serving not only the country but also international markets, revealing prices.

16.1.6 For both livestock and machinery, valuation will be by comparison. What have other animals or machines like the one in front of you been sold for? What are the similarities and differences? A dispersal sale will illustrate the range of values for different animals in the same herd, from a good, young, freshly calved commercial dairy cow to an older, dry cow.

16.1.7 Produce is more typically reported with prices for relatively homogenous commodities meeting standard specifications. While there are a limited number of major buyers, the markets are such that supply and demand interplay freely, though now on a more global than local basis. Local prices for such internationally traded produce will also be affected by currency movements. However, only a relatively small share of world production crosses borders, sometimes magnifying the effects of national surpluses, shortages or trade restrictions. With prices accessible for much produce from wheat to hay, the valuer's task is more often about assessing volumes and understanding quality.

16.1.8 For a few major crops, there will also be forward prices, agreed now to the delivery of a stated volume meeting a given specification at an identified future date. It is then for the farmer to meet that commitment. If, for some reason such as crop failure in drought or flood, he does not have that volume when it is due to be delivered, the necessary volume will need to be made up by being bought in to fulfil the contract with the circumstances likely to make that expensive. However, selling part of an expected crop forward where there is an acceptable price is often a useful means for a farmer to manage the sale of part of a year's crop. The remainder is then sold when the farmer chooses. Choice in this matter turns on the farmer having adequate and legally compliant storage whether on the farm or, increasingly, in pooled storage. That alternative can be cheaper than building storage and may, according to the facility and terms, offer flexibility in blending stocks and both skill and scale in marketing.

16.1.9 A few but material items of produce – notably clamp silage and manure – have most value on the farm, posing challenges for their valuation while some others, such as hefted sheep (sheep that know their boundaries on an open hill and are acclimatised to local ticks), may have a greater value in their location.

16.2 Livestock

16.2.1 The agricultural valuer should have a knowledge of the main breeds of dairy and beef cattle and sheep. Even if never asked to value them, this is part of the general knowledge and conversation of the countryside, and not only in pasture areas. As well as what is put through markets, a good variety representing their breeds should be on display at leading agricultural shows, notably the Royal Welsh Show, where other species such as pigs, poultry and goats can also be seen with owners often willing to talk.

16.2.2 Cattle – Cattle have now almost entirely been separated between dairy breeds and beef with only a few farmers having dual purpose animals. Surplus calves from dairy farms may go on for beef production and that choice may influence the selection of the bull used by the dairy farmer.

16.2.3 In all cases, all cattle (and also buffalo and bison) must have cattle passports for them to have value. An animal without a passport is unsaleable.

16.2.4 Dairy Cattle – Dairying has become heavily focussed on the Holstein Friesian breed developed to be an efficient producer of high volumes of milk, commonly aiming for 10,000 litres per cow each year and often achieving more. There may be growing experimentation with other breeds, whether to achieve a particular milk composition or for some "New Zealand"-type systems where animals kept out of doors, often on lighter land or in drier areas, may need to be smaller than the ever growing Holstein Friesian to protect pastures.

16.2.5 The markets for milk include sales for liquid milk processing and distribution for producers nearer urban areas (including some who are producer-retailers) while those further afield are more limited to buyers of milk for processing as cheese, butter, and other milk products including milk powder. Some milk producers are themselves makers of cheese, yoghurt, ice cream and other products, enabling them to keep the sale value of the milk inside their business and adding value for the sale of the final product.

16.2.6 Information for valuations of dairy cattle will mostly come from dispersal sales where a decision has been taken to sell the whole or a significant part of a working herd. Dispersal sales are commonly well reported in the farming press. Few good dairy animals ordinarily go for sale in a livestock market.

16.2.7 Typical issues in considering dairy cattle are:
- breed
- age – older animals are nearer the end of their milking life
- lactation information as to annual yield, the composition of the milk (average quality milk might be 4.1% fat and 3.3% protein) and the stage in the current lactation. Holstein Friesians, often with short lives, may be of less interest after three lactations when they may have a higher cell count.
- the physical condition of the cow:
  - its conformation, with dairy cows, distinguished from the blocky shape of a beef animal, looking like a wedge whether viewed from the side, the rear or from above. The udder should be spacious and well formed with good milk veins, four well-spaced teats and no warts
  - its health, with mastitis, lost quarters of the udder and lameness as common conditions
- its calving status, as a cow in calf or freshly calved is likely to be more valuable.

16.2.8 Beef cattle – A wider range of breeds and of crosses between them will be more commonly found with beef cattle than for dairying, with all generally aiming for well-fleshed carcases with good conformation and good proportion of lean, saleable meat, all maximising the final sale value. Within that constraint, the choice of breed will be a matter of the farmer's personal preference, taking account of farming system and the character of the farm, especially in areas requiring hardier and more thrifty breeds.

16.2.9 As already noted, the classic beef conformation is a solid rectangle, well fleshed, firm with a long back and well-developed hind quarters and loins rather than a large belly, so emphasising the higher value cuts.

16.2.10 Typical issues in considering beef cattle are:
- breed
- conformation
- condition
- age

16.2.11 Pedigree and high value cattle – Despite the growing interest in animal genetics, formal pedigree status is not necessarily a marker for high value. Pedigree societies with herd books maintain breed lines and typically hold breed sales, supporting both those that are committed to the breed in question but also a wider pool of genetics for livestock.

16.2.12 Parts of the dairy sector see high performance commercial animals achieve very high values when some pure-bred pedigree lines do not. Semen selection for both dairy and beef animals and embryo transfer, more usually for dairy, can be ways of creating value.

16.2.13 High value bulls and rams – Some specialist breeders focus on producing these animals that can achieve remarkable prices at specialist sales, being bought for the genetics (and perhaps reputation) that they can bring to other farmers' herds and flocks. This is to be seen as a specialist endeavour and, as with other walks of life, success revealed at a sale is rewarding in more than financial ways: in personal satisfaction and enhancing reputation.

16.2.14 Sheep – The great majority of flocks use cross-bred sheep for the commercial production of lambs, typically for slaughter but sometimes as store lambs for others to fatten. A smaller number of flocks are run as pedigree animals to produce breeding stock.

16.2.15 With a now limited UK market, exports to the continent, whether France, the Low Countries and Germany or the Mediterranean, with their differing tastes, account for over a third of lamb production. That, with changing demand at home, has moved markets away from fat lambs to leaner carcases and specific breeding lines for this, such as the rise of the Beltex sheep with larger hind quarters. There will still be local variations and traditions, especially in the area requiring hardier stocks as with Swaledales in the Northern Pennines and Herdwicks in the Lake District, and where demand can be built for rare breeds or specific provenance.

Nonetheless, the larger commercial demand is for carcases provided to a standard conformation that can be consistently handled in a slaughterhouse and give the required meat.

16.2.16 Valuing finished lambs is largely a matter of their weight and the market price for that conformation.

16.2.17 Valuing store lambs typically considers:
- breed
- condition
- weight
- a sense of how the finished lamb market is moving.

16.2.18 Valuing breeding ewes will typically see consideration of:
- breed and type
- age, tackled by assessing the ewe's teeth: as, with only bottom teeth, a ewe
   - up to 12 months, has its suckling teeth
   - at a year, it has two large teeth
   - at two years. four
   - at three years, six
   - at four years, eight

but then loses them and becomes broken mouthed, possibly then with only cull rather than breeding value
- a good udder to rear lambs well
- the good health needed to rear good lambs.

16.2.19 There will be a particular value for a ewe with unweaned lambs at foot, seen as a single valuation unit, referred to in some districts as a couple with single couples for one lamb and double couples for two.

16.2.20 Milking sheep (or goats) might be viewed on the same points as dairy cows.

16.2.21 For the valuation of the benefit of hefting and acclimatisation of hill sheep and of bound sheepstocks in Scotland see Chapter 29.

16.2.22 Pigs – Pigs are predominantly kept indoors, often in enclosed high health status units, but also outdoors on light land within an arable rotation. While there are markets for meats from traditional breeds such as Tamworths and Gloucester Old Spots, this is very much a minority, and the general market is for a standard product with a great focus on the genetics and production systems behind that. Relatively few are traded on open markets mid-life, though some systems see pigs move between more specialist producers at stages in their lives, as after weaning. Nonetheless, there may be occasions when there is a need to value, including a divorce or partnership dissolution.

16.2.23 Valuing breeding sows will start with an assessment of confirmation (well-fleshed but not fat, good feet, good legs, an adequate number – say 14 – of well-placed teats) and health with no sign of mastitis or other infection.

16.2.24 For a sow in farrow, note:
- the farrowing date
- the number of litters already produced
- the boar in question.

A sow close to farrowing will typically be worth more than one only recently served. Over a sow's life the balance will shift from the expectation of more litters to her cull value. The sow and her litter of unweaned pigs might be a single valuation unit with consideration of the health, number and evenness of size of the piglets.

16.2.25 A stock boar should have good health, sound feet and legs and not be fat or lazy.

16.2.26 The nearer store and fattening pigs are to final slaughter, the more the value will be a matter of weight and market price with a judgment of leanness and conformation. Further back in the chain (those between weaning and 60 kilogrammes), more attention might be paid to breed, size, conformation and health with valuation by comparison.

16.2.27 Other livestock – The valuations of other farmed animals, from trout to alpaca, are specialist topics for which the valuer would be advised to turn to a specialist.

16.2.28 One recognised phenomenon is where an enterprise becomes a sudden fashion, as has been seen with angora goats and ostriches. Breeding livestock, as the constraint on numbers, can become valuable very quickly. This creates a bubble which can then burst when the market reaches saturation and late purchasers are often financial losers.

16.3 Machinery

16.3.1 Machinery, as the fundamental tools for undertaking farming, may range from a combine harvester or exceptionally powerful tractor with a field train to an aged implement or a quad bike. Values may need to be found for any of these, working from comparison in the market and knowledge of machinery. On some occasions, it might be right to bring in a specialist machinery valuer.

16.3.2 Some machinery may be very large, heavy or, like a grain drier, have to be fixed to property to work. That may make it necessary to determine whether:
- it is a fixture which runs with the land or a chattel that can be sold independently;
- if it is a fixture, whether it is a tenant's agricultural fixture saleable under the tenant's statutory rights or can otherwise be removed legally.

16.3.3 For machinery, as much as any other topic, the importance of patient description, forcing a close inspection of the machine, is clear. That may often be supported by a checklist for condition and also necessary legal documents, as appropriate, such as registration documents and MoT certificates. Tractors will have a record of the hours for which they have worked, often a better measure of use than age.

16.3.4 United Kingdom road traffic law imposes its own requirements on road vehicles and so, especially for tractors, self-propelling equipment and trailers, there are legal standards to be met, with non-compliance making the machine much less valuable or unusable on the road.

16.3.5 On leaving the EU, the United Kingdom carried forward a quantity of EU legislation on machinery safety standards, often additionally implemented in domestic regulations covering what can be put on the market or used in work, electrical safety, machinery safety, provisions for lifting equipment and other matters. CE markings are needed to enable sale into the European Economic Area Single Market.

16.3.6 Close consideration must be paid to the condition of the equipment and the signs of how well it has been maintained. Common features to check are PTO guards, machinery guards, tractor cabs and roll bars. While it is often easy to see damage, it is not always so easy to spot that something is not there.

16.3.7 Sometimes it may be difficult to establish who owns a machine. Not only might that be an issue within families, but problems can occur where items have been stolen or are subject to an outstanding Hire Purchase (HP) agreement where the apparent owner in possession of the item will in law be only its hirer until he has made the final payment to purchase. The internet now means that a check can usually be made for a machine's hire purchase status.

16.4 Growing crops

16.4.1 Approaches – These may need to be valued in various circumstances from the end of a tenancy or stocktaking to a sale contract, compulsory purchase or business dissolution. While conventionally viewed on the basis of costs adjusted for risk and circumstances, as a crop matures there is a balance between that approach and considering the crop's prospective value as a harvestable proposition but recognising the remaining costs and risks. There is a market for the sale of some standing crops, such as maize, that a farmer might buy, cut and take for fodder.

16.4.2 Starting with cost – This assesses a crop on the basis of the cost the farmer has committed to the crop with seeds, fertilisers, cultivations and other acts of husbandry. Those are the costs that someone taking to that field would have saved while gaining the advantage of a crop that might then have been well established at the right time and so better placed to yield well at harvest. This would anyway be the basis for almost all stocktaking and was the case under the 1986 Act but is now no longer prescribed for England and Wales by it. Scotland, working until 2019 on a common law basis, also had no legal requirement for this. However, it is almost always likely to be the basis used in practice for young crops subject to adjustment for the circumstances.

16.4.3 However, that is not in itself a measure of the value of that crop and so needs adjustment where there is a clear difference. Drought, flood, rabbits or other factors may have rendered that expense less effective or useless, requiring its downwards adjustment or removal.

16.4.4 Equally, over time that crop would come to be recognised as established. Historic tenancy practice tackled that for autumn sown crops on a spring tenancy (so needing a valuation on a spring termination) to have that accumulation of cost supplemented by an additional enhancement payment capped at the rental value of the land in question. That would also apply to grass or clover seeds sown, whatever the termination date, provided no crop had yet been taken from it.

16.4.5 Once it has been established what has been used, the costs of seeds, fertilisers and sprays can either be taken from the farmer's records or from knowledge of the marketplace that anyone who might take hypothetically take to that crop would be in.

16.4.6 The cultivations from initial preparation of the land and seed bed through to the valuation date need to be set out. Since 1945, the CAAV has published for each year typical costs of a range of agricultural operations such as ploughing, drilling and spraying, taking account of the overhead costs of the machinery used as well as operating costs such as fuel. Where the farmer has done the work, these should be used as a basis that is recognised (including by HMRC for stocktaking) and often specified in contracts, adjusted as necessary for circumstances such as higher wear and tear on sandy or stony soils or where there are smaller fields. Where a contractor has done the work, such as spraying, the paid invoice for that work is then a direct cost of the crop.

16.4.7 Looking at value – The nearer an average or good crop is to harvest the more it will be looked at in that light as a commercial prospect. There will still be costs and acts of husbandry (including harvesting) and risks from heavy rain, hail, flood, fire or disease.

16.4.8 This requires a prudent assessment of yield, recognising how hard it can be to see a good grain yield until it is harvested. For root crops, sample digging can allow a view to be taken on yield and quality. A market price can then be applied to that yield – where available, a forward price for the likely harvest date might be appropriate.

16.4.9 The outstanding costs required for the crop to reach harvest then need to be set out. This might not only be harvesting and, for straw crops, baling but also spraying.

16.4.10 A figure for risk should be deducted from that, judged on the circumstances and the time remaining to harvest. That is the commercial risk that a buyer would assume and, in reality, might overlap with the profit the buyer would want from the venture.

16.4.11 Multi-annual crops – Orchards, vines, asparagus, soft fruit and other crops may also need assessment but with the prospect of future crops, not just the one in the year in hand, and the life of plants involved.

16.4.12 That may variously mean:
- having a view of the life of those plants
- applying and writing off the cost of establishing the plants, sometimes substantial over the likely life of the plants
- recognising that newly established plants may be some time from yielding or their peak yields – vines might not give useful grapes for three years and reach sustained production levels in five years.
- the annual costs of managing the plants and the crop.

16.5 Produce

16.5.1 The main items being considered here are hay, straw and silage, harvested on the farm and kept for livestock on the farm.

16.5.2 Harvested combinable crops have a ready sale market with valuation turning on an assessment of volume, weight and quality to which the current market price can be applied. The same approach can be taken to stored roots crops.

16.5.3 In assessing the weight of stored crops from a calculated volume the following figures are offered as approximate guides to densities for different crops in m^3/t; divide the volume by the following figure to give the weight. Valuers may need to adjust according to their knowledge of the circumstances.

Hay			Bulk grain in store at 16% moisture	
Small bales	6.0		Wheat	1.3
Big bales	8.0		Barley	1.4
Unsettled	10.0		Oats	1.9
Oilseed rape	1.4			
Barley Straw				
Small bales	11.0		Beans	1.2
Big bales	16.0		Peas	1.3
Wheat straw			Potatoes	1.6
Small bales	13.0		Sugar beet	1.8
Big bales	18.0		Onions	2.0
Grass silage				
18% DM	1.3			
30% DM	1.6			
Maize silage - clamp	1.3			

16.5.4 Hay – A distinction is drawn between:
- meadow hay taken from old pastures and seen as suitable for cattle and sheep
- seeds hay from rye grass, timothy and cover mixtures, more often used for horses.

Good hay should smell sweet, be leafy and, cut young, not be fibrous.

16.5.5 For many years, hay has been baled, using either:
- conventional small bales, with some 50 to a tonne, convenient for an individual to handle and so might be baled in this way for sale in small volumes to individual horse and animal or pet owners or for use on small farms
- big bales that can be handled by farm machinery, whether:
  - round bales of perhaps 150 to 200 kilogrammes according to how they are made
  - rectangular bales that, if made to a high density, can be up to two thirds of a tonne each.

Valuation is a matter of assessing the weight of the hay present, applying current values for the hay as it is and adjusting for any factors that might reduce that value.

16.5.6 Small bales are easier to assess as a sample of bales drawn from various places in the stack can be more readily weighed to find an average weight and then counted, allowing for the possibility of there being irregularity, such as a void, in the structure of the stack. Where it is not possible to count the bales, a weight can be assessed by determining the volume of the stack and then dividing it by the density:
- very compact – 6–7 m^3 per tonne
- medium – 8–9 m^3 per tonne
- unsettled – 10–11 m^3 per tonne.

16.5.7 Weathered bales on the exposed side of the stack should be disregarded as may be at least half the bales in the bottom layer as they might be damp. There could be considerable wastage where bales have been left out in the open, even if sheeted.

16.5.8 With big bales, assessing an average weight may mean having three or four taken to a weighing facility. A total weight of the big bales can then be determined.

16.5.9 Current values can then be applied to the hay, allowing for:
- its quality
- its convenience for use
- how well-stacked and protected it is.

For a tenant right valuation, bales not reasonably needed for the holding might be disregarded.

16.5.10 Straw – Straw, while a relatively high volume, low value commodity, is now the subject of national and even international trade. Areas that have specialised in arable faming typically have a surplus of straw, and livestock areas need to import it. This becomes more extreme in times of straw shortage, often in dry years when fodder may also be short, when prices can rise substantially. Arable farmers may also choose to incorporate straw into their soil where it is not worth baling.

16.5.11 Of the different crops from which straw might come:
- wheat straw is usually used only for bedding
- barley and oat straw is used for feeding, though winter barley and oat straw is more brittle and so of lesser value than spring barley straw.

Straw should not be dusty, damp or weathered.

16.5.12 Straw for farming use is increasingly handled in large bales, again either
- rectangular (sometimes called "big square bales") with the classic Heston bale (1.2m x 1.2m x 2.4m) weighing just over half a tonne, and smaller sizes proportionately less
- round; again sizes vary, but a standard size is 1.2m in diameter and 1.2m wide.

Small straw bales are still made for the same reasons as small hay bales.

16.5.13 Valuation is again a matter of assessing weight and applying a value per tonne. Straw is lighter than hay, with perhaps 70 small bales to a tonne. Where bales cannot be counted, common density figures are:
- heavier bales – 10–11 m^3 per tonne.
- medium bales – 12–13 m^3 per tonne
- light bales – 14–16 m^3 per tonne.

16.5.14 The general principles then are those for assessing hay. Wrapped straw bales may suffer less outside than hay, but once exposed will see considerable deterioration.

16.5.15 Silage – For a full exploration of this, see CAAV Numbered Publication 183, *Silage: A Valuer's Guide* – silage, mostly made from grass but also from maize (for its starch contents) and other green crops, can be found both baled, for which there is a market with each bale containing only a small volume, and in large volumes in the clamp, where it is rare for it to be sold. Especially on a dairy or beef farm in the autumn, a large clamp of silage can have a significant value for the stock of winter feed that it represents.

16.5.16 Grass silage should be a light green to greenish-brown colour, have a smell between pleasant and faintly acidic and be firm. Made by a natural process of fermentation, dark brown or black silage is overheated; olive green or dark green is underheated with a rancid or musty smell. Where secondary fermentation occurs, it will have turned the silage brown with a fishy or musty smell.

16.5.17 The quality of baled silage turns first on its wrapping. Wrapped silage stores better than bagged silage. Where either wrapping is badly punctured, the whole bale may be ruined.

16.5.18 Clamped silage, if well-made, with its large volumes to surface area ratio, is protected by the floor and walls of the clamp and the sheeting put over it, helping preserve it from natural decay. Where the walls have been sheeted there may be even less wastage, but there may still be wastage in the top six inches and the slope at the mouth of the clamp.

16.5.19 Assess the volume of the silage in the clamp. That can then be converted into weight once the density has been calculated by taking and weighing a sample volume.

16.5.20 When considering clamp silage, there is only rarely market evidence for its value. Due regard should be had to any sales, though they tend only to occur in times of fodder shortage, but must then consider the costs of haulage and the risks of deterioration from moving it.

16.5.21 Unfortunately, the value of wrapped or bagged bales is often not directly applicable as:
- it is rarely analysed, requiring the market to allow for that risk
- there are often greater losses where a bale is damaged
- the market price will typically assume that the purchaser removes the bales with the haulage costs of moving a bulky product
- its smaller quantities and associated flexibility place it in different market.

16.5.22 Samples of clamp silage should be taken and analysed, taking care for those samples to be representative of the clamp as it may contain more than one cut. The analysis will give useful information about:
- digestibility (the "D" value – look for it to be between 60 and 70)
- ash content indicating soil contamination – look for this to under 10%
- fermentable ME (metabolisable energy) – look for this to be above 9.9MJ/kg corrected dry mater and not below 9.
- pH levels with acidity between 3.8 and 4.3 acceptable and values below 3.8 being too acidic.

16.5.23 With that lack of market evidence, the approach uses a "basket of feeds" to calculate from the values of common feeds the unit costs of the energy and protein that the silage has been analysed to provide. This reflects the basic economic insight that where two goods give the same benefit, the buyer will choose the cheaper one, substituting it for the other until the prices equate. Here the main benefits are in units of metabolisable energy and protein, which the farmer can obtain from a range of sources, and so will seek to formulate a feed ration that optimises those benefits for the lowest cost. The approach here finds what else the farmer might have to pay for what the silage will give. Incidentally, that also helps indicate that silage is not a free food but comes at the cost of good management of the pastures as well as of making the silage.

16.5.24 The two baskets of feeds used to calculate average values for energy and protein are:

- for energy – feeding barley straw, good hay, barley, wheat, beet pulp and maize gluten
- for protein – Brazilian soya, rape meal, beans, peas and distiller's grains.

16.5.25 Using commodity values as reported in the farming press, the formulae produced from work by ADAS for the CAAV to find the value of a unit of metabolisable energy and a unit of consumable protein are:

- for energy

$$ME = \frac{(\text{Average cost of energy feeds} \times 39) - (\text{Average costs of protein feeds} \times 14)}{213}$$

- for protein

$$CP = \frac{(\text{Average cost of protein feeds} \times 12.9) - (\text{average costs of energy feeds} \times 15.6)}{213}$$

Note – the values of 39, 14, 12.9 and 15.6 are fixed, as the underlying arithmetic is seen as unlikely to change.

16.5.26 The resulting financial values for metabolisable energy and consumable protein are then applied to the figures from the analysis for the silage's energy and protein content, adjusting for its dry matter with this formula uniting the two values to give a single figure:

$$\frac{(\text{Unit Price ME} \times \text{ME\%} \times \text{DM\%})}{100} + \frac{(\text{Unit Price CP} \times \text{CP\%} \times \text{DM\%})}{100}$$

16.5.27 In addition to those noted earlier, adjustments are then made for such factors as
- the ammonia level with discounts where this rises above 10%
  - by 1% for each % up to 15%
  - then by 2% for each % up 25%
  - thereafter by 3% for every 1 % of ammonia
- starch content, recognising its extra energy, with the formula

  Silage value corrected for starch = silage value x $\dfrac{100}{(120 - \text{starch \%})}$

- density which varies with the corrected dry matter percentage.

16.5.28 As with any valuation, but particularly for one constructed on this more indirect basis, the valuer should at the end step back and consider whether the result found is the value payable for that silage. Would a farmer be reasonably expected, in the current fodder market and other economic circumstances, to pay that money for the feed value the silage offers?

16.5.29 For bagged silage, sample at least 1 in 20 of the bales and apply the same approach to the composition found. However, high dry-matters percentages should be treated with caution. There is more direct market evidence of the sales of bales which are themselves easier to sell on.

Basic example of a valuation of clamp silage made from grass

The silage has been analysed to show values of:

ME	10.6MJ/kg
Dry Matter	23%
CP	13%

No direct market evidence is found to be available, so the valuer moves to the basket method (perhaps any a cross check for adjustments to be made for any market evidence).

Step 1 – Obtain the values per tonne of the components of the baskets of feeds

Energy feeds		Protein feeds	
Barley straw	£25	Brazilian soya	£210
Hay	£80	Rape meal	£143
Barley	£72	Beans	£88
Wheat	£76	Peas	£96
Maize gluten	£87	Distiller's grain	£119
Beet pulp	£87		
Average	£71.16	Average	£131.20

Note – Those figures are dated ones to encourage readers to look at the values current at the time of reading.

Step 2 – Assess the unit values of ME and CP, applying the average feed values to the formulae given previously.

$$\text{ME} = \frac{(£71.16 \times 39) - (£131.20 \times 14)}{213} = \frac{(£2{,}775.24 - £1{,}836.80)}{213} = £4.40/\text{unit of ME}$$

$$\text{CP} = \frac{(£131.20 \times 12.9) - (£71.16 \times 15.6)}{213} = \frac{(£1{,}692.48 - £1{,}110.09)}{213} = £2.73/\text{unit of CP}$$

Step 3 – Calculate the feed value of the silage as analysed using these unit values

$$\text{Feed value} = \frac{£4.40 \times 10.6\text{MJ} \times 23\% \text{ DM}}{100} + \frac{£2.73 \times 13\%\text{CP} \times 23\%\text{DM}}{100}$$

$$= £10.72 + £8.16 = £18.88 \text{ per tonne of silage}$$

Step 4 – Apply any further relevant adjustments – **not considered here**
Step 5 – Review the answer – **is it a sensible price in the circumstances at the time?**

Further reading

Silage: A Valuer's Guide (CAAV, 2006)
Guide to the Costings of Agricultural Operations (CAAV, published annually)

17 Environmental valuations

Note – This is a developing area of practice in which significant experience is still being gained and for which the mechanisms for transactions are still emerging. Nonetheless, it calls on the agricultural valuer's traditional skills of practical appraisal and perception of what will give rise to value.

17.1 Introduction

17.1.1 With the growth of concern about climate change, biodiversity and other environmental matters having an increasing importance in driving policy – as, for example, in DEFRA's 25 Year Environment Plan and the Clean Growth Strategy – insights from environmental economics are being used to develop tools to take this into account in markets and so in values. With agricultural policy and land management among the significant means seen to deliver change here, rural professional work is now seeing more consideration of these matters, and they are coming bear on values and decisions.

17.1.2 Much of the argument turns on tackling the point that businesses and people can, through their actions, impose wider costs on society because no cost is imposed on them for doing so. One current example, already a substantial issue in Northern Ireland, and coming forward with DEFRA, is ammonia from farming. Agricultural activity is held to be responsible for 88% of ammonia in the air, notably from slurry management. That both deposits nitrogen on important environmental sites and adds to air quality problems downwind, helping pollute air in cities with public health consequences. Farmers may be completely unaware of those effects of an unseen emission and have paid no cost for them. The policy question becomes one of how such issues might be taken into account in decision making on farms and elsewhere, whether through regulation, taxation or some form of market incentive.

17.1.3 At the same time, there is growing evidence of private sector activity finding tools to resolve environmental issues in transactions with farmers and land managers. Classic examples include:

- New York saving itself the cost of new water treatment plants by coming to agreements with dairy farmers in the Catskills mountains to reduce nitrate output
- Nestle agreeing with farmers on ways to protect the catchment for Evian water

and now many schemes being managed by water companies in the United Kingdom to influence land management to reduce harmful run-off into water courses that would be much more expensive to remove. Examples include agreements to limit the use of metaldehyde to control slugs, managing pastures to limit clostridium, buffer strips to shield water courses and support for improved slurry and farm water management.

17.1.4 One early development in this area was with the EU Habitats Directive requiring the provision of compensatory habitat by development that was publicly desirable but adversely affected the environment. That has led to work at Steart Point in Somerset to replace habitat loss arising from work elsewhere on the Severn Estuary and at Sunk Island in Yorkshire to compensate for habitat lost by a new terminal at Immingham. The cost of that was then an economic factor in that development.

17.1.5 That theme was picked up by England's Natural Environment White Paper in 2011, with biodiversity offsetting for some development, and now more comprehensively by the government's proposal in England that all development should result in a net biodiversity gain (initially of 10%), whether on-site, off-site or through contributions to local government to do such work. That would tend to deter development of more sensitive sites by making that more expensive and promote improved biodiversity. Off-site agreements would be facilitated by the allied proposal for conservation covenants to be part of English law.

17.1.6 Everything in those two paragraphs is to be resolved by contracts between parties, with environmental factors now part of their commercial considerations. That drives values, encourages innovation and new approaches and requires advice by rural valuers. What was previously just damage with no crystallised cost can now be a factor in decision making.

17.2 Natural capital and payments for ecosystem services

17.2.1 That approach has led to a larger structure of argument founded on the concepts of natural capital and payments for ecosystem services. While these and related concepts have differing origins, they offer analogies applying the perceptions of economics and accounting to promoting environmental improvement.

17.2.2 "Natural capital" is presented as the stock or capital of nature or the environment,

> that part of nature which directly or indirectly underpins value to people, including ecosystems, species, freshwater, soils, minerals, the air and oceans, as well as natural processes and functions.
> (Natural Capital Terminology, Natural Capital Committee, August 2019)

with the implied premise that it is better to have more, rather than less, of it. That may then be considered in component parts of natural capital assets, probably identified as is apt to the case in hand.

17.2.3 Loosely allied to it, "ecosystem services" are more the flow of beneficial things from natural systems and have been defined as the benefits provided by ecosystems

> that contribute to making human life both possible and worth living.
> (UK National Ecosystem Assessment)

They are then conventionally discussed in terms of different types of services, which may be fully in the marketplace, partly so or not at all:
- *Provisioning* – products obtained from ecosystems, such as food, fresh water and timber
- *Regulating* – benefits obtained from the regulation of ecosystem processes, such as climate regulation, hazard regulation, disease and pest regulation, regulation of water, air and soil quality
- *Cultural* – non-material benefits that people can obtain from ecosystems, such as recreation and tourism, aesthetic experiences and cultural heritage
- *Supporting* – ecosystem processes such as soil formation and nutrient cycling. As this category is needed to support the other three types of ecosystem services, the focus is usually on the first three.

17.2.4 However, and in practical terms, the discussion of ecosystem services on rural land becomes more about what might be done by management to provide services to society and nature, replenishing the stock of natural capital and reducing environmental harm. Again, in practical terms that returns to the questions of whether this creates cost or value and how that drives decisions as well as the relationship with what might instead be required as a baseline by regulations.

17.3 Public goods

17.3.1 The third and equally elastic concept of "public goods" comes from economics but is now more generally used in policy debate to mean things that society wants which are not delivered through markets but which could be delivered by a policy of "public money for public goods". That has been articulated in the post-EU agricultural policy being developed for England and the Sustainable Farming Scheme to come forward in Wales. In this context, food is not seen as a public good because it is traded in the marketplace.

17.3.2 The possible public goods being considered for England under the Agriculture Act are set out as:
(a) "managing land or water in a way that protects or improves the environment;
(b) supporting public access to and enjoyment of the countryside, farmland or woodland and better understanding of the environment;
(c) managing land or water in a way that maintains, restores or enhances cultural or natural heritage;
(d) managing land, water or livestock in a way that mitigates or adapts to climate change;
(e) managing land or water in a way that prevents, reduces or protects from environmental hazards;
(f) protecting or improving the health or welfare of livestock;
(g) conserving native livestock, native equines or genetic resources relating to any such animal;
(h) protecting or improving the health of plants;
(i) conserving plants grown or used in carrying on an agricultural, horticultural or forestry activity, their wild relatives or genetic resources relating to any such plant;
(j) protecting or improving the quality of soil." (s.1)

17.3.3 Interventions using public money released from the inherited Basic Payment Scheme are expected to be developed to achieve some or all of those goals, creating incentives for changed land management and farming. At the same time, potential changes in trading arrangements may also prompt new decisions.

17.4 Payments to change behaviour

17.4.1 For an offer from the state or a private company to change behaviour by voluntary agreement, the payments and terms must be:
- enough to attract an intended seller (depending on the service wanted, there might be one specific seller or several possible ones) who might consider the costs or lost opportunities associated with the change as well as the hope of profit
- be worthwhile to the buyer so that deal securing what the buyer wants is still beneficial.

Both will be informed by what they perceive others agreeing (where they see evidence of it) and their insights into the position of the other, as well as the relative strength of their motives.

17.4.2 That is the art of finding the price for that transaction, not its value to either or its value in the marketplace. Parties may well need advice on this, but it may be little different from a neighbour wanting to buy garden land from a farmer or negotiations over a restrictive covenant.

With that understanding, a valuer advising on such matters will be able to apply conventional valuation approaches with the differing bases of value and methods to these developing issues.

17.5 Transactions prices for agreements

17.5.1 Where an ecosystem service is already substantially traded – such as farm produce or timber – then it is usually easy to apply the concept of market value to securing a change in its supply. With the well-established understanding of market value, that would be the expectation of what would be agreed between hypothetical parties in a functioning and unstressed marketplace.

17.5.2 There are already some initial signs in some land transactions of hope value based on the potential of land for biodiversity offsetting and net gain agreements. That would simply be an aspect of market value.

17.5.3 That might also be applicable as markets in some other ecosystem services or public goods become more developed and transparent so that it is then evident what is being generally agreed for what services. In many cases, that may be tailored to particular circumstances and with evidence offered by failures to agree as well as from settled transactions. That is no more than typical practice in the property market.

17.5.4 Perhaps more often for what will typically be discussed as a payment for ecosystems services agreement, the concept of fair value may be more relevant – that is, the value that is fair between two specific parties, such as for a settlement of property between family members with differing interests.

17.5.5 Indeed, on occasion marriage value (or other forms of special value) may be relevant, perhaps especially where an agreement over the management of an additional area disproportionately enhances the biodiversity value of an existing site – biodiversity value tending to increase with scale. That might be similar to situations where landlord and tenant are negotiating a surrender of the lease or the tenant buying the freehold.

17.5.6 With the undeveloped or individual nature of markets for most arrangements that might now be seen as transactions in ecosystem services or public goods, those valuation bases of fair value and special value are perhaps more often in play than market value. However, all these lead to the **transaction price** between two parties for the actual agreement to provide and buy the services.

17.5.7 One way being used to reveal transaction prices by some buyers is reverse auctions. These see the potential suppliers bid the prices at which they are willing to sell their ecosystem service(s) with those bidding up to an undisclosed reserve price being accepted. Examples from water companies seeking changes to land management have been joined by the Woodland Carbon Guarantee scheme, which uses an online auction to determine the price at which Woodland Carbon Units (in tonnes of CO_2) would be sold to the government.

17.6 Worth of an ecosystem service to a buyer (individual, business, charity or society)

17.6.1 Most discussion of the valuation of ecosystems services is, though, less about such transaction prices needed to effect change than about the value of an ecosystem service to a party who might then take a decision about seeking that service. Very often, this is a value to society in general, for which the state might act in seeking agreements for the desired services. It might also be a value to an environmental body or a private company wanting or needing that service.

17.6.2 With that understanding, this applies the valuation basis of **worth** (sometimes known as investment value): that is, the value of something to a specific party, so framing the price at which that party would or would not bid for something.

17.6.3 Once found, that worth then helps that party and its advisers
- know the maximum it might wish to pay for an ecosystem service; and
- make choices between competing options with their likely purchase or implementation costs on the basis of value for money.

With such assessments often described in terms such as a natural capital valuation, that second point is important as this helps with the understanding that this is less about actual values than about decision making. The value itself may have no objective existence or be crystallisable as such, but, in a world of scarce resources, it helps frame choice and has been used to that effect by bodies as the Forestry Commission. If there are two projects with an equal cost, but a natural capital valuation shows one to be twice as beneficial as the other, the choice may become easy.

17.6.4 As many of the more important services for this have no direct markets at present, much effort has been put in by environmental economists and others into finding ways to reach the value, some of them more speculative or remote than others. What they do not produce is a bankable value but rather a means of informing decision making.

17.6.5 Such approaches generally try to assess:
- what might the ultimate beneficiary or beneficiaries be willing to pay for the supply of new ecosystem services or actions to avoid habitat degradation (i.e. maintaining the supply of existing ecosystem services)?
- what are the costs imposed by not acting say on water management or air quality?

208 *Valuations*

It can be difficult to quantify the value of the delivery of some ecosystem services, particularly regulating services such as air quality or biodiversity and habitats.

17.6.6 Approaches to estimating the values of natural services often use methods relying on people's preferences, as they might express them in a marketplace, including:

- *revealed preferences* – as with market data and so limited by where this is available whether in actual or surrogate markets.
- *stated preferences* – survey data from questioning people about hypothetical situations with the problems of working with the differences between what people say and how they might really behave.

17.6.7 Other approaches to estimating values could include:

- *Surrogate market valuation* methods use indirect expenditure data such as the time and/or money spent by visitors to a national park to indicate the value they place on it.
- *Hedonic pricing* might assess how much people are willing to pay for an amenity when choosing a property.
- *Avoided damage costs* or *replacement costs* are those associated with not having the service. Values might be placed on carbon sequestered, the costs of poor air quality on earnings or the heath service, or of flooding on downstream property
- *Contingent valuation* draws on asking people how much they would be willing to pay to preserve something – even something that they might never see but prefer to know that it survives or is in good condition.

17.6.8 In the absence of data for the subject service, a benefits or values transfer method might be used. This is the use of comparables from arguably similar situations. This becomes more possible as a greater volume of this work is done and so far as confidence can be gained in how to understand its reliability. It is likely to require significant skills in adjusting for differences between situations.

17.6.9 The conventional valuation hierarchy for this work appears, in descending order, to be:

- market data for the specific ecosystem service
- revealed preferences, using actual situations for the specific service
- stated preferences, using hypothetical situations for the specific service
- benefits transfer (comparables) using evidence from potentially equivalent services, requiring further judgement and adjustment.

This is similar in style to the three-level approach to evidence applied by accounting standards under which almost all property valuation is in the lowest level, Level 3, because of the reliance on judgment and adjustment.

17.6.10 While comparables may be gleaned from accumulated experience and contacts, several websites are beginning to offer information that may allow some sense of comparison. These include the Ecosystem Service Valuation Database (previously the TEEB Valuation Database).

17.7 Agreeing the reward – possible transactions in the market

17.7.1 With the potential buyer of the services and change now better informed, the transaction still needs to be agreed.

17.7.2 Aside from the terms of the agreement, the price paid for the supply of ecosystem services may need to take into account:
- *sellers' opportunity costs* – thus, not only the impact on earnings from returns foregone for current operations (for example, from agricultural production) but also the loss of potential opportunities in the future from the use of the land as economic possibilities change. As an example, conversion to woodland might be seen as hard and expensive (if not impossible under regulation) to reverse, so reducing the opportunities to revert to farming if prospects there improved or for use for a diversification or other possibility that might become viable;
- *start-up and ongoing maintenance costs* – what are the costs involved in delivering the actions required for the ecosystem service? This might be particularly relevant for 'asset-building' schemes which focus on restoring an area's ecosystem services;
- *transaction costs* – to cover, for example, the costs of developing and settling the agreement (advisory and legal costs as well as any SDLT/LBTT/LTT that might apply to any covenants);
- *operational costs* – establishing the baseline for monitoring change, training, developing a monitoring framework and providing third party assurance;
- *taxation* – usually most relevant to the seller of the services, this could include:
  - any taxation of the transaction
  - taxation of income arising under the agreement, whether falling within the statutory definition of the trade of farming, counting as trading or property income and or having any sideways loss relief consequences
  - any extent to which the transaction might be treated as a part disposal of the land
  - the buyer's ability to recover any VAT that the seller must charge
  - often importantly for many concerned with rural land management, any change to the seller's eligibility to reliefs from capital taxation, most obviously whether the management required to deliver the service:
    - affects the seller's qualification for Business Property Relief from Inheritance Tax, as where it alters the application of the "wholly or mainly" test of investment activity
    - for Agricultural Property Relief from Inheritance Tax, it means that the land is no longer either agricultural land or pasture, or is not used for agriculture for the necessary period before death

- for Agricultural Property Relief, whether it alters the qualification of a house as a "farmhouse" whether by changing the activities undertaken from the house or affects the application of the "character appropriate" text
- affects the seller's qualification for Business Assets Disposal Relief (formerly Entrepreneurs' Relief) from Capital Gains Tax on a disposal by increasing the element of the business deemed to be non-trading.

- *costs of alternatives to the buyer* – for example, for improved drinking water quality or cleansing wastewater, comparing the cost of building a water treatment plant with that of investing in natural ecosystem service-based filtration; and
- *the balance of supply and demand for the service in question in that location* – it may not matter very much where carbon is sequestered (save that some sites may offer more potential than others) but a specific location may be important for a habitat, biodiversity or flood attenuation while buyers will tend to seek the lowest-cost suppliers of desired services or where, as with biodiversity, value often increases disproportionately with scale.

17.7.3 Buyers of ecosystem services may wish to ensure that existing services are preserved (maintenance of good practice) but are also likely to want "additionality", the improvement of existing services or creation of new services. The payments for those two outcomes might differ once the factors mentioned earlier are taken into account. While achieving improvement might often require a higher payment, not paying for maintenance might lead to loss of the service, whether by degradation over time for lack of care or outright reversal (as with ploughing out reversion land). It should be noted that some maintenance work may be to meet standards required by regulation and so perhaps not merit payment.

17.7.4 Longer-term value – While for many the initial challenge will be to assess the price for transactions there is also the question of whether or not the service offered, and particularly the contractual approach adopted, has any longer-term impact on the value of the host holding, and whether that is either positive or negative. Where the transaction was achieved by a freehold sale or long leasehold, then that residual impact will be limited, save to the extent that having that scheme as a neighbour may impact on the retained land.

17.17.5 However, where the transaction involves a management agreement or similar continuing arrangement with a routine payment, then that income stream is there to be valued in just the same way as the rent from a farm building let as a workshop. How might that relate to the value of the land for agricultural or amenity purposes where those are still available? If the ecosystem service payment is doing no more than replace agricultural income on a "profits foregone" basis then there is no additional value;

although that begs the question as to why the farmer would enter into such an arrangement when contracts are likely to be long-term. Where there is a clear surplus over the income available from traditional farming, then there may be additional value. Whatever the actual answer, as with other diversifications explored in Chapter 14 its recognition will turn on a range of factors while one farmer's "useful additional income"' may be another's "waste of good land" and that may change with the times. In a developing market, it may be a while before the potential for value is recognised here not least as for some it will never replace "proper farming" and other many arrangements may appear small to financial investors.

Part 3

Valuations for taxation, compulsory purchase, utilities and communications

18 Valuations for capital taxes

18.1 Introduction

18.1.1 Valuations for taxation may be required for both capital and revenue tax purposes and for some valuers these are a significant part of their workload. In the revenue tax sphere, valuations are most likely to be required for stocktaking purposes for Income Tax (covered in Chapter 19 next), but other taxes such as ATED and non-domestic (business) rates may also prompt the need for valuations. In the capital taxation sphere, the focus of this chapter, the most common valuations are for Inheritance Tax, but they may also be required for Capital Gains Tax, or Corporation Tax where a company disposes of an asset and SDLT on property transactions. In both cases, valuers may be instructed to produce valuations either to inform tax planning before transactions take place or after the event. A natural reluctance to incur fees amongst some clients may see valuations delayed until the last possible moment and it is not uncommon for valuers to find themselves negotiating with HMRC, in the person of the District Valuer, to reduce a tax liability which could have been avoided had advice been sought earlier.

18.1.2 This highlights a particular element of taxation work for rural valuers; valuation will be at the core of the activity, but it is also part of the valuer's role to advise to what exemptions and reliefs may apply in practice to the particular holding and business. This is commonly the case with Inheritance Tax, particularly Agricultural Property Relief (APR), but applies across the spectrum of tax work.

18.1.3 Previous editions have included a sample probate valuation for valuers. This reflects a time when valuations might be presented in very different formats, depending on the purpose of the valuation. The increasing trend for standardisation, whether from internal quality assurance regimes within firms or external influences, including the expansion of the RICS Red Book to include fiscal and other valuations previously excluded, now means most valuers adopt a similar template for most valuations, including those for taxation. However, there are specific requirements for tax valuations which are explored further in the following sections.

18.1.4 UK Guidance Note 3 (UKGN 3) of the Red Book, *Valuations for Capital Gains Tax, Inheritance Tax and Stamp Duty Land Tax*, covers valuations for taxation. This points out that the statutory definition of Market Value for various taxes differs from the "standard" valuation under the Red Book. To that end, whilst the basis of valuation will be Market Value this will be subject to the variations in the Guidance Note. The Guidance Note focuses on the interpretation of the market and the participants, vendor, purchaser and potential special purchaser involved in assessing the value under a hypothetical transaction. As the note observes the differences between the statutory interpretation of value and Market Value are commonly very small, to the extent that they are unlikely to have any impact on the value. However, there are some circumstances where the difference may be very marked or where the assumptions implicit in statute may introduce a value where logic suggests one would not exist.

18.2 Tax valuations generally

18.2.1 Basis of value – The statutory definition of market value required by s.160 of the Inheritance Tax Act 1984 (IHTA) is necessarily adopted by UKGN 3 in these words:

> . . . the price which the property might reasonably be expected to fetch if sold in the open market at that time, but the price must not be assumed to be reduced on the grounds that the whole property is to be placed on the market at one and the same time.

18.2.2 Taken at face value it is difficult to imagine how that can produce a significantly different valuation to the definition of Market Value in VPS 4 of the Red Book. Much is made of the fact that these will be hypothetical transactions, but other than where valuation is focused on apportioning known facts then it is always dealing with a hypothetical transaction. The various tax definitions have been tested in case law over the years so that valuers should base their valuations on a number of assumptions:
- the sale is hypothetical; decisions in *Crossman v IRC*, *Duke of Buccleuch* and *Lynall v IRC* all held that the issues which might have affected an actual sale were irrelevant;
- however, whilst the sale is hypothetical the market to be considered is the actual market at the valuation date (*Walton*);
- both vendor and purchaser are hypothetical (unless the purchaser is regarded as a special purchaser (*Walton*)), willing to trade, but prudent not desperate;
- the property would be prudently lotted (*Buccleuch*) and offered by the most appropriate method on the open market;

- the arrangements have been made for the sale at the valuation date and adequate publicity has taken place;
- however, where the property would not normally be taken to the market without work being done (or in the case of *IRC v Gray* restructuring the legal interests) the property must be valued as it is at the valuation date making allowance for those shortcomings;
- the valuation should reflect the bid of any special purchaser (provided such a purchaser who is willing and able to proceed can be shown to exist) thus providing a distinction between valuations for tax purposes and Market Value

18.2.3 Taken literally, this creates an uncomfortable mix of the actual and the hypothetical but, in practice, this only creates significant difficulties where there are unusual circumstances which would preclude the property being taken to the market. In these circumstances, the disparity between the client's valuer and the District Valuer may be significant, which will demand particular care by the valuer in managing the case as explored further later on.

18.2.4 **Untradeable assets** – Whilst the property is real sometimes the asset to be valued is something which would seldom or on occasions never be traded on the open market.

18.2.5 The two classic examples of such assets are:
- non-assignable tenancies with the vast majority of both AHA and FBT tenancy agreements being non-assignable and
- undivided shares in property.

The decision in *Crossman v IRC 1937* specified that the value for tax purposes is that which a purchaser would pay to stand in the shoes of a transferor, including an agricultural tenant, even where there was a bar preventing that happening in practice.

18.2.6 The principles of the valuation of agricultural tenancies are explored further in Chapter 32. The issue is brought into particularly sharp focus in tax terms where there are intra-family tenancies, not least the traditional arrangement where the parents in a farming family let as landowners to a family partnership involving themselves and one or more children.

18.2.7 The question was addressed in two notable tax cases, *Baird* and *Walton*.

18.2.8 Whilst, in contrast, it is possible to trade undivided shares in property in the open market, it is very unusual for that to happen. The impact of the shares has been considered in cases including *Wight v Moss* which dealt with joint owner-occupiers with half shares in the property and *Charkham v CIR* which involved multiple minority shareholdings.

18.2.9 The question is, what discount from a proportionate share of the full value of a property would someone pay to buy a part share in it when the remainder of the property is owned by other people who will still be there? In *Wight v Moss*, the tribunal decided that where there were two joint owners, each with a half share in the property and the owner of the share not being valued remained in occupation, the appropriate discount should be 15%. This approach should be adopted unless the other owner has chosen not to be in occupation and the purpose for which the trust which created the interests can no longer be achieved in which case the discount should be 10%. Where there are joint owners but no right of occupation the discount in *Youlden* was 10%.

18.2.10 The situation in *Charkham* was rather more complicated; the case involved the valuation of minority undivided shares in two central London properties at five different valuation dates and the discounts ranged up to 22.5%. There were particular circumstances in this case, not least the prospects for development of one of the properties. However, the Tribunal Member concluded that it was reasonable to assume both that the trustees would act in the general interest of the beneficiaries and, notwithstanding that, any purchaser would discount their bid for the shares to reflect the lack of direct control. The value of a shareholding of less than 25% is directly only in the income and indirectly should those controlling the company ever decide to sell or liquidate it.

18.2.11 Case management – In most cases there are two stages to tax work: the initial valuation and the subsequent negotiation with the District Valuer (DV). That makes it particularly important for valuers to ensure that the client understands the extent of work involved at each stage, not least so that the client is not surprised when they receive a second fee invoice for negotiations with the DV, having already paid for the valuation. The Valuation Office's workload is substantial, and it may be between 12 and 24 months before the DV considers a valuation. Clients may face some difficulty if they assumed that the passage of time meant matters were concluded and they are then faced with a tax bill which they did not anticipate.

18.2.12 This highlights two other issues which apply to tax work, commonly, but not exclusively, valuations and advice associated with Inheritance Tax.

18.2.13 Farming clients in particular will have been coached by previous generations that the answer to minimising tax liabilities is to keep values low, leading to some intriguing redistribution of assets so that in the past it was not unusual when inspecting a farm for an Inheritance Tax valuation to be told that the surviving widow rather than her deceased husband had owned all the farm equipment. The current combination of Agricultural Property Relief and Business Property Relief means that much of a farming family's assets will be relieved from Inheritance Tax and, given that those valuations would become the acquisition cost for the beneficiaries, there may be a very strong argument against pursuing low values.

18.2.14 Conversely, some taxpayers may also be aware of the benefit of reliefs and may assume, for example, that where a property benefits from 100% Agricultural Property Relief from Inheritance Tax there is no tax to pay. As we shall see in the following section, this is not necessarily the case and, whilst valuers may be aware of the subtleties involved, it would be wrong to assume either that their client shares that knowledge or that other professionals will have explained that to the client. Therefore, it is important for the agricultural valuer:

a) to liaise with the other professionals involved and ensure that that clients are aware of the potential liability for taxation;
b) in assessing that potential liability not only to consider the situation based on the valuations and advice you have provided but also the view the DV might take, particularly where there are significant questions about relief, whether a farmhouse qualifies for Agricultural Property Relief from Inheritance Tax (see section 18.3 later on) or how far Principal Private Residence (PPR) Relief might extend for Capital Gains Tax (section 18.4 later on); and
c) to gain a sufficient understanding of the situation as a whole to recognise the relevance of any changes in valuation arising from negotiations with the Valuation Office; there is little attraction for the client in his valuer spending hours fighting over the last £20,000 of value on an unrelieved property for Inheritance Tax where that value would be covered anyway by the Nil Rate Band.

18.2.15 In many cases, the valuations proposed by the DV and the client's valuer will be very close, so that the difference between the negotiated settlement and the original valuation submitted by the Valuer will be minimal. However, there are circumstances where this is not the case and it is important that professionals and client together plan for this eventuality. The agricultural valuer is at the core of this process, being the professional who will negotiate with the Valuation Office.

18.2.16 Positive husbandry – A client who assumes that the tax reliefs for farming come with being a longstanding member of the farming community may need to be reminded of the need to be a farmer who farms, summarised here as being able to show positive husbandry of land. This summarises the concept of agricultural activity as crop production, pasture management and, where relevant, animal husbandry rather than the property maintenance of gates, fences, hedges and ditches. This has been illustrated in a series of cases for several taxes, including *McCall* and *Charnley* for Inheritance Tax and *Allen* for Capital Gains Tax. This can be tested for both the aging and failing farmer who fades out of activity and where the landowner is coming to arrangements with others over the farming of the land. For the valuer, this leads to the need to appraise the situation and the parties' issues and motives to advise on an appropriate answer, whether that be a practical way in which the client can remain active with positive husbandry or helping an understanding that this is not realistic.

18.3 Inheritance tax

18.3.1 Valuation – The basis of valuation for Inheritance Tax is set out in Section 160 of the Inheritance Tax Act 1984. Aside from the general valuation issues set out earlier in this chapter, the principal issue with Inheritance Tax is that the value to be assessed is that of the asset in the hands of the donor the moment before the transfer is made, whether by lifetime gift or on death. Again, this may sound self-explanatory but there are some practical implications. Thus, in a family tenancy situation, where Father is the Landlord and Son the Tenant under an Agricultural Holdings Act tenancy, then were Father to gift his interest to the Son this would create a holding with vacant possession with significantly greater value. However, for Inheritance Tax purposes the value to be assessed is the Landlord's investment.

18.3.2 Valuations for Inheritance Tax may be required for a lifetime gift but most commonly on death, which used to be referred to as probate valuations, but which the Red Book authors would rather were referred to as Inheritance Tax Valuations. Valuations of agricultural assets on death will include the deceased's interest in freehold property, live and dead farming stock, entitlements to payment under the Basic Payment Scheme and one would assume its post-2019 successors.

18.3.3 There is a temptation to limit the detail in a valuation for Inheritance Tax on the basis that the client's family will know the assets far better than the valuer ever will. However, a comprehensive description will help to justify the approach taken to the valuation and answer many of the questions which may otherwise be raised by either the Valuation Office or the snappily named Inheritance Tax, Trusts and Pensions department at HMRC, previously known as the Capital Taxes Office.

18.3.4 Freehold properties – However, the valuation of freehold property should be as comprehensive as possible, fully describing the property in detail with a plan and accompanying schedule and details of tenure and occupation. Where the property is let the report should include full details of any tenancies, including any improvements which can be disregarded in assessing the value of the deceased's estate, where let.

18.3.5 Where there are properties with potential development opportunities, then these should be noted, along with the constraints on development or, in the case of a barn close to the farmhouse for example, the adverse impact the development would have on other property. This is an area which has gained in relevance with the relaxation in National Planning Policy in respect of the conversion of traditional buildings; however, that relaxation has not been reflected by many local planning authorities and often the conditions required to secure development will dissuade all but the most enthusiastic from pursuing development. In such cases, it is not unreasonable to assume that the hypothetical vendor will be equally discouraged.

18.3.6 The prospect of more extensive development can be challenging for both valuers and taxpayers alike. Unless the position has already been crystallised at the date of valuation by the grant of planning permission there will be an element of conjecture and hope about any value. These challenges are addressed in greater detail in Chapter 14 on valuations for development, but in Inheritance Tax situations, in particular, the critical issue is the situation at the date of death (or gift) not three months later when the valuation is being prepared. The prospect of development can change overnight and even where planning permission has been granted there can be significant differences in value depending on the s.106 obligations (s.75 in Scotland and Article 40 of the Planning (Northern Ireland) Order 1991) and Community Infrastructure Levy or planning Tariff charges in England.

18.3.7 Valuers may be required to provide valuations both for Market Value and Agricultural Value (see later on) and it is important that the valuation makes clear which basis is being used.

18.3.8 Live and dead farming stock – Again, it is important to provide a full description of both live and dead farming stock and, except where livestock would naturally be grouped together for sale, a pen of store lambs for example, stock and machinery should be valued individually as should produce and where relevant growing crops. All livestock should be described by breed, sex, age and type. Additional information should be provided for dairy cows including yield, number of calvings, last calving date and service details. Full details should also be provided for farm machinery including make and model number, age and in the case of tractors and other self-propelled machinery the number of hours and the condition of tyres all of which have an impact on value. The valuation of live and dead stock is explored in greater detail in Chapter 16.

18.3.9 Agricultural Property Relief – The two principal reliefs from Inheritance Tax for farmers and landowners are Agricultural Property Relief (APR) and Business Property Relief (BPR). Conditional exemption for heritage property will also be important for those eligible to claim it and indeed there are some agents who specialise in this area of work. However, and in contrast to APR and BPR, Conditional Exemption is not something that a valuer would normally be required to comment upon when undertaking a valuation.

18.3.10 In contrast, solicitor's instructions to produce Inheritance Tax valuations will commonly include a request to the valuer to provide commentary on the extent to which the property is likely to qualify for relief and the value involved.

18.3.11 Agricultural Property Relief is available at 100% of the property's "agricultural value" (see later on) where the land is:
- owner-occupied
- on a tenancy granted since 1 September 1995 (which is not simply synonymous with FBTs and includes AHA tenancies let since that date whether by succession or surrender and regrant)

- where possession can be recovered within 12 months, extended by ESC F17 to 24 months so fully covering a normal Notice to Quit period
- some exceptional pre-1995 lettings where "double discount" relief is available.

APR is available at 50% for property remaining on a tenancy that was let before 1 September 1995. With Business Property Relief relieving the assets of qualifying farming businesses (see later), the essential role of APR is to offer relief (albeit often partial) to qualifying farmhouses and, by giving relief on let land, to ensure equality of taxation between choices over whether land should be let or farmed in hand.

18.3.12 The principal provisions for Agricultural Property Relief are contained in sections 115 to 124C of the Inheritance Tax Act 1984. The definition of property which might be eligible for APR is set out in s.115(2):

> ... agricultural land or pasture and includes woodland and any building used in connection with the intensive rearing of livestock or fish if the woodland or building is occupied with agricultural land or pasture and the occupation is ancillary to that of the agricultural land or pasture; and also includes such cottages, farm buildings and farmhouses, together with the land occupied with them, as are of a character appropriate to the property.

18.3.13 The critical issue is that agricultural land is required first before other associated property, including buildings, farmhouses and cottages, can benefit as tested in the case of *Williams v HMRC 2005* where it was held that the intensive poultry buildings on the site dominated the land such that the occupation of the buildings was not ancillary to the land.

18.3.14 There are very few cases where agricultural land would not qualify for APR and most farm buildings will also qualify. The challenge can be with farmhouses and farm cottages. Whatever the asset there are three tests:
- is the property agricultural?
- is it occupied for the purposes of agriculture?
- is it of a character appropriate?

18.3.15 At first glance it can be difficult to distinguish between the first two tests. The first test turns on the definition in s.115(2) as set out earlier and has been tested in cases including *Higginson, Rosser 1* and *McKenna*. In *Higginson* and *McKenna* the dominant nature of the property was not agricultural, whilst in *Rosser* the bulk of the land previously owned with the farmhouse (39 acres out of 41) had been gifted to a different ownership. The question of common occupation rather than ownership has been tested more recently in the *Hanson* case, where it was held that common occupation would also satisfy the requirements of s.115(2) and common ownership was not essential. However, the doctrine which emerged from *Rosser* and, before it, *Starke*, to retain sufficient land with the house to secure APR on the farmhouse has remained with those professionals involved in this work ever since. That said, for those families with some uncertainty over eligibility for full APR the introduction of the Residential Nil Rate Band, with its certainty of relief on a qualifying house, may provide a better alternative. That would change this dynamic as the relief is only fully available where the value of the estate is below £2 million on death, thus prompting the opposite approach in those cases, giving away land to get below the Residential Nil Rate Band threshold.

18.3.16 The second test is set out in s.117 of the IHTA and brings the timing in to play. To qualify for relief the transferor must either have:
- owned and occupied the property for the purposes of agriculture throughout the two years prior to the date of transfer; or
- owned the property for seven years and it had been occupied for the purposes of agriculture throughout the period (thus can be let for that period).

18.3.17 It is important to note that the periods in question are both immediately prior to transfer. Thus, a transferor may have lived on and run the farm for 30 years but if that occupation ended say 6 months before the date of transfer the taxpayer will no longer qualify for the relief. This issue was tested for a cottage in the *Atkinson* case where, after a somewhat quixotic decision in the First-Tier Tribunal the Upper Tribunal confirmed the position. This may cause particular difficulty where an elderly farmer falls into poor health and moves perhaps to hospital or a nursing home to spend their last days. To date, HMRC have shown due discretion where the move was relatively short, say for a final illness, but the matter has not yet been tested other than in *Atkinson* where the taxpayer had not lived in the property for a number of years.

18.3.18 Whilst timing is an issue, the other challenge is the extent to which the property was genuinely occupied for agriculture. Thus, in *McKenna*, where the farming was undertaken under an arrangement which demanded little input from the taxpayer who lived in the farmhouse it was decided that the house was not occupied for the purposes of agriculture. For APR, the "farmhouse" is the dwelling from which the day to day farming is conducted. This may also be an issue where buildings are in non-agricultural use. Where occupied for a diversified business by the farmer then Business Property Relief will replace APR; however, this may not apply where the building is let to a third party.

18.3.19 The third test, whether the asset, most typically a farmhouse, is of a character appropriate. There have been two principal challenges; in simple terms where there is a large house, is the house too grand for the farm or with more modest houses is it still out of character if there is only a small farm? This has been tested in a succession of cases of which perhaps the most noteworthy, and the one most commonly remembered by students, is *Antrobus (Lloyds TSB (Personal Representatives of Antrobus)) v Inland Revenue Commissioners 2002*. In this case, involving a 126-acre farm in Worcestershire, the Special Commissioner set out five tests before concluding that the farmhouse, a substantial Grade II* detached country house was of a character appropriate.

18.3.20 The same issue was addressed in *McKenna*, again with a large country house, where an additional test explored the commercial relationship between the profitability of the farming which might be carried out on the holding and the value of the property. In the event, the conclusion was that, on the *Antrobus* tests, had the manor house passed the first two tests to be a farmhouse, it would not then have been of a character appropriate.

18.3.21 The challenge of commerciality was tested again at the other end of the scale in the *Golding* case with a small detached house, in a dilapidated condition, on only 16 acres, which was found to be of a character appropriate. In this case the judge considered both the extent of commercial activity, concluding that it was reasonable given that Mr Golding was 80 when he died in occupation of the holding he had had since 1940 and his previous business activity, and the broader issue of APR and smaller farms. He also noted the condition of the property which was in disrepair. The question of commerciality has been the focus of some discussion, but *Golding* would appear to have addressed that for the time being.

18.3.22 Taken together these and earlier cases give considerable scope for technical analysis and debate with much legal ink spent on this subject. When all those arguments have been distilled, it is clear that the question of character appropriate is more than the sum of the parts of the individual tests. On occasions it is difficult to escape the conclusion that the only real analysis is that Miss Antrobus and Mr Golding lived like farmers and the McKennas did not.

- **Appropriateness** - Is the house appropriate in size, content and layout to the holding?

- **Proportionate** - Is the house proportionate to the farming activities?

- **The elephant test** - You may not be able to describe a farmhouse which is of an appropriate character but you know one when you see it.

- **The educated rural layman** - Would the educated rural layman regard the property as a house with land or a farm?

- **Historical association** – How long has the house in question been associated with the property and is there a history of agricultural production?

Figure 18.1 The *Antrobus* Case's Five Tests for "Character Appropriate"

18.3.23 Agricultural value – Agricultural Property Relief applies only to the Agricultural Value of the property and this is an issue which has come under far more focus in recent years, particularly since the matter was tested in the second *Antrobus* case. Prior to that generations of valuers acting for taxpayers were used to arguing that there was no difference between Agricultural Value and Market Value. However, whilst that will still apply in most cases for the land and most buildings valued, the situation is rather different with dwellings and possibly other assets. The valuer may be instructed to offer an opinion of Agricultural Value as well as Market Value but, even where not asked, the valuer should both ensure that the basis of valuation is clear and also, in the context of case management, draw attention to circumstances where there may be a distinction between the two.

18.3.24 Agricultural value is defined in s.115(3):

> ... the agricultural value of any property shall be taken to be the value which would be the value of the property if the property were subject to a perpetual covenant prohibiting its use other than as agricultural property.

18.3.25 The first *Antrobus* case having established that the farmhouse was of a character appropriate, the taxpayer's representatives and the Valuation Office could not then agree on the impact of s.115(3) on the value of the farmhouse. In the event the Tribunal found that if anything the provisions of s.115(3) were more restrictive than an agricultural occupancy condition and the value should be that which would be bid by a farmer occupying the farmhouse to farm the property. On that basis and in the absence of sales evidence from the taxpayer, it adopted a discount of 30% for that very specific case.

18.3.26 Adopting that precedent, the Valuation Office will regularly seek to distinguish between Market Value and Agricultural Value for farm dwellings, although the extent of that discount will turn on the circumstances of the case with the potential for a settlement to be above or below that 30% figure, and sometimes for the two values to be the same where, for example, there is a very commercial farmhouse in close proximity to the farm buildings or there is an Agricultural Occupancy Condition.

18.3.27 One of the issues in *Antrobus II* was the lack of evidence of sales which might be used as a proxy for s.115(3). Sales subject to agricultural occupancy conditions may offer such evidence, although it is worth observing that s.115(3) does not limit the population of bidders, rather the use of the property, whereas an agricultural occupancy condition can only be satisfied by approximately 2% of the population who might qualify as being employed in agriculture. An alternative source of evidence is sales subject to user covenants. A number of vendors including local authorities and traditional institutions sell property subject to restrictions on use which are often sufficiently similar to offer some evidence relevant to s.115(3). Agricultural valuers would do well to retain evidence of such sales as well, given the judgement in *Antrobus* II, to keep a list of bidders for any agricultural property to record the extent to which farmers or lifestyle buyers were part of the market for any property.

18.3.28 Business property relief – Business Property Relief is set out in s.103–114 of the IHTA and in general is more the province of accountants and solicitors than valuers. The agricultural valuer may be asked to comment on the use of particular properties and to provide evidence to inform a decision on the nature of a business, although this is seldom to the same extent as required for Agricultural Property Relief. However, it is important for an agricultural valuer to appreciate the principles of BPR, not least as it may apply to arrangements for land farmed through various contracting arrangements.

18.3.29 BPR is available for relevant business property where the business has been owned for at least two years at the point of transfer. It is assessed after Agricultural Property Relief and is generally at 100%, although rates of 50% apply in some partner and trustee circumstances.

18.3.30 The relief applies to business assets with the critical test, beyond whether the assets are used in a business in the first instance, being whether that business qualifies for relief or whether it is a business which consists *wholly or mainly* of making or holding investments, in which case it will not qualify for relief. Following some caravan park cases, this issue was tested in a farming context in *Farmer* where there were both business and potentially investment activities and four tests were established:

- what was the overall context of the business?
- what was the capital employed?
- how was the time spent by directors and employees?
- how was the turnover split?

18.3.31 In that case it was concluded that the business did not wholly or mainly consist of investments and hence BPR was available. In contrast to APR, there is no question of apportioning between elements eligible or ineligible for relief as demonstrated in the *Hertford* case, where the main part of the house was opened to the public and qualified for BPR and that benefit then also extended to the private apartments in the house.

18.3.32 The same principle was tested in the *Balfour* case, sometimes referred to as *Brander*. This involved a mixed landed estate in Scotland extending to approximately 1,900 acres including three let farms, two in hand farms, 26 let houses, business premise parkland, woodland and sporting rights. Again, the test was whether or not this estate comprised wholly or mainly of investments. In this case, the business was considered as a single entity as, although there were two entities, they were seen to have a common guiding mind. Evidence was adduced of value, revenue and most particularly of the demands on the owner's time of the different parts of the business and, with the view that capital values were not relevant as the estate was not to be sold, the decision was that the active business elements outweighed investments and BPR was available.

18.3.33 This decision excited considerable interest amongst advisors and commentators with the simple conclusion being that an agricultural estate involving a substantial number of let properties could still benefit from BPR. Whilst a very noteworthy judgement, it was not the carte blanche that some commentators would have it to be and each case will turn on its merits. It has, however, emphasised the importance of in-hand business as part of the wider portfolio, which has doubtless influenced some owners' decisions on whether or not to let land.

18.3.34 One less successful case for the taxpayer was *McCall* which highlights the other key area where BPR is particularly relevant. In contrast to APR, the relief is not limited to Agricultural Value and so will also apply to premium values, such as where development is involved. In this Northern Ireland case, some 33 acres of land had been let for many years on conacre with some management work undertaken by the taxpayer's son-in-law, but the bulk of the husbandry done by the occupier. There was no doubt that Agricultural Property Relief applied to the Agricultural Value of £165,000. The difficulty for the taxpayer was that the land was zoned for development with a value of £5.8 million; BPR would have relieved the full development value. However, in this case the lack of activity on behalf of the taxpayer meant that the land was so held to be wholly or mainly an investment and hence BPR did not apply and there was a large liability to tax.

18.3.35 The decision in *Charnley* shows the counterpoint to *McCall* with an elderly farmer, Thomas Gill, taking in graziers still found to have had a qualifying business because of his positive husbandry. As the grazier testified to the Tribunal, Mr Gill was "farming his land using my stock". Qualifying for BPR not only gave full relief to full value of the land and also the farm machinery, but the facts for that then gave APR in the farmhouse as well as the land – Thomas Gill's farming activity was done from that house.

18.4 Capital gains tax

18.4.1 Valuation – Valuations for Capital Gains Tax can be amongst the most interesting areas of agricultural valuers given that valuations may be required both for the present day and the date of acquisition together with apportionment of value between constituent elements of the property.

18.4.2 The principal legislation is enshrined in the Taxation of Chargeable Gains Act 1992 as amended by a succession of chancellors of both political hues, feeling the urge to interfere with the application with significant amendments in the 1998, 2008, 2010 and 2016 Finance Acts. As a consequence, there are now four rates of taxation and what for many years had been a non-tax for farmers is increasingly relevant.

18.4.3 In principle, tax is charged on the difference between the proceeds of sale of a property less the cost of acquisition making allowance for expenses and the costs of any investment. Where an asset, often such as farmland, has been owned continuously since 31 March 1982, its value at that date is substituted for the earlier cost of acquisition. Property inherited since March 1982 will usually have its base value as at the date of inheritance (unless holdover relief was elected for on a gift). The basis of valuation is set out in s.272 of the 1992 Act:

(1) In this Act "market value" in relation to any assets means the price which those assets might reasonably be expected to fetch on a sale in the open market.

(2) In estimating the market value of any assets no reduction shall be made in the estimate on account of the estimate being made on the assumption that the whole of the assets is to be placed on the market at one and the same time.

The complication for many valuers will come where the property was acquired prior to 31 March 1982. In those circumstances the value may be rebased to that date and for many practising valuers that will be well before they were in practice and for some before they were born. That minority of valuers who were practising in 1982 and can remember the details of the market will feel they have some advantage, but for everyone else and for those joining the profession they will have to hope that their firm has good evidence of a market very distant from where we are now: before, for example, anyone had thought of residential barn conversions.

18.4.4 A variety of reliefs are available, two particularly noteworthy ones being that neither transfers between spouses nor transfer on death are chargeable transfers for CGT purposes. Amongst the more important other reliefs are that on the taxpayer's Principal Private Residence (which tenant farmers can often apply to another dwelling), the gain from which is 100% relieved from tax and Entrepreneurs Relief (renamed Business Assets Disposal Relief in April 2020), Rollover Relief and the closely allied Holdover (or Gift) Relief. These are revisited later on, but the relief on the Principal Private Residence requires the valuer to apportion value where a principal dwelling is sold alongside other property, an activity further complicated by the changes in the Finance Act 2016 Autumn Statement, which created a distinction between tax rates for residential and other property.

18.4.5 **Apportionment of value** – A valuer instructed to produce valuations for a farm for Capital Gains Tax may find himself required to produce a matrix of values to reflect the different exposure to tax for constituent elements of the property.

18.4.6 Thus, for example, a farm of 200 acres comprising a farmhouse (the taxpayer's principal residence), a second farm dwelling (let to a farmworker) and farm buildings and land is sold in the open market as a whole in 2020, having been acquired by the taxpayer in 1980. The valuer instructed to provide a Capital Gains Tax valuation will be required to provide separate valuations in 1982 for the farmhouse and curtilage, in the vast majority of cases up to 0.5 hectares of land, the farm cottage and its curtilage and the land and buildings. He will also be required to apportion the 2017 composite sale value between those constituent parts.

18.4.7 The impact of those valuations and consequently the respective potential negotiating positions of taxpayer's valuer and Valuation Office will be influenced by the application of relief and tax rate to those individual elements; thus:

Asset	Farmhouse	Farm Cottage	Land and Buildings
Tax Rate	28%	28%	20%
Relief	Principal Private Residence (100%)	~*	~*
Effective Rate of Tax	0%	28%	20%

*Subject to any impact from Entrepreneurs' Relief – see later

18.4.8 The sales and acquisition costs and any investments made in the property will also have to be apportioned or applied, but this is generally done by the client's accountant.

18.4.9 In some instances the basis of apportionment can be challenging, particularly between the property eligible for PPR and the remainder. A recent case, in a development scenario rather than on a farm, is *Oates* where the property comprised a dwelling house, which was eligible for PPR, a former ice cream factory and surplus land. Planning permission for residential development was granted over the whole shortly after the date of apportionment. A number of methods of apportioning value were explored with a straightforward pro rata apportionment based on acreage being dismissed. Ultimately the decision used the existing use value of the constituent elements of the site to apportion the proceeds which reflected the development value. That said, in most cases the apportionment of farm values should be simpler, albeit the same challenge may occasionally arise as in *Oates* where there is a particular marriage value between the constituent parts.

18.4.10 Reliefs – Most agricultural valuers of a certain vintage will have encountered the impact of Rollover Relief where purchasers fortunate to have sold their farm or part thereof for development (or, indeed, sold any other business) will be significant bidders for land to secure Rollover Relief. The relief has routinely been blamed by commentators, some of whom have practical experience of the land market, for distorting markets. Rollover Relief enables the taxpayer to defer the tax on the gain from the sale of qualifying business assets by reinvesting the proceeds of that sale in replacement business assets, one of which is farmland, doing so within a limited period. Thus, the taxpayer can roll the gain over until such time as a sale is not followed by a replacement purchase or where death or Entrepreneurs' Relief is available to remove or reduce the tax on the gain. This enables businesses to reinvest in themselves rather than find the tax inhibiting business change by taking value out of the business.

18.4.11 There are detailed rules to be satisfied of which timing is perhaps the best known. The general rule is that the replacement purchase must be made between one year before and three years after the disposal, although HMRC have some discretion over this timing.

18.4.12 Whilst Rollover Relief applies to disposals, Holdover Relief, sometimes referred to as Gift Relief, is available for gifts. Once again, the relief serves to defer the gain, preserving the old base value. There are various qualifying assets, although in this case the range is broader than Rollover Relief including both let farms and farmland and Heritage property.

18.4.13 While earlier generations benefitted from Retirement Relief long since abolished, Entrepreneurs' Relief was introduced in 2008. This provides that gains made on disposal of all or part of a business or disposal of assets following the cessation of business will only attract CGT at 10% rather than the 20% which would otherwise apply. The amount of relief was extended in 2010 so that it applied to the first £10 million of lifetime gains. However, anxiety over potential abuses of the Relief led this to be reduced to £1 million of gains in the 2019 Budget, so back to the level when first introduced in 2008.

18.4.14 The assets have to be held in the business for one year before qualifying making it more relevant for tax planning where potential development land is involved, albeit with more limited benefit following the 2019 changes. However, it is critical that the requirements for sale or cessation of a business are satisfied. Thus, the sale of part of a farm where there is no change in the business will not qualify for Entrepreneurs Relief, an issue tested unsuccessfully by the taxpayer in *Russell* where a third of the farmland involved was sold but the business continued otherwise unchanged.

18.4.15 The attractions of Entrepreneurs' Relief had rather overtaken the historic attractions of Rollover Relief for some taxpayers who saw the merit in paying a tax rate of 10% and then having the benefit of the rest of the proceeds available untrammelled by the constraints of Rollover Relief. However, the 2019 changes may make Rollover Relief more attractive again for some of those families, potentially increasing the demand for land in an already undersupplied market. A chancellor's lot is not a happy one; they legislate to remedy apparent abuses in one part of the tax system only to create unforeseen consequences elsewhere.

18.4.16 The impact of Principal Private Residence Relief (PPR) on the apportionment of value has been explored earlier. However, the extent of the land to be included with the house has been more strictly applied in recent years. Statute provides for the relief to apply to up to 0.5 hectares of garden or other land should this be required for the reasonable enjoyment of the property. Thus, with larger country houses it may be possible to demonstrate that additional land should be included. The test is strictly interpreted based on what the market would require and tests of historical association between the house and adjoining land and comparable transactions in the market.

18.4.17 The principles have been tested in both the *Longson* case involving the sale of a house and 18 acres where the claim for additional land including stabling was ruled out, and *Henke v HMRC*. In this latter case, Mr Henke acquired land without a dwelling then constructed a house which became the principal residence and later secured planning permission for two further dwellings. When the two plots were sold, the taxpayer claimed PPR on the plots on the basis that this land had been acquired in a single lot with the plot on which the private residence was built. The taxpayer failed to produce any expert evidence and lost. It is perhaps interesting to note that relatively few valuers seem to be asked to offer an opinion on the extent of land 'required' with a Principal Private Residence.

18.4.18 Part disposals – One further area where the valuer may be called upon is when there is a part disposal of a property; in that case the acquisition cost needs to be apportioned between the part of the property sold and that retained. The value of the part disposed can be assessed by a classic A/A+B formula as follows:

$$\text{Acquisition cost of whole} \times \frac{\text{Proceeds of the property disposed}}{\text{Proceeds of the property disposed} + \text{Value of the retained property on disposal date}}$$

18.4.19 This situation can occur routinely in agricultural disposals, whether the sale of a field from a larger holding, or the sale of a barn for conversion to a dwelling. In this latter case the formulaic approach is likely to give a higher acquisition cost than valuing the building separately, particularly given that the agricultural value of traditional buildings, which is likely to be the basis of value when the property is acquired, is relatively low. Different rules apply for small part disposals where the proceeds are less than £20,000 and can be deducted from the base value.

Further reading

Concise Rural Taxation (Charles Cowap, published annually)
Entrepreneurs' Relief (CAAV, 2012)
Surrender and Regrant of Agricultural Tenancies: A Review of Issues (CAAV, 2010)
The Taxation of Rural Dwellings (CAAV, 2016)
Valuation of Agricultural Tenancies (CAAV, 2009)

19 Agricultural stocktaking for income tax

19.1 Introduction

19.1.1 It is uncommon for a set of annual farm accounts to report on an entire production cycle completed from start to sale, benefiting from no activity in the previous year and with no spending or effort for the following one. As a result and under fair accounting, the accruals basis requires that where and when there are sales, full account must be taken of the relevant work and payments made in a previous year. Similarly, a value for work and spending undertaken in one accounting year to create sales in future years needs to be found and carried forward to be matched against those sales, rather than be a cost for the accounts of the year when they were incurred, potentially distorting those accounts.

19.1.2 The United Kingdom's self-assessment basis for taxation requires the taxpayer to submit accurate accounts and, with accounting standards, that expects the costs of what is sold to be matched against the income from it. That means there will almost always be a need for a record of:
- the stocks that have been brought forward into the year to be matched against income in that year or later year
- the stocks to be carried forward from this year to be set against income in future years.

This needs to be on a fair and consistent basis; undervaluing is as dangerous as overvaluing.

19.1.3 Taking the simple example of cereals cropping and a typical arable accounting year-end date of 30 September it is likely that:
- there will be grain from that summer's harvest still in the grain store and not yet sold on 30 September
- work and sowing to establish the following crops may already have been done that September.

Both need to be recognised in the accounts. If they were not recognised with values, then:

- the sale of grain in the following year would have no costs to offset it, overstating the profit
- the costs of establishing the following year's crop would be a cost in the year they were done, understating the profit.

19.1.4 As a very simple illustration, a business buys and sells items but has not sold all of them at its accounting year end:

Bought 10 items for £50	£500		
Sold 6 items for £100	£600		
Apparent profit	£100		
But it still has a stock of 4 unsold items			
To be added (carried forward)	£200	A fair profit of	£300

19.1.5 The £200 of cost carried forward is then there to be set against the £400 of intended sales of the remaining four items when that is achieved. That enables the profit really achieved in each year to be stated correctly.

19.1.6 While it might be that in a stable and well-ordered farming system these effects might typically tend to cancel each other out, it becomes more relevant where there is change, as:
- at the start or end of the farm business
- on any change in the farm business, as if it rented in more land and so increased turnover
- where outside factors change the timing of what is done, such as a change in the usual pattern of crop sales or where bad weather delays the establishment of the new crop
- with a change of farming policy, perhaps from arable to livestock.

19.1.7 While this feeds into the management and tax accounts for a farm business, this assessment is best undertaken by an agricultural valuer. It is typically more often called for by arable than livestock farmers (partly because of the exclusion of production animals under the herd basis and HMRC's acceptance of deemed costs for most livestock).

19.2 The approach

19.2.1 Important guidance is given by both financial reporting standards and HMRC's *Business Income Manual* and its *Help Sheet 232* (still known by many as *BEN 19*, its previous reference). Departures from the principles set out in these documents may require careful justification.

19.2.2 A fundamental constraint is the desirability of consistency in practice, so that successive years are assessed on the same basis without changes that might influence or distort the results and so enable easier comparison between years, aiding business review and management.

19.2.3 With very limited exceptions, mostly concerning publicly quoted limited companies and clients that have chosen a "mark-to-market" (or "fair value") basis, farm stocktakings are generally assessed on the lower of net cost or the net realisable value of the work. In practice, that means that a stocktaking is an assessment of costs incurred in the activity up to the year-end date (also the balance sheet date).

19.2.4 Arable stocktaking – Thus, for crops in ground and harvested crops, it will be the costs of:
- the seeds, fertilisers, sprays and other inputs committed to that crop by that date
- the farming operations involved, with the costs of the machinery and labour in land preparation, drilling and other activities, whether done in-house or by contractors. While contractors' costs will be evidenced from invoices, the costing of in-house operations is typically done using the CAAV's *Costings of Agricultural Operations* published annually and giving typical costs per acre worked for the use of machinery (with its depreciation, repairs and other costs), fuel and labour.

19.2.5 This is essentially an assessment of the "direct" costs of a crop and so, unless land has been specifically rented in for the crop in question (as perhaps for potatoes), the rent of land is not taken into account, that being seen as an overhead.

19.2.6 For sole traders and partnerships, as unincorporated businesses, the business owner's labour ("proprietorial labour") is excluded from the assessment as that is a matter of business risk. However, the cost of retained labour is included as is the labour of directors for limited companies.

19.2.7 On the same principles as for livestock discussed later, in those cases where the attributable costs cannot be identified, HMRC allows harvested crops to be assessed at a "deemed cost" of 75% of their market value. However, it is generally likely that cost can be established and, once actual costs have been used, it is not usually possible to revert to deemed costs.

19.2.8 The accountant may require depreciation to be shown separately so that it can be removed as an adjustment for the tax accounts and replaced by capital allowances. The CAAV Costings allow that to be done.

19.2.9 Livestock – HMRC has recognised that it can be harder to identify the costs of rearing livestock and has, as a result, allowed the use of "deemed costs" for certain species (and no others) where they were home-reared or purchased when young:
- cattle at 60% of market value
- sheep at 75% of market value
- pigs at 75% of market value.

Taxpayers cannot alternate between deemed and actual costs; if for whatever reason a move is made to actual costs they must then be applied as a matter of consistency.

19.2.10 Actual costs are to be used in all other situations including animals of qualifying species that were purchased when mature.

19.2.11 Breeding livestock, if not on the herd basis, can be assessed by writing their value down to cull value over their production life.

19.2.12 Net realisable value – The net realisable value (the value after the cost of selling the product) will only be material if it is clear at that balance sheet date that the crop is of little or no value. That might, for example, be the case if a crop, once sown, is then washed out by flooding or failed in a drought.

19.2.13 Fair value – While rarely met for farming in the United Kingdom, international accounting practice (as set out in International Accounting Standard IAS 41) has tended to favour using market values for "biological assets" and "agricultural produce" rather than costs for stocktaking, even though this is in practice more volatile between years and advances unbanked notional profits for taxation.

19.2.14 The valuer is asked to come to a reliable estimate of the fair value, net of disposal costs, working from the available evidence, though allowing a default to a costs basis.

19.3 The process

19.3.1 The valuer should establish the division of work with the client's accountant for clarity in the instructions and to avoid the risk of omission of work or double counting. This is important as, strictly, the tax accounts adjustments for the herd basis (excluding the value of production animals where an election for this basis has been made) and depreciation are made by the accountant.

19.3.2 With that preliminary, the letter of engagement, typically drafted by the valuer, can record the terms from the instruction, any assumptions that are to be made and limit liabilities including the way in which the valuation may be used.

19.3.3 Undertaking the stocktaking requires the methodical review of what is being grown and what has been grown, reconciling:
- areas of cropping, with operations undertaken for each crop, direct costs applied, likely yields, crops in store and records of sales or likely on farm feeding, etc
- numbers of animals with birth, deaths, purchases and sales, etc

with the evidence of activity from invoices, sales documents and other sources as well as stocks of grain or consumables (such as fertilisers, chemicals, fuel, feedstuffs, machinery parts and potato boxes) on the farm. Does the picture as established make coherent sense in the circumstances?

19.3.4 For RICS members, this is a matter covered by the Red Book.

19.3.5 With much of the information necessarily coming from the client and the client's records, that will need to be noted in the valuation report.

Further reading

Guide to the Costings of Agricultural Operations (CAAV, published annually)
Guidance Notes for Agricultural Stock Valuations for Business Purposes (CAAV, Third Edition, 2012)

20 Valuations for business rates and council tax

20.1 Introduction

20.1.1 Business rates – Non-domestic rates ("business rates") are levied in each part of the United Kingdom as a tax on an assessed rental value of each non-domestic property, payable by the occupier as tax on occupation and used as part of the funding of local government. While agricultural property is generally exempt, that is subject to definitions and periodically tested by changing practices. Some other uses are fully or partly relieved. Nonetheless, the basic principle is that all property is, by default, rateable and so has to be valued.

20.1.2 While farmers, having won full exemption in 1929 on the basis that farmland is more like production plant and machinery than premises, that principle of universal rating means that rates are likely to be met as a necessary cost of much farm diversification or other non-agricultural activity on estates or in the countryside. Very often, the scale of enterprise may mean that the relevant form of small business rate relief applies to lift the possible charge. In a striking inversion of the system, rating assessment, the basis for aid to be paid to small businesses in the early days of the 2020 Covid-19 pandemic, meant that where farmers did not otherwise have a rating liability – historically a reason for some avoiding diversification – they could not get the grants.

20.1.3 The law and practice of rating is of long history, steadily becoming more formalised in the nineteenth century and subject to much statutory amendment and case law since, as well as changing business operations (as shown later on for renewable energy). However, its essential principles are:
- to identify the taxable property (the "hereditament")
- identify its rental value on a statutorily assumed lease and assumptions about the condition of the property as at a standard date.

20.1.4 The actual tax is then levied at a (more or less) standard percentage ("poundage") set each year in each country in the United Kingdom. The scale of money raised through this system (and distinctively from physical retail property challenged by the internet) means that business rates are increasingly contentious.

20.1.5 Agricultural land and buildings, fish farming and (save in Scotland) sporting rights are generally exempt (see later on). There are then reliefs available, expressed differently in each country, but notably for small businesses with lower rateable liabilities and charities. There is typically only a short-term relief of a few months for property that is empty.

20.1.6 However, land drainage rates are levied on farmland and buildings within internal drainage board areas as a contribution to the water management works such as pumping and flood defence that benefit farmland as well as other uses.

20.1.7 Domestic property – Domestic property in England, Wales and Scotland is taxed under Council Tax, according to the band of sale values into which the property was placed as at the valuation date in each country. This also funds local government.

20.1.8 The valuation is on certain assumptions about the property's condition.

20.1.9 Particular issues arise over mixed ("composite") hereditaments, where a property is both residential and non-domestic, whether the other part is subject to business rates (as with, say, a public house) or exempt as with a farm.

20.1.10 In Northern Ireland, domestic property continues to be subject to the previous system of domestic rates, working from a rental value for each dwelling.

Business rates

> *Note* – *This summary is with reference to the law in England with the main statutory provisions being in the Local Government Finance Act 1988 supplemented by other legislation and much case law. The law and practice in other parts of the United Kingdom are essentially similar, though some specific differences will be mentioned in some cases. Nonetheless, care should be taken on this point.*

20.2 The hereditament

20.2.1 This is the unit for rating assessment, whether that be an area of land (even if underwater), a building or part of a building, a sporting right, a mine or a right to use land for advertising. A network of fibre optic cables has been found to be a hereditament. As this is about what qualifies to be a separate item in the rating list, case law shows that this turns on it being an item of rateable property which is:
- capable of definition so that the area of land to be occupied is identifiable, even if there are no physical boundaries.
- a single geographical unit with its parts contiguous or with a functional connection between them, albeit recently revised by statutory changes

in England to reverse a decision that two separate floors in a building were separately rateable
- capable of separate occupation, but if several such contiguous properties are in one occupation they will be one hereditament
- put to a single use or with a sufficiently strong functional connection between differing uses.

20.2.2 Plant and machinery – This is a valuation of the property, not of any possessions in it, but can raise sometimes complex issues over plant and machinery. The effect of the detailed Plant and Machinery Regulations (separately provided in England and Wales, Scotland, and Northern Ireland) is that these are, in principle but by no means always, disregarded. However, the latent value of the property being able to install them can be considered (*Kirby v Hunslet Union Assessment Committee*).

20.2.3 The drafting of these regulations, having preceded the emergence of on-farm renewable energy installations, has more recently caused issues for them, partly by the chance of the timing of valuations during the development of the sector and more notably for hydroelectric plant, particularly in Scotland. Relying more on fixed engineering works that fall to be rated, small scale hydro-facilities have seen much higher rating assessments than other technologies of equivalent output, which are more predominantly equipment and machines that are outside rating.

20.3 Rateable value

20.3.1 The rateable values on which rates are charged are set for all properties across the whole of a country as at a common date (the "antecedent date") so that they are all assessed as at the same time, even if not brought into rating until later as if it had not then been built. That can be an issue where new property types and uses develop between valuations, such as on-farm renewable energy or data centres. With each re-valuation, values may move between regions or types of property according to market movements.

20.3.2 In England and Wales, the valuation process is undertaken by the Valuation Office Agency's district valuers with this forming a large share of the Agency's work. The Agency has a very large bank of rental information collected from property occupiers which is now much more searchable as a resource in its work. The 2017 Valuation List for England, intended to be regularly reviewed, is publicly viewable on the Agency's website.

20.3.3 In Scotland, this is handled instead by the Assessors. Rating in Northern Ireland is done by Land and Property Services.

20.3.4 The ratepayer may seek professional support from a valuer to consider the merits of a proposed valuation and handle any appeal. Some valuers specialise in this area of work.

20.3.5 The basis for that assessment is defined by statute as

> an amount equal to the rent at which it is estimated the hereditament might reasonably be expected to let from year to year on these three assumptions –
>
> (a) the first assumption is that the tenancy begins on the day by reference to which the determination is to be made;
>
> (b) the second assumption is that immediately before the tenancy begins the hereditament is in a state of reasonable repair, but excluding from this assumption any repairs which a reasonable landlord would consider uneconomic;
>
> (c) the third assumption is that the tenant undertakes to pay all usual tenant's rates and taxes and to bear the cost of the repairs and insurance and the other expenses (if any) necessary to maintain the hereditament in a state to command the rent mentioned above.
> (Local Government Finance Act 1988, Sch 6 Para 2(1))

20.3.6 So that all valuations are ultimately on a common basis, this assumes:
- a hypothetical tenancy with particular obligations on the tenant, whether or not there actually is a tenancy
- a hypothetical tenant, rather than the actual occupier
- a hypothetical landlord
- a rent negotiated at arm's length

and looks for the rent that might reasonably be expected to be paid if:
- the property were let from year to year, so with an implied prospect of continuing but able to exclude expected long-term changes that might affect the rent for a long-term fixed tenancy
- the property is in a state of reasonable repair, save for those that a reasonable landlord would consider uneconomic. This does not require any improvements to be assumed.
- the tenant pays the tenant's rates and taxes
- the tenant pays for the repairs, insurances and other matters needed to keep the property in a condition to command the rent found. It seems assumed that the property has been put into reasonable repair before the hypothetical lease commences.

That may, therefore, very well not look like the rent actually being paid for property.

20.3.7 In terms recognisable from other valuation provisions, Lord Denning described the process of finding the rateable value:

> The rent prescribed by the statute is a hypothetical rent . . . which an imaginary tenant might reasonably be expected to pay to an

imaginary landlord for a tenancy of this [hereditament] in this locality, on the hypothesis that both are reasonable people, the landlord not being extortionate, the tenant not being under pressure, the [hereditament] being vacant and to let, not subject to any control, ... the period not too short not yet too long, simply from year to year.

(*R v Paddington (VO) ex parte Peachey Property Corporation Ltd*)

20.3.8 Case law holds that the hereditament is to be valued as it stands at the date of valuation with all the facts as they were then, irrespective of possible future changes. That means:
- disregarding possible structural changes
- assuming the mode of use applying at the valuation date (though this may be limited by development control restrictions)
- assuming as noted earlier a state of repair commensurate with the property, not its actual state of repair.

This physical state is at the date the valuation list takes effect, typically two years after the antecedent date as at which the valuation is assessed.

20.3.9 The requirement that the rateable value be that at which the property could "reasonably be expected" to be let necessarily looks to using all the evidence available when preparing the valuation.

20.4 Valuation methods

20.4.1 Fortunately given the variety of properties to be assessed, no method of valuation is prescribed, save for certain statutory undertakers while informal formulae are in place for particular properties, including schools and universities.

20.4.2 The great majority of properties met by the agricultural valuer would be assessed by conventional comparison with other properties. More particular properties with less direct market evidence available may be assessed in the basis of the profits or the contractor's methods. In practice, the great weight of comparable valuations helps anchor the "tone of the list" when considering other valuations.

20.4.3 Where there is little or no rental evidence for a type of property of use, but:
- its value lies in the profits to be made using it, then the valuation may be based on the profits method
- where there is less direct resulting profit, the valuation may work from its capital value or its construction cost which is then decapitalised (the contractor's method) at the risk of confusing cost and value.

20.4.4 There will generally be a good sense of the market for the typical lettings likely to be found in rural areas. More individual or specialist properties, such as public attractions, are likely to require more consideration as these are often operated in hand rather than let.

20.4.5 Rental comparison – It may be that a rent is already being paid for the property, but that may have been agreed some years ago or have terms very different from those assumed for rating valuation. Nonetheless, it is evidence to be considered alongside the rents paid for similar properties, always looking to compare like with like, using evidence that is as contemporary as possible for the valuation date.

20.4.6 The comparison will typically be by reference to a standard unit, often a square metre, for each component type of the property. Those types may be divided by:
- different uses such as storage rather than an office area
- different commercial values, distinguishing front areas (zone A) of a shop from rear areas that are still retail but generally seen as of lesser value.

20.4.7 Account is not to be taken of rents that are subject to statutory control, for long leases, between connected parties, fixed by tender, set as interim rents under the Landlord and Tenant Act 1954, on the sale and leaseback of a property, which include goodwill or were otherwise part of a larger transaction.

20.4.8 Adjustments are likely to have to be made for differences:
- between the properties being considered, with differences in size and composition generally handled by the units used for comparison, but other differences need to be considered on their facts
- from the terms required as for repairing obligations, where there is a premium or a rent-free period, service charges, phased rents or where the rent includes the rates.

20.4.9 The profits method – This is the method used where the value of a property lies in its ability to trade, sometimes with the benefit of a regulatory licence, and there may be little or no relevant rental evidence. It may also offer a cross check on the outcome of rental evidence.

20.4.10 The underlying principle is that the tenant is only motivated to hold such a property to be able to trade in the specified way and will view rent as a fraction of the profit expected to be achieved, sometimes in different ratios for differing profit levels (as is seen in some livestock market leases).

20.4.11 This relies on preparing typical trading accounts for a tenant of average competence and standing as at the valuation date, showing:
- annual turnover
- the working expenses, including repairs, in achieving that turnover.

There may then be an allowance for the tenant's return on enterprise and tenant's capital (in that way allowing for the existence of the tenant's fixtures that make trading possible). The evidence for this may come from the actual trading accounts preceding the valuation date, adjusted for all relevant factors and where they include trade on other properties.

20.4.12 The fraction of the remaining profit assessed from those accounts to be seen as rent would vary according to the balance of the contributions in earning that figure from the tenant's skills, energy and investment rather than the site and premises.

20.4.13 The contractor's method – Most often used for properties that are specialist, rarely let and often the results of engineering work, this finds the required rent (rateable value) by decapitalising the cost of the land and buildings. This method, with all the risks for value of a cost-based approach, is intended to find the annual costs to the occupier of providing the premises directly. It may be rejected if it can be shown that a tenant would not pay a rent on this basis.

20.4.14 The first step is to establish the cost of constructing the property, whether a building, reservoir, dry dock, a sports stadium or other premises, as at the valuation date.

20.4.15 As the work will rarely be new, an adjustment is made to that current cost of construction to produce the "adjusted replacement cost" or the "effective capital value", allowing for physical condition, current utility and any element of obsolescence.

20.4.16 The underlying land has then to be valued as a cleared site usable only for this purpose.

20.4.17 Those two values, for the land and the construction, are then decapitalised with an interest rate found from the real rate of interest adjusted for factors such as the additional cost to a typical borrower, depreciation and repairs. Statutory decapitalisation rates are set for some types of property.

20.4.18 There may then be adjustments for wider factors, such as the economic prospects for the sector or, if not already considered, the limitations of the site.

20.4.19 The outcome is then subject to the review required in all valuations, especially those involving so many steps, of whether the result found is one that could actually be expected to be paid.

20.4.20 Use of other rating assessments – As rateable values are determined, so they may be seen collectively to create patterns suggesting where the values for other similar properties might lie. This is not direct evidence of rental values but is seen as creating a "tone of the list", if only because they show what rating valuers have accepted and so can be held against them.

20.5 Agricultural, forestry and sporting exemptions from business rates

20.5.1 The principle is that all property is rateable unless otherwise provided. Agricultural land and agricultural buildings, with fish farming and forestry, as well as sporting rights outside Scotland, have the benefit of statutorily defined exemptions from rating, varying slightly in each part of the United Kingdom.

20.5.2 That does not necessarily mean that all activities that might be thought of as agricultural are exempt, requiring care in more specialist cases or where practices have changed. The definitions in rating law do not follow the definitions of tenancy law or planning law, meaning that apparent similarities can mislead. Even when it seems to give a similar answer, it may be by a different route. Thus:
- not all farmland is "agricultural land"
- not all farm buildings are "agricultural buildings".

20.5.3 Agricultural land – The first exemption is for five types of agricultural land – here excluding buildings – which are:
- land used as arable, meadow and pasture ground only
- land used for a plantation or a wood or for the growth of saleable underwood
- land exceeding 0.10 hectare and used for the purposes of poultry farming
- anything which consists of a market garden, nursery ground, orchard or allotment
- land occupied with, and used solely in connection with the use of, a building which (or buildings each of which) is an agricultural building for rating purposes.

20.5.4 Any other land will be rateable and there are five specific rateable exceptions from those types of agricultural land:
- land occupied together with a house as a park
- gardens (other than market gardens)
- pleasure grounds
- land used mainly or exclusively for purposes of sport or recreation
- land used as a racecourse.

20.5.5 The main exemption for agricultural buildings covers buildings occupied with agricultural land and used solely in connection with agricultural operations on that or other agricultural land. That means they must be both:
- occupied with agricultural land; so where that is not true, they are rateable unless they fall into the categories described later on
- only used for agricultural operations (not secondary processing, sales, etc) on that or other land, so allowing a farmer who is an agricultural contractor to retain the exemption while farming other people's land in addition to the home farm. There is a very limited de minimis exception, commonly taken as 5% of activity.

That means that where a building is storing farm produce but no longer in conjunction with agricultural operations, it is rateable.

20.5.6 The buildings that are be exempt when not occupied with other agricultural land are:
- market garden and nursery ground buildings
- livestock buildings
- buildings for bees
- buildings in use by syndicates (such as partnerships and unincorporated co-operatives) and bodies corporate (often co-operatives).

20.5.7 Exemptions are provided for land and buildings used for fish farming.

20.5.8 Sporting rights are exempt in England and Wales and those held with land are exempt in Northern Ireland. Scotland has removed the exemption.

Residential property

20.6 Council tax and domestic rates

20.6.1 Tax is also raised on the occupation of residential property, including farmhouses, with:
- Council Tax levied in England, Wales and Scotland on the basis of capital values as at a standard valuation date (1991 in England and Scotland; uprated in Wales to 2003) with the charge based on the band of capital values into which a property falls.
- rates in Northern Ireland, based on rental values assessed in a similar way to business rate, with the last valuation being 2005, and the rate poundage set then applied to each value.

20.6.2 In some rural cases, there are questions as to how much land is within the curtilage (the natural boundary) of the dwelling and so part of the Council Tax assessment.

20.6.3 Council tax and composite hereditaments – This, typically for farmhouses, is the main area where an agricultural valuer would meet "composite" hereditaments, where the farm is the unit of occupation. The agricultural land and buildings are exempt from rates, but the house is assessable for Council Tax.

20.6.4 To arrive at the value of the dwelling for Council Tax, the capital value of the whole hereditament is established and then apportioned between the residential and agricultural parts. That might usually then require identifying the capital values for both parts if sold separately and then applying the ratio between those two figures to the united value. In practice, this often saw the farmhouse move down one or two bands.

20.6.5 In Northern Ireland, the rental value of a farmhouse is discounted by 20%.

Land drainage rates
20.7 Land drainage rates

20.7.1 Drainage rates are payable in England and Wales under the Land Drainage Act 1991 by the occupier of agricultural property in an internal drainage district. In these districts, bodies, usually internal drainage boards, act locally to manage flood risk with the maintenance, improvement or new works to watercourses or structures. The drainage rates funding this work are levied separately on agricultural property which would not otherwise pay rates. The occupiers of other property contribute through general local taxation.

20.7.2 The rateable value of farmland and buildings is currently assessed as its rental value under the provisions of the Agricultural Holdings Act 1986, but the 2020 Environment Bill would give powers for that basis to be replaced.

21 Compensation for compulsory purchase

21.1 Introduction

21.1.1 Compulsory purchase is the exercise of a power granted by law to take property, including houses, buildings and land (residential, commercial, agricultural and other) from its owners and occupiers for purposes approved by the law. Those purposes include defence, public health (as for water and sewage), energy, transport (roads and railways, ports and airports) and urban regeneration where it is typically hard to acquire the necessary property by private purchase alone. With much of these either needing rural land or connecting urban areas, a large proportion of compulsory purchase work concerns rural landowners and valuers.

21.1.2 With the magnitude of that challenge to private property rights, strict procedures are provided with opportunities to test the case for expropriation. That may be taking forward a Compulsory Purchase Order (CPO), General Vesting Declaration (GVD), a Development Consent Order (DCO for Nationally Significant Infrastructure Projects (NSIPs)) or, as with HS2, specific legislation. While this chapter does not focus on those procedures, they allow the negotiations that may enable the valuer acting for a client to achieve changes to the detail of the scheme and potentially helpful accommodation works that may be more helpful to a continuing farm than simply maximising the compensation that is the subject of this chapter.

21.1.3 Where the property is taken, the law provides for the payment of compensation that, broadly speaking, covers the value of the expropriated property and associated losses, damage and the costs associated with resolving that. Assessing this is a substantial job for the valuers on both sides.

21.1.4 The basic principle applied in the United Kingdom is that of equivalence, so that affected person is left no worse off than before and is to be no better off, however much their property may have been taken against their wishes and across all their plans. The basis for the assessment is market value, even though the assumption of a willing seller might be more hypothetical than usual.

21.1.5 Very few of those involved in using compulsory purchase powers to take property from its owners and users (and, indeed, few of those acting for those affected) actually have practical experience of being subjected to it. That calls for some imagination as to the position of someone threatened by the loss of home or business and, as often in rural work, with living and farming alongside the project as it proceeds across their land.

21.1.6 That market value is to be assessed in a hypothetical "no-scheme world" so that no account is taken of the changes that the scheme requiring the compulsory purchase brings. It is as though the intended project would never happen.

21.1.7 The equivalent chapter in the fourth edition began with the statement ". . . the past 50 years have seen a dramatic increase in claims arising on compulsory purchase . . ." The same could equally be said of the 12 years since that edition and, if anything, the pace of increase has accelerated. Major transport infrastructure schemes have generated much of this work with significant investment in road and rail alongside landmark regeneration projects such as the Olympics facilities in London.

21.1.8 That was set to increase further over the next ten years with it being said that there could be a tripling of the professional work on compulsory purchase over the coming years. The various phases of HS2, taken together a long enough project to fully occupy some careers from graduation to gold watch (or time implant as it may be by then), and other major road and rail links were only part of this. Post-Covid, the emphasis on housing alongside infrastructure as twin engines for economic recovery are likely to see even more compulsory purchase activity. This will place even more strain on a system which arguably should have been overhauled generations ago, but now unlikely to be the focus of wholesale review with everything else on successive government agendas over the next decade.

21.1.9 This presents an opportunity for valuers in both the urban and rural spheres who are interested in or willing to engage with compulsory purchase. Some practitioners retreat at pace when encountering opportunities for this work, particularly if they have an established client base and workload elsewhere, complaining that the returns are insufficient reward for what they see to be the increasing frustrations of compulsory purchase work. However, for the younger valuer at the start of their profession, willing to pursue this area of skill, there is a substantial body of work in front of them in all but the most remote parts of the country. Those who make it to a senior level can earn substantial rewards, particularly if they become sufficiently skilled and respected to deal with high value work and appear at Tribunal.

21.1.10 Those interested in this work will find far more comprehensive points of reference in the subject bible, *Compulsory Purchase and Compensation* by Barry Denyer Green and, for Scotland, Jeremy Rowan-Robertson's *Compulsory Purchase and Compensation: The Law in Scotland*, both supported by the excellent CAAV numbered publications, *Professional Fees in Compensation Claims* and *Good Practice in Statutory Compensation Claims*.

21.1.11 The title of the latter is deliberately chosen. There is increasing concern at the conduct of compulsory purchase negotiations and this publication is part of a concerted effort to improve behaviours. The RICS has also updated its guidance note in this area seeking to offer guidance and support to all surveyors involved including those feeling under pressure from clients exercising statutory powers in a way which may make them uncomfortable. Valuations for compulsory purchase are covered by the Red Book with all the obligations that places on Chartered Surveyors in terms of their behaviours. Again, those interested in this area of work may consider joining the Compulsory Purchase Association, a multi-disciplinary body including leading surveyors and lawyers practising in this area.

21.1.12 This is a technical and much litigated area of practice; indeed enthusiastic practitioners will soon become familiar with a plethora of Victorian cases, prompted by canal, railway and urban sanitation developments which still offer precedents for today's litigation. Further guidance will be found in those sources referred to in the previous paragraph. This chapter seeks to introduce the rural valuer to the key issues involved and at least provide a framework for wider research.

21.1.13 In many cases, valuers, particularly those acting for the claimant, will be introduced to a scheme fairly late in the process, although it should be part of a rural valuer's routine to be aware of proposed infrastructure or regeneration work which might impact their area. Other valuers, acting for acquiring authorities, whether in-house or as retained consultants, will be involved at an earlier stage, not least in referencing schemes (identifying the owners and occupiers of property which will be impacted) and in producing Land Cost Estimates to support infrastructure proposals.

21.2 Legal framework

21.2.1 The basic legal framework is that established by consolidating legislation in 1845. In Scotland, these statutes remain direct points of reference, but elsewhere those principles (and often wordings) have brought forward in later statutes. Everywhere, they have been amended. The result is a wide variety of overlapping statutes for the compulsory purchase arena so that, whilst they are still very relevant, the practising valuer needs to have regard to more than just the "Six Rules" which, for England and Wales, are set out in section 5 of the Land Compensation Act 1961, although they will still feature heavily in the rest of the chapter.

21.2.2 The legal framework differs between parts of the United Kingdom, though that is more in its form and some detail than in principle.

21.2.3 England and Wales – The fourth edition of this book referred to six principal statutes for England and Wales, namely:

250 *Valuations for taxation, utilities, etc*

- Land Compensation Act 1961 – with its six rules, augmented as we shall see by the Neighbourhood Planning Act 2017
- Compulsory Purchase Act 1965 – providing for Severance and Injurious Affection
- Agricultural Miscellaneous Provisions Act 1968 – with compensation to a tenant for disturbance subsequently incorporated into the Agricultural Holdings Act 1986
- Land Compensation Act 1973 – introducing in Part 1 statutory rights for those affected by works but who have not had land taken, as well as more on Severance and Injurious Affection
- Planning and Compensation Act 1991 – amendments to the rules on Notices to Treat, certificates of alternative development, planning assumptions and interest payments
- Planning and Compensation Act 1994 – introducing basic loss and occupiers' loss payments replacing the previous farm loss payment under the 1973 Act and it could also have mentioned the Land Acquisition Act 1981.

21.2.4 Further piecemeal amendment has seen the addition of two substantial pieces of legislation with some marked new developments in this area.

21.2.5 The Housing and Planning Act 2016 – This Act deals with a wide range of issues, including amendments to the 1961, 1965 and 1973 Acts mentioned earlier as well as the Compulsory Purchase (Vesting Declarations) Act 1981 and the Acquisition of Land Act 1981.

21.2.6 Amongst other things, it provides for:

- powers of entry for acquiring authorities to enter onto land and property for surveying purposes, extending beyond the land to be acquired to include other land and subject to at least 14 days' notice
- new rules on timing in respect of advance payments under the 1973 Act to accelerate the process which had become delayed by acquiring authorities asking for further particulars
- amendments to timing and procedures in respect of Compulsory Purchase Orders (CPOs), notices to treat and General Vesting Declarations (GVDs)
- compensation for claimants for expenses or losses where a Notice to Treat is withdrawn
- technical provisions for procedures where land is severed by a scheme and the claimant wishes to serve a notice requiring the authority to purchase the severed land
- easements to be overridden where they interfered with works subject to the payment of compensation
- related housing development to be included in a Nationally Significant Infrastructure Project (NSIP) Development Consent Order in England
- a model claim form for compensation and advance payments as a guide to claimants to ensure that they provide sufficient information for a claim to be properly handled by the acquiring authority.

21.2.7 Old hands at compulsory purchase on the claimants' side might be heard to argue that there was nothing wrong with the standard of claims submitted, seeing the problem to lie more with the resources available or the acquiring authority's willingness to address them. Whilst many practitioners may have provided comprehensive detail to support claims, that was certainly not always the case. There is no doubt that some agents acting only occasionally in this area have been surprised when a simple request to *"pay my client the money"* was met by a response seeking some justification beyond the valuer's professional opinion. There is certainly no excuse for, at least, making sure that the relevant information on the model form is provided; practitioners can find the current versions of the claim form, government guidance on compulsory purchase and Crichel Down Rules and development consent orders on the compulsory purchase and national infrastructure sections of the government website respectively.

21.2.8 The Neighbourhood Planning Act 2017 – Close on the heels of the 2016 Act, this too is far reaching in some of the changes introduced. While mainly focused on planning issues, the principal changes in respect of compulsory purchase involve:

- the treatment of minor tenancies, of particular relevance to tenants of commercial properties with short-term leases (potentially including agricultural tenants with short-term Farm Business Tenancies)
- repeal of Part 4 of the Land Compensation Act 1961 which had allowed for a claimant to make a subsequent claim where planning consent was granted for additional development of the land taken any time up to ten years after the compulsory purchase
- changes to the timing for the publication and service of notices confirming a CPO
- extending powers to city mayors, the Greater London Authority and Transport for London
- codifying the "no-scheme world" assumption to provide direct legislative reference to a principle otherwise best encapsulated in the decision in *Pointe Gourde*
- a new regime for temporary possession with far reaching powers, although yet to be implemented.

The last two elements are the real heart of the compulsory purchase part of the Act or, arguably in the latter case, the teeth and are considered in more detail later on.

21.2.9 Before leaving the 2017 Act, it is worth reflecting briefly on the repeal of Part 4 of the 1961 Act. This legislation gave the claimant the chance to revisit their claim in the event that planning permission was granted for further development of the land acquired within ten years of the Order. The repeal of that legislation, in the face of representations from landowning and professional bodies, means that the claimant's valuer has to have even more regard to hope value, already a problematic area for compulsory purchase, when considering any claim for land taken.

Whilst it was not an overused provision, it provided a safety net which may have assuaged some claimants' anger over compulsory purchase. Removing it has taken away a future risk for local authorities in particular, but it is unlikely to help make affected landowners feel any more comfortable, or more importantly, amenable and so may be something of an own goal for the acquisition process.

21.2.10 In practical terms and some circumstances it might reinforce the attraction of a private treaty negotiation "in the shadow" of compulsory purchase where the vendor is free to add restrictive terms to protect their future interest based on a commercial judgement.

21.2.11 The consultation process around the 2016 legislation also considered several proposals for further change, leading to revised rateable value limits and statutory blight provisions, but two, which would be of particular interest to claimants, remain on hold:

- raising the statutory rate of interest from 0.5% below the base rate (with a floor now set of 0%) possibly to 8%; and
- reversing the Basic and Occupier loss payments for let land where at the moment the landlord receives an additional 7.5% (subject to limits) and the tenant 2.5% so that the larger payment would go to the occupier.

21.2.12 Aside from general statute law, there has always been project specific legislation enabling infrastructure development, most notably in the railway sector; HS2 has hybrid Bills for each stage with their own distinct processes. As well as enabling the relevant orders for compulsorily acquisition, these Bills may also provide for specific compensation schemes. Long running projects, such as HS2 and before that HS1 (the Channel Tunnel Rail Link), often include special schemes for householders and, to a lesser extent, businesses adversely impacted to reach a settlement with the acquiring authority before any orders for possession are served. These schemes are intended to assist those facing particular challenges when the scheme has frozen normal life: for example, more senior residents wishing to downsize or move for other reasons who can no longer sell their homes at pre-scheme values.

21.2.13 Wales – The Welsh Parliament now has the authority to enact planning legislation which, as in England, could be a host for compulsory purchase legislation, but this is an area where there is a degree of uncertainty and as yet no distinct Welsh legislative code.

21.2.14 Scotland – Scotland has a parallel code of its own legislation, retaining the underlying 1845 statutes of the Land Clauses (Scotland) Act and the Railways Clauses Consolidation (Scotland) Act, rather than, as in England, subsuming them in later enactments. They are then supplemented by much other legislation, but principally:

- the Acquisition of Land Authorisation Procedure (Scotland) Act 1947

- the Land Compensation (Scotland) Act 1963
- the Land Compensation (Scotland) Act 1973
- the Town and Country Planning (Scotland) Act 1997

and also, as elsewhere, much case law from the courts.

21.2.15 A wide ranging consultation was launched by the Scottish Law Commission in 2014 with its report published in 2016, but the Scottish Government had still not published a response by 2020. In the meantime, the Land Reform (Scotland) Act 2016 has given communities the powers to take neglected land.

21.2.16 As in England, Scottish government guidance on compulsory purchase is available on the government website under the compulsory purchase section. Whilst intended to inform government agencies, it is helpful to claimants and their agents in setting out the procedures and as a useful reference point in the event that agency actions do not seem to accord with the guidance.

21.2.17 In Northern Ireland – Distinctively using Vesting Orders rather than Compulsory Purchase Orders, the principal legislation in the province is provided by:

- the Lands Tribunal and Compensation Act (Northern Ireland) 1964
- the Planning and Land Compensation Act (Northern Ireland) 1971
- the Land Acquisition and Compensation (Northern Ireland) Order 1973
- Land Compensation (Northern Ireland) Order 1982
- the Planning Blight (Compensation) (Northern Ireland) Order 1981.

The Land Acquisition and Compensation (Amendment) Act (Northern Ireland) 2016 introduced Basic and Occupiers' Loss, assessed in a rather different way to England and removed Farm Loss Payments.

21.3 The process

This section concentrates on the situation in England and Wales. The processes in Scotland and Northern Ireland are substantially similar, but reference should be made to the relevant legislation and guidance.

21.3.1 Statutory routes – The legal background, reviewed briefly earlier, provides two principal routes for an acquiring authority to acquire an interest in land. Nationally Significant Infrastructure Projects are covered by their own rules and procedures.

21.3.2 The two alternatives for schemes which are not NSIPs will be influenced by the nature of the scheme and, in particular, whether the authority is interested in taking possession to do work itself. If the scheme is for a local transport improvement, for example, where the authority will want to enter the land to do the works, the original Notice to Treat/Notice of Entry route which allows for early possession will be preferred. In the alternative where the authority will not be involved in work on the land but is acquiring it for sale to a third party (for example, in a regeneration scheme), a General Vesting Declaration ensuring early transfer of title without the authority needing to take possession may be more appropriate.

21.3.3 In many cases, the claimant's valuer will only become involved once the CPO has been confirmed, notices served on the interested parties and fixed to the land. Alternatively, agents may be very fully engaged where a hybrid bill is involved as explored further in the following section.

21.3.4 Once a Compulsory Purchase Order has been made the acquiring authority has three years to serve a Notice to Treat or make a General Vesting Declaration, the two alternative routes for schemes which are not covered by National Infrastructure provisions.

21.3.5 By negotiation – The acquiring authority could also seek to acquire the land in question by negotiation in the shadow of a CPO, either securing the Order, or having a reasonable certainty of doing so, approaching the owner informally in advance of serving notices. In one of the authors' experience, this route may be particularly effective where:

- there is a single owner, or very few owners involved
- the proposal is well trailed in local plans and most likely therefore in the local press, so it is no surprise to the owner
- there is a degree of added value in the land anyway, so the "no-scheme world" is less relevant
- the acquiring authority is small so that:
  - generally, the client and professionals involved have a reasonable awareness of each other, the client normally being well briefed on the scheme; and
  - the expertise required to secure the CPO would have to be bought in.

In those circumstances, the acquiring authority can avoid the delays and costs involved in securing a CPO, which would be particularly disproportionate for a small number of claimants. That, in turn, allows it to be less parsimonious in the approach to the claimants which can lead to a swifter and more cost-effective settlement for the authority. The author has seen such an approach work particularly well in schemes ranging from a mixed residential, cottage hospital and nursing home site to an extension for a secondary school.

21.3.6 Notice to treat and notice of entry – This is the "traditional" route, involving the acquiring authority serving an initial notice, the Notice to Treat, on the landowner, which fixes the interest in the land that is to be compensated, preventing any subsequent improvements or other actions having an impact on value. It is to be hoped that the client receiving the notice will be alive to the issue because notice should be served on their behalf, registering the nature of their interest in the property involved and setting out an initial claim, usually within 21 days.

21.3.7 The acquirer then has the right to take possession following service of a Notice of Entry giving at least three months' notice, a considerable improvement on the previous 14-day minimum period.

21.3.8 Whilst the process is compulsory, it is based on a 'normal' land transfer and thus the claimant has to convey the land separately to the acquiring authority.

21.3.9 General vesting declaration – This alternative version, introduced in 1981, has the advantage for the acquirer that the GVD itself automatically transfers the interest in the land to the authority at a date specified in the declaration. As set out earlier, this makes it particularly attractive where the authority is acquiring the land to resell and is not interested in taking possession.

21.3.10 The authority does not need to serve a preliminary notice of intention two months prior to executing the GVD; the purpose of the GVD and other detail will now be in the Declaration. However, this has not shortened the overall timetable; the two-month period has been added to the previous 28-day minimum time between GVD and transfer so that vesting date for transfer must now be at least three months after the execution of the GVD.

21.3.11 Nationally Significant Infrastructure Projects (NSIPs) – For England, the Planning Act 2008, amended by the Localism Act 2011, classed a variety of larger scale transport, water, waste and energy projects as NSIPs and provided Development Consent Orders (DCOs) as the process to grant approval for them and associated land acquisition. This has been done as part of an attempt to reduce some of the delays and costs which have dogged major infrastructure investments, for example the Sizewell B nuclear power station. One notable exception is large onshore wind farms which were excluded in 2016.

21.3.12 The application for a DCO is made by the developer of the proposed scheme and this can involve considerable work for valuers employed in those firms which act for acquiring authorities and developers. One of the requirements will be accurate referencing and an attempt to make contact with potentially affected parties, which can often be rebutted. This can involve substantial blocks of work for referencing teams in relatively tight timescales as this work will naturally tend to be delayed until there is at least a degree of clarity over the likely route. Approval of the grant of a DCO rests with the relevant Secretary of State, advised by the Planning Inspectorate.

21.3.13 The DCO can be seen as equivalent to the Compulsory Purchase Order, although the provisions for elements of compensation are assessed under the 2008 Act rather than earlier compulsory purchase legislation. Once the DCO is granted, the developer can proceed by either the Notice to Treat/Notice of Entry or General Vesting Declaration routes described earlier.

21.3.14 The Housing and Planning Act 2016 enabled developers using compulsory powers under a Development Consent Order to acquire potential residential development land on the basis of "geographical proximity" provided that:
- such a related housing development has now more than 500 dwellings
- the right will be constrained by planning policy, particularly for example in a protected environment
- local policies on affordable housing will apply.

The housing proposals will be judged as part of the overall DCO proposal and subject to the same requirements for prior consultation

21.3.15 As part of the compulsory purchase, the compensation assessment will be on the *no-scheme world* basis so that, if the potential for the housing development is contingent on the NSIP, the developer will be able to acquire the land at market value on the basis of its existing use, without that additional residential development value. Some landowners, anyway not willing sellers, may see this as the acquirer (unlikely to be a natural housing provider), taking unreasonable advantage of its powers to its financial benefit.

21.3.16 Devolution and compulsory purchase – The procedures for a CPO are very similar in Scotland to England and Wales with the same timetables for either Notice to Treat or General Vesting Declaration. The situation is different in Northern Ireland with compulsory acquisition through a Vesting Order as set out in the Local Government Act (Northern Ireland) 1972. As the name suggests, this is essentially equivalent to the GVD regime with the interest in the land transferring at the date of vesting.

21.3.17 Devolution and national infrastructure – Similar arrangements apply in Scotland and Wales, although with different rates of progress:
- in Scotland, where the government has established an Infrastructure Commission to develop a 30-year Infrastructure Plan
- in Wales, where Developments of National Significance (DNS) and the associated procedures are similar to NSIPs although these do not extend to conferring compulsory purchase powers which the developer will have to seek separately

21.4 The valuer's role

21.4.1 The role of the valuer extends beyond simply assessing, submitting and negotiating a claim. Whether acting for acquiring authority or claimant, this is an area which requires both high levels of technical knowledge, particularly an appreciation of the asset types involved and the relevant legislation, and well developed "soft" or people skills.

21.4.2 At the heart of the process, the two protagonists – and despite best efforts it is difficult to see them very often as partners in this venture – are:

- a landowner or tenant, whether corporate, public or third sector but commonly in the rural world where so much infrastructure work takes place, private individuals; and
- the acquiring authority, whether local, national, directly democratic, arms' length agency, a public/private hybrid or commercial company

21.4.3 However, it is their particular positions which are the problem, and where the careful hand of a competent and assiduous, and very often experienced, valuer on the tiller is essential; years served are not always a proxy for greater wisdom than youth.

21.4.4 The landowner
- like most of us, will not deal well with compulsion
- will normally not want to sell anyway and will see the compulsory purchase exercise as a smash and grab by the acquiring authority, stealing the family silver at the very least
- in a farming environment, may well have a lifetime's affinity with the area and, specifically, with the property to be taken
- when and if they rationalise matters, may conclude, not unreasonably, that as they do not want to sell the property the payment should reflect that and compensation be set at a level at which they would be foolish not to sell

21.4.5 Their agent, who would hope to have their trust and respect to deal with matters effectively on their behalf, will have to deliver the chastening message that
- in most cases, there is not much they can do other than grin and bear it, and keep a note of the time they spend doing so for the later claim (although read on)
- whatever the local cost benefit (or, as the client will see it, disbenefit), of the scheme the CPO or DCO means that it has been accepted as being in the public interest
- as it is unlikely that the work will go as quickly or smoothly as any timetable they have seen may suggest, they should be ready to live with the scheme for some time
- and, worst of all, whilst their logic for a ransom price might not be faulted, it is unfortunately overtaken by the legislation which treats them as a willing seller, happy to part with the land at market value, or less where development may arise as a function of the scheme

21.4.6 Managing client expectations is a key part of compulsory purchase work, particularly but not exclusively, for the claimant's valuer. Ultimately how they approach matters may have significant impact on the nature of the negotiations, the respect with which they are treated by contractors and in some cases the level of compensation. As we will see later on, the owner redoubtably defending their farm from even a square yard (metres do not work in such emotional conflicts) being taken may do far worse in compensation terms when the authority accedes to his demands than his neighbour who lost a small area of land.

21.4.7 There may be challenges for the authority's valuer as well, not least in being the advocate of their unwelcome client's attentions. Major CPO schemes need a planned and strategic approach and, whilst the engineering costs of infrastructure schemes will nearly always dwarf the land acquisition budget, it is the valuer who knows how to approach owners and the challenges of acquisition to best effect. That role is not always recognised, not least as the project teams are normally led by engineering or strategic directors in the authority or the consultant team for whom the acquisition of property rights is something of a foreign country.

21.4.8 That frustration can be particularly apparent in the very largest projects where the drive for standards of conformity at all levels, and a fear of delegation, can create their own problems.

21.4.9 Given this challenging environment, the key is for valuers to rise above the conflict and to treat each other, the parties, other professionals and contractors with respect. There will inevitably be disagreements, but they must be dealt with pragmatically and with cool heads. There is no merit and much to lose in valuers trying to channel the animosity and frustrations of their clients when engaging with their opposite number. The approach to be taken is best expressed in the CAAV guidance particularly in the chapter "The Principles of Good Professional Practice in Compensation Claims". The clue is very firmly in the title and any valuer acting in this area should have those doctrines very much at heart.

21.4.10 Whatever role they take, valuers will hope to be drawn in at a sufficiently early stage to make a meaningful contribution. Acting for the acquiring authority, this may on occasion be no more than making sure all the necessary work is done to enable proper negotiation of a claim. However, that is of itself critical to cost. For those who have not tried it, the author can report from experience that it is no fun at all trying to defend Part 1 claims (of which more later on) when your client, instructing you after the event, did not take any pre-scheme noise readings.

21.4.11 Acting for the claimant, valuers will ideally be retained sufficiently early in the project to be involved in any consultation on the scheme, particularly where there are some variables and room for negotiation. Early engagement can create a more tolerable outcome for their client, where often design measures mitigating the impact of the scheme are more important than the ultimate level of compensation. Those measures are much easier to achieve before the design is completed than when the swing shovel arrives on the farm.

21.4.12 Early stages – Depending who the valuer is acting for they may be involved in a range of activity before land is taken and construction starts.

This may seem more obviously applicable to the acquiring authority's valuer but, provided instructed early enough, it is more likely to be the case where the claimant's valuer is keeping aware of proposed projects and is, for example, referring to them on their website. There is much for the claimant's valuer to do as well beforehand.

21.4.13 On the acquirer's side, valuers may be involved:

- in providing initial valuation and strategic advice in advance of land cost estimates, particularly relevant where regeneration projects are involved and there may be a number of different schemes under consideration
- land referencing, that is collecting data on all potentially affected interests in land, a mammoth task where there is a linear scheme, such as overhead electricity lines and all the optional routes need to be referenced
- preparing Land Cost Budgets, normally with numerous iterations
- engaging in preliminary discussions with landowners and occupiers and their agents, identifying key acquisitions and where appropriate pursuing negotiations in the shadow of CPO.

21.4.14 Land cost budgeting is a particularly important element of regeneration projects where the viability and consequently sometimes the extent of the scheme may depend on property costs. That is seldom the case with infrastructure budgets, where engineering costs will be key, but the exercise is still important, both in setting broad financial parameters and in identifying potential difficulties through the review of scheme drawings and land interests which the exercise will involve.

21.4.15 The budgets are often required at short notice and with inadequate information, particularly in the early stages of a project. There is usually an indication when someone from Highways in the client council makes contact with an apparently innocent question over the value of farmland in the district. Seldom is that not a precursor to a request for a land cost budget and it is important for the valuer, whether in-house or external, to prepare the ground. As we shall see multiplying the area of land taken by the average price per hectare or acre will fall far short of the likely compensation cost (and, in passing, valuers should beware of the capacity of non-specialists to mix those values and measures up to dramatic effect).

21.4.16 This can be a particularly irritating exercise for the CPO specialist who will feel the inadequacy of information far more than the general valuer. To that extent, this should be a team exercise; it is impossible to predict the precise impact of a scheme from a plan as weather, timing, unforeseen archaeological remains and inaccurate digger drivers will all contribute to the real outcome in a way which cannot be foreseen at this stage. This is not the time to spend long hours fretting over the precise figure for Injurious Affection and, most particularly, disturbance. Rather, the important thing is to capture the key foreseeable elements of each claim and provide accordingly. Nor is this the time to be too conservative; ideally the budget can be produced as a range of figures but, as the project looms closer, there is no point in setting unrealistically low budgets which may be attractive at the time when the project's viability is in doubt but far harder to meet later.

21.4.17 The claimant's valuer, if retained sufficiently early, may also be involved in a range of activities beyond the initial challenge of explaining the likely progress of the project and acclimatising the client to the principles of compensation. These will involve trying to understand the client's property and business, not least who the owners are of the interests that are actually affected, to better appreciate the impact of the scheme. There may also be informal discussions with the acquiring authority, whether over fundamental issues such as the route or the extent of a scheme or more detailed matters such as accommodation works and timing of entry.

21.4.18 There may also be some involvement in more formal representations on behalf of the client with the forum depending on the nature of the scheme:
- where a CPO is involved, affected parties may make representations on the draft CPO which will be considered at the Inquiry by a Planning Inspector
- the same basic process applies with a DCO, where again there will be an Inquiry
- where a hybrid bill is involved, HS2 for example, affected parties can petition against the scheme when the Bill is going through Parliament

21.4.19 The valuer will need to be skilled in assembling arguments on the practicality of the particular elements of the scheme which affect their client, considering impacts and the cost/benefits and practical implications of any alternative schemes which may be offered in mitigation. Where a hybrid Bill is involved the petition must be in a prescribed form and there are precise timetables. Guidance will be found on the Bill website and it is important to check on the requirements for each Bill rather than rely on past experience and assume the same provisions will apply.

21.4.20 Engaging with the other valuer is important and again covered very clearly in the CAAV publication, *Good Practice in Statutory Compensation Claims*. There is much to be said for that engagement being:
- as early as possible – early warning of potentially damaging issues for a business offers more scope for the acquirer to mitigate the impact
- as open as possible – acquiring authorities and their agents will probably have little or no knowledge of the claimant's particular business, complaining after the event that they have ruined a crop of onions by preventing irrigation where that could have been avoided, will be of little use if the acquirer's agent was not told of this in the first instance.

21.4.21 Where a DCO is involved the acquirer is obliged to consult with affected parties. Unfortunately, the formal requirement for the process sometimes seems to militate against its usefulness, perhaps because the effort to engage with hostile parties is just seen as another item on the tick box of NSIP requirements. That is not symptomatic of all schemes and this is where experienced valuers, given sufficient freedom to act by their acquiring clients, can bring real benefits for both parties, avoiding the worst impacts for the claimant and enabling smoother access for their client.

21.4.22 Again, the valuer's diplomatic skills will be required. Affected clients may view the acquirer's agents as the last person they want on the farm, but there is nothing to be gained and much to be lost by allowing the acquiring authority to proceed in ignorance of those issues which are important to the claimant.

21.4.23 This will be further tested as and when access is required for preliminary surveys; the general reluctance amongst landowners to allow access led to the changes in the 2016 Act. Different rules apply depending on the nature of the scheme, but survey access is often a point of dispute, one recently tested in *Sawkill* which was decided in favour of the acquirer. That said, many schemes still offer a licence fee for access, seeing the relatively small sums involved as a much more effective way of securing access and co-operation than using purely statutory rights.

21.4.24 Before construction – Once the relevant order is confirmed and notices served or a GVD executed, the claimant's valuer should submit an initial claim based on the model claim form. There is not much time for this as the form should be returned within 21 days. Where the claimant has instructed a valuer well in advance, the form can be prepared at least in draft prior to the trigger service of the notice. In any event, it is most unlikely that sufficient detail will be available to complete the form, but that should not prevent submission; rather an incomplete form should be submitted for future addition.

21.4.25 Advance payments – Service of the notice also means that the claimant is entitled to an advance payment of 90% of the acquirer's estimate of the compensation due. This is unlikely to include all the items of disturbance because the compensation for that cannot really be assessed until after the event. However, it should include land and property taken and at least an estimate of injurious affection and severance.

21.4.26 These latter issues, challenging at the best of times, can be particularly difficult to assess where land is acquired in phases under a series of notices, as may be the case with a linear scheme such as roads. Agents on both sides may find it difficult to assess the cumulative effect of the full land take partway through the process. However, that does not seem to warrant the approach, adopted by some acquirer's agents at the time of writing, that this is too difficult to assess and thus there should be no payment at all. It should be possible for the agents, recognising that there will be some injurious affection, even if as yet unclear, to reach a conservative estimate against which an advance payment can be made.

21.4.27 For the affected client, who will be incurring costs both for professional fees and adjusting the business in the face of the impact of the scheme, neither of which may be fully recoverable until sometime well into the future, 90% of a conservative estimate is more welcome than nothing at all, and considerably less offensive.

21.4.28 During construction – Hopefully, both valuers will be involved with site meetings before works commence together with the contractor(s) involved and the owners and occupiers. This should deal with practicalities during the work including:

- access to the site for contractors, including off-site issues such as road closures which might affect a milk company collection for example
- access across the site, where relevant for the occupiers
- service interruptions and temporary solutions, although these may be difficult to identify precisely at the time
- critical dates/times and access issues for the occupier which might be target weeks, for harvest or silaging, or key times of the day on a dairy farm
- biosecurity, with the claimant's agent ensuring that their client's producer contracts requirements are very clear to the contractor, both to avoid breaching them or, failing that, so that the consequent claim is clearly founded
- fencing and security, particularly where compounds or material dumps are going to be sited on the farm; these can include some valuable equipment and materials which can be an invitation to those inclined to help themselves, that may attract some unwanted attention to the farm.

21.4.29 Whilst still some time off, this will also be a good time to discuss restoration works, removal or otherwise of temporary fencing and the like, not least as this may be the first time that the contractor has been involved in direct discussions. The meeting should also deal with communication with contact details exchanged. Whilst it is fairly obvious that the claimant will want the contractor and acquirer's contact details and vice versa, it will also be helpful for the contractor to know other people who may be entering on the farm and working near the site, such as routinely used agricultural contractors. Just to add to the complexity, there will likely be some data protection points here that are perhaps more likely to be honoured in the breach than the observance.

21.4.30 Where the claimant's valuer has been instructed in good time, then this will be far easier and much of the meeting may simply confirm previous discussions; otherwise there may be a steep learning curve. Whatever else, the valuer must meet the client in good time in advance, not half an hour before the meeting, so that they can fully appreciate the issues and understand their client's priorities. As in most things, the client will have some "need to have" issues and some "nice to have" ones. It is essential that the valuer knows which are which.

21.4.31 That meeting will be much easier if the acquirer and contractor has provided detailed plans and specifications well in advance of the works. While that used to be the case, in the modern design and build environment (who knows which comes first) it seems increasingly common for this information to be provided at the last moment, or even at the meeting. Modern mobile communications may be extraordinarily advanced but not to the stage that having scheme drawings of a scheme impacting your client's land being emailed to your mobile phone as you drive to the site meeting is a substitute for a plan received a week in advance.

21.4.32 That can be a source of frustration for both agents and, having acted for acquirers and claimants alike, securing early and reliable detailed designs can be a considerable challenge. As there is unlikely to be time for multiple pre-commencement meetings, it is better to postpone if the data is unavailable, albeit that may meet with resistance from contractors and acquirers who are likely to be on a tight programme of deadlines.

21.4.33 Where land is to be restored, the acquiring authority should make a record of condition in advance of the works. It should be as detailed as possible not least, as in contrast to a record for a tenancy serving as a benchmark for maintenance, this is intended to identify how the land should be restored when some of those features have been destroyed by the works. Clearly, claimant and agent will wish to check this record; it is helpful for all if it can be agreed. Alongside the record, it is very useful to keep a schedule of reinstatement, identifying what is to be done, particularly where this is not simply direct replacement of what was there before, by whom and by when.

21.4.34 In an ideal world, the scheme then proceeds smoothly with limited involvement for agents. That is not always the case. Whilst the claimant has an obligation to mitigate their loss, including avoiding unnecessary site visits by their agent, there is every prospect that further meetings or visits will be required, particularly where contractors are on site for more than one season. These may be routine visits, checking on progress and the impact on access routes, reinstatement of land to be returned and other matters for which it may be helpful to agree how many routine meetings the acquirer will be willing to include in any claim for costs. However, there will often be unforeseen visits where accidents have occurred or urgent changes are required. Whatever the urgency, the agent should advise the contractor of their visit and, preferably, the acquirer as well. Many contractors will now have particular site protocols and access arrangements which should be respected. Where a number of colleagues might be involved on the same scheme for different claimants, it may be cost-effective to deal with these in one visit.

21.4.35 Whatever the purpose or nature of the visit or meeting, thorough notes should be kept. Some specialist agents have detailed project management software to manage all the data involved, the accessibility of which is commonly in direct inverse proportion to the valuer's years of service. It is also important that notes are exchanged with the other party and essential that agreements are documented. What may seem obvious to the claimant's agent, dealing with only one client, will easily be lost in all the different affected farms and parties that the acquirer's agent is dealing with unless they are prompted by a note from their opposite number.

21.4.36 There is a complication here. Up until the start of the works, most of the contact will have been between claimant, their agent and the acquirer's agent and possibly engineer, all referring to plans and speculating on impacts. Once the contractor is on site, the dynamic changes with direct contact between claimant and contractor's staff because they are both on site. There is a difficult balance to be struck. It would be ridiculous for the contractor to have to go through an Agricultural Liaison Officer to the agent and then to the farmer to confirm that it is alright for them to take down a gate for two hours, but the agent will need to know if that becomes two days and gives rise to a claim because stock got out. There is no fixed way around these issues; the best thing for both agents is to be open about them at the site meeting so that all present can understand the issues involved, particularly the claimant, as it is worth remembering that, whilst the expert on the farm, he will be the least experienced on infrastructure work.

21.4.37 Contractor/farmer arrangements over temporary matters are common and very practical. However, there is a broader question over privity of contract, revisited again in the next chapter, as these remarks apply equally to utility as to other infrastructure and CPO work. The starting point should be that the acquirer is responsible for compensation for the loss and damage suffered by the claimant. The Lands Tribunal case *Donovan v Dwr Cymru* confirmed that even though the contractor had gone outside of the notice area and caused damage giving rise to a claim they were there solely as Dwt Cymru's agents and thus the utility could not avoid its statutory responsibilities to the claimant.

21.4.38 However, there is also the principle that someone instructing a contractor is not responsible for the contractor's negligence or other torts. Some acquirers seem to be relying more on this to lay off liability onto the contractor. That leaves the claimant either a frustrated spectator whilst the two parties argue over the finer points of contract and common law or, simply and more often, rebuffed by the acquiring authority. Whilst that argument may develop soon after the incident which gave rise to the claim occurred, that is no excuse for the claimant's agent holding back. The best reaction is to submit the relevant claim, or the notice that the claim will be made, to the acquirer as soon as possible so that the real injury is not lost in the wider debate.

21.4.39 At the end of construction – In the same way that the parties met on site before works commenced, they should also meet on completion to confirm:

- satisfaction or otherwise with the way any land returned has been left
- outstanding works or snagging (checking for minor faults to be rectified), with arrangements over completion and costs
- any potential signs of future difficulties; for example, flooding in adjoining areas might suggest the drains have not been properly reinstated
- timing of the removal of any remaining plant and equipment; in the electricity sector, for example, scaffolding may be left on site well after completion of the works pending the scaffold contractors' return to remove the towers
- confirmation of maintenance liabilities both in the short and long term; a forestry contractor might be responsible for beating up newly planted landscaping in the first couple of years but not thereafter
- that the acquirer will provide plans of the works as completed.

21.4.40 Whilst there should be some matters which can be signed off, there are usually matters which will be saved for checking in the future with drainage problems and subsidence being issues that may be very difficult to detect until well after completion of the scheme. Where major works are involved with very heavy equipment or very substantial excavation, impacts to retained land may still be emerging ten years after works have been completed.

21.4.41 Preparing, submitting and negotiating the claim – Part of this work will already have been started and hopefully agreed and an advance payment made before the works have completed. Further, much data will already have been compiled to inform the claim, for example where particular incidents such as flooding or escaping stock have given rise to a loss.

21.4.42 The claimant's responsibility for mitigating the claim extends to fees. That should not extend as far as the claimant's agent giving up a reasonable argument at the first hurdle, but it does demand a degree of reasonableness of approach on both sides. Where that is not reciprocated, more justifiable time and costs will be involved. The agents are there to represent their clients but also to reach an appropriate and equitable settlement on behalf of, and arguably sometimes despite, their clients.

21.4.43 It is up to the claimant's agent to construct and submit the claim providing sufficient evidence for the acquirer's agent to understand, assess, negotiate and ultimately agree the claim and the appropriate quantum. This may involve subsequent site meetings to assess particular issues. For RICS members, CPO claims are now covered by the Red Book with all the requirements for process and evidence that involves where reporting such items.

21.4.44 Beyond that many acquiring authorities will have very specific reporting and authorisation requirements with delegated authority a thing of the past so that the signing off of their valuers' reports and the associated claims will go through a number of processes. It may be unfair to suggest that chains of authority lengthen proportionately to the length of the infrastructure or scale of the scheme, but some of the arrangements are Byzantine in nature. This is particularly annoying when evidence, already provided to a preliminary authorisation committee, is requested by the next tier above. That said, there is no merit for the claimant's valuer in venting their wrath on their opposite number who will be equally frustrated; rather the only three things one can do are:
- supply as much evidence as possible in the first instance
- work hard to manage the client's expectations and keep them informed; and
- make a note of your time for the fee invoice.

21.4.45 Fees – This can be a contentious and occasionally confusing issue not helped by one or two preconceptions and some lazy language used by professionals acting on both sides, although generally that has improved in recent years.

21.4.46 The general presumption amongst clients, which is wrong, is that the acquiring authority will pay the claimant's fees as of right. Depending how that is interpreted that could leave the prospective client to believe:
- that fees will be dealt with directly between his agent and the acquiring authority and he will not be involved in any way; or
- he may have to pay his professional fees, but they will not be invoiced until the end of the process and he will be reimbursed almost straight away with minimal impact on cash flow.

21.4.47 He is unlikely to be expecting, for example:
- that fees for preliminary consultations with his agent prior to the scheme becoming fully developed may not be borne by the acquiring authority
- that he will be liable for his total professional fees, but the acquiring authority may query or challenge the extent to which these are reasonable, both in terms of time and the level of charge
- he is obliged to mitigate his claim so he will not be recompensed by the authority if he calls out his agent to inspect more than has been agreed or is reasonable.

21.4.48 It is his valuer's responsibility to explain the situation to him and ensure that he provides clear and explicit terms of business so that there is the least possible room for debate in the future. At the same time, the valuer should confirm his proposed level of fees with the acquiring authority's agent to ensure that there is no difficulty later in respect of hourly rates. This can lead to some challenging conversations about the level of fees proposed and the extent to which the claimant's valuer may, in their opposite number's opinion, be either too senior or simply too expensive. The former can be countered on the basis that the experienced valuer will be able to draw on that experience to deal with matters more rapidly than a less experienced colleague or in the case of the small practice or sole practitioner they are the only resource available and it should not be for the acquirer to dictate through fees the claimant's choice of agent.

21.4.49 There may be some difficulty here with some acquirers comparing the level of fees that they are paying their consultants with the higher figures charged by the claimant's agent, but that is a false comparison. The acquirer's agent acting on a larger scheme will generate thousands of hours work with economies of scale that systemisation can bring. The claimant's agent will be billing a fraction of that time and waiting until the end of the project to be paid, at least for the disturbance part of the claim, which can often absorb a disproportionate amount of time.

21.4.50 Further and better guidance can be found in the CAAV publication *Professional Fees in Compensation Claims* as well as the RICS Professional Statement and the VOA *Land Compensation Manual* found online.

21.4.51 Fees are increasingly claimed on a time basis, the old practice of using Ryde's Scale having fallen by the wayside amongst most claimant's agents. However, some authorities and utilities still seek to enforce the payment of fees on the basis of Ryde's Scale plus a percentage uplift; whether in so doing they are pursuing anti-competitive practices and breaching the competition legislation remains to be seen.

21.4.52 Given the tendency to charge on a time basis the claimant's valuer needs to keep accurate and comprehensive time records, not just recording time spent but also the nature of the activity, who else was involved in any meetings and associated expenses. The keeping of accurate records has been a key element in those cases on fees which have gone to the Lands Tribunal, notably *Poole v South West Water, Matthews v Environment Agency* and *Newman v Cambridgeshire County Council.*

21.4.53 The CAAV annual Fees Survey provides data on the levels of fees being charged as well as the basis of those charges for different work and different levels of practitioner.

21.4.54 The valuer has a contract with the claimant and, ultimately, the fee, if properly managed, will be part of the claim against the acquirer as it is accepted as a necessary part of the disruption and loss to the claimant arising out of the scheme. Thus, where the acquirer disputes the fee, he is disputing the overall level of the claimant's claim. It is not directly an argument with the valuer but with the claimant whose claim it is. That means that only the claimant can pursue the matter through Tribunal if they choose so to do. Whilst few cases get to Tribunal, the basic principles are the same as any other job:

- the valuer should have clearly set out instructions with the client
- these should ensure that the client is clear as to his liabilities
- they should be compliant with the relevant professional regulations and relevant legislation
- the valuer should be able to provide sufficient evidence to support the fee invoice when asked.

21.5 Potential heads of claim

21.5.1 There are a range of potential claims depending on the nature of the holding, the nature, particulars and extent of the scheme which the compulsory purchase is supporting and the interests of the parties.

21.5.2 The legislation supporting and, in some cases, circumscribing the claim has developed over time as explored in Sections 21.2 and 3 and varies across the four administrations, although with England and Wales essentially under the same legislation at present.

21.5.3 That said, whilst the detail will change, the claims will fall under a series of Heads of Claim:
- for the owner/occupier:
    - payment for the land and property taken
    - severance – the impact on the retained land where the works sever the retained land (as where fields are now left on the other side of the new bypass from the buildings)
    - injurious affection – other adverse impacts on the retained land from the scheme
    - disturbance – claims for items other than those reflecting the value of the land
    - Basic and Occupier's Loss Payments – statutory payments available in addition to the claim (these are set statutorily and vary around the United Kingdom)
    - fees and interest – the claimant's reasonable professional costs explored in the previous section
- for the tenant:
    - the value of the tenancy over land taken
    - the tenant right or waygo payment that would be available on the land involved
    - disturbance
    - Occupier's Loss Payment
    - severance and injurious affection
    - fees and interest
- for the landlord:
    - land taken
    - severance
    - Injurious Affection

- Basic Loss Payment (statutorily set compensation due to the landowner)
- fees and interest

21.5.4 In each case, the claimant will be entitled to an advance payment of 90% of the acquirer's assessment of the claim, subject to the difficulties in assessment discussed previously.

21.5.5 In addition, looking forward, valuers may have to consider the particular circumstances around Temporary Possession and, for NSIPs, the implications of housing development under the Housing and Planning Act 2016. In some circumstances, Blight will apply.

21.5.6 Whilst not strictly an item of claim, the issue of accommodation works will also form part of these negotiations to the extent that:
- all parties should be clear and agreed over the works to be done and who will undertake the work, whether main contractor, subcontractor or claimant and, if the claimant, the arrangements for procurement and payment
- the efficacy and impact of any accommodation works will be reflected in the final negotiated settlement.

21.5.7 Further, some projects, particularly the larger scale transport works, such as HS2, will have their own specific provisions for special compensation schemes to support residents and others affected by the proposals. The rules can be complex and occasionally overlapping. Claimants' agents will need to study the schemes carefully to check whether clients are likely to be eligible.

21.5.8 This chapter has concentrated in the main on the impact on people who have land and property taken by the scheme as the valuer will commonly be instructed by someone who has had part of their farm taken under a CPO. However, there are also provisions for those who have not had land taken but who are still impacted by the scheme, most notably Part 1 of the Land Compensation Act 1973. While increasingly the province of specialist firms who operate solely in this area on a national basis, there is no reason why a local valuer should not do this work.

21.5.9 The nature, extent and scale of the claim will vary depending on the nature of the property and the nature and impact of the scheme. Compiling a claim is far from simple, perhaps one of the reasons that some valuers fight shy of this work. However, and behind the technicalities, a number of core themes are involved which are consistent through the full range of valuers' work:
- what would the market make of this in finding a value? In particular, how would it react and what adjustment would it make to values reflecting the impact of the scheme on the claimant?
- what is the cost of remedying this loss?
- what practical actions can be taken to mitigate the impact of the scheme?

21.5.10 The application of those principles and the legal basis of the claims is explored further in the next section.

21.6 The basis of the claim

21.6.1 Introduction – The basic principle involved is that of equivalence with the concept being that the injured party will be put in the same position as if he had suffered no wrong, normally but not exclusively through financial compensation. This was established in statute in the Land Clauses Consolidation Act 1845 and has been explored in a number of cases from the nineteenth and twentieth centuries including *Livingstone v Rawyards Coal Company, Horn v Sunderland Corporation* and, the most recent, *Director of Buildings and Land v Shun Fung Ironworks*.

21.6.2 In the last case, Lord Nicholls expressed the position very plainly:

> No allowance is to be made because the resumption or acquisition is compulsory; and land is to be valued at the price it might be expected to realise if sold by a willing seller not an unwilling seller. But subject to these qualifications, a claimant is entitled to be compensated fairly and fully for his loss. Conversely, and built into the concept of fair compensation, is the corollary that a claimant is not entitled to receive more than fair compensation.

21.6.3 Going back to the start of this chapter, that is unwelcome news for the claimant who is an unwilling seller and will start from the premise that he should be treated that way.

21.6.4 For generations of valuers, the statutory reference has been to the six (really five plus one) rules in section 5 of the Land Compensation Act 1961 that:

(1) No allowance shall be made on account of the acquisition being compulsory:
(2) The value of land shall . . . be taken to be the amount which the land if sold in the open market by a willing seller might be expected to realise:
(3) The special suitability or adaptability of the land for any purpose shall not be taken into account if that purpose is a purpose to which it could be applied only in pursuance of statutory powers, or for which there is no market apart from the requirements of any authority possessing compulsory purchase powers:
(4) Where the value of the land is increased by reason of the use thereof or of any premises thereon in a manner which could be restrained by any court, or is contrary to law, or is detrimental to the health of the occupants of the premises or to the public health, the amount of that increase shall not be taken into account:

(5) Where land is, and but for the compulsory acquisition would continue to be, devoted to a purpose of such a nature that there is no general demand or market for land for that purpose, the compensation may, if the Upper Tribunal is satisfied that reinstatement in some other place is bona fide intended, be assessed on the basis of the reasonable cost of equivalent reinstatement:

(6) The provisions of rule (2) shall not affect the assessment of compensation for disturbance or any other matter not directly based on the value of land:

21.6.5 The basic rules, taken from the 1961 Act, are almost identical in Scotland (Land Compensation Act (Scotland) 1963) and Northern Ireland (The Land Compensation (Northern Ireland) Order 1982).

21.6.6 The no-scheme assumption in England – For England, a new rule 2A was added by the Neighbourhood and Planning Act 2017 as an attempt to codify the no scheme world assumption previously developed by case law, saying:

(2A) The value of the land referred to in Rule 2 is to be assessed in the light of the no-scheme world principle set out in Section 6A

21.6.7 In seeking to codify the *no scheme world* the draftsmen were given something of a poisoned chalice to the extent that the principle seems to have been reasonably well understood from *Pointe Gourde;* one of those key cases remembered alongside *Stokes v Cambridge* and, latterly, *Antrobus* by both practitioners and students alike. Notwithstanding that the legislation sought to clarify the situation by adding c.6A to the Land Compensation Act 1961, whether that works in making for a clearer understanding or not will be judged over time.

21.6.8 The new section 6A then provides (the full text is available in the statute or the CAAV publication) that:
- any increase or decrease in value as a consequence of the scheme or the prospect of the scheme for which the land is to be acquired is disregarded
- in so doing, the following five rules (*the no-scheme rules*) – in practice a set of assumptions – should be applied assuming that:
  - Rule 1) The scheme was cancelled on the relevant valuation date
  - Rule 2) No action has been taken by the acquirer for the purposes of the scheme
  - Rule 3) There is no prospect of the scheme or any other scheme to meet the same need being carried out as a statutory function or through CPO powers
  - Rule 4) That no other scheme would have been carried out as a statutory function or using CPO powers if the scheme was cancelled on the relevant date

- Rule 5) If there was a reduction in the value of the land as a result of:
  - the prospect of the scheme
  - the fact that the land was blighted (as defined in Sch 13 to the Town and Country Planning Act 1990) that is to be disregarded.

21.6.9 Whether this definition adds anything to the understanding of this area remains to be seen and, indeed, the Courts have yet to test it and the place of precedents such as *Pointe Gourde*.

21.6.10 The no-scheme world more generally: *Pointe Gourde* – The underlying principle, considered in many cases, is seen as crystallised by the decision in *Pointe Gourde*. Pointe Gourde Quarrying owned land in Trinidad which was required for a naval base. The land contained a substantial limestone deposit which was particularly well suited for the construction of the base. The company argued that that deposit would attract a premium in a market where the demand for stone would increase as a consequence of the building of the naval base and so claimed for the additional value. However, this claim was disallowed on the basis that there would be no added value without the scheme.

21.6.11 The *Pointe Gourde principle* has since been applied in a wide range of cases exploring the interpretation of what constitutes the "scheme", in which *Waters v Welsh Development Agency* is an early test of the value of land for offsetting and the interaction between the principle and the interaction between the principle and statutory disregards.

21.6.12 The doctrine of equivalence applies to disturbance alongside all the other elements of the claim, as confirmed in *Hughes v Doncaster Metropolitan District Council*. However, there is a distinction in the treatment of the claim as indicated by Rule 6 of s.5 of the 1961 Act. Whilst other elements of compensation are based on market value, that for disturbance is based on the value to the owner. This difference was considered in *MWH Associates v Wrexham CBC* where the Upper Tribunal considered the distinction between claims for diminution in the value of a landfill site (Rule 2 and so market value) and loss of profits (Rule 6 and so value to the owner). The Tribunal noted two important distinctions:

- that for the former claim the relevant facts are those applying at the valuation date whereas for loss of profits subsequent events are also relevant, thus echoing the *Bwllfa principle;* and
- for the former claim the reaction of the market is the important thing and the intentions of the owner are not relevant, whereas those intentions are essential to the claim for disturbance (in this case loss of profits)

21.6.13 They should be considered in the round: does the overall settlement of compensation for those items, having taken account of any accommodation works to mitigate the impact of the scheme, put the claimant back where he was before the scheme took place? Parts of the claim, such as for land taken, may be susceptible to direct comparables, but there may there may be less market evidence from which to draw analysis for a severance claim. That said, the valuer can stand back and think "what would the market pay for this property with and without the road, railway, secondary school or runway?". That should be the test of equivalence.

21.6.14 Armed with those twin sets of rules for equivalence and the no-scheme world, the valuer can now apply their mind to the assessment of compensation. However, there are various difficulties which remain, some informed by statute and others by precedent.

21.6.15 Valuation date – The date for which the valuations in the claim are to be assessed is the earliest of:
- the date when the acquiring authority enters and takes possession of the land where the notice to treat procedure is used, or the date the title of the land vests in the acquiring authority when the general vesting declaration procedure is followed.
- the date values are agreed
- the date of the Upper Tribunal's decision on compensation.

21.6.16 Normally, valuations deal with information and evidence available at the time, but the question arises with compulsory purchase to what extent should account be taken of post valuation date events. Valuers will occasionally refer to the *Bwllfa* principle as supporting the use of evidence which arose after the valuation date. However, the case *Bwllfa and Merthyr Dare Steam Collieries v Pontypridd Waterworks Company* dealt with compensation for disturbance rather than land taken. The claimant had sought compensation for the fact that the water company had prevented them from extracting coal at a time when the price was rising. The question arose whether the price to be used in assessing profits foregone was the price when the extraction was stopped or the higher price over the period of loss. The House of Lords decided that the subsequent information could be used, although acknowledging that this might not be the case in respect of land taken.

21.6.17 For assessing the value of land taken, the distinction is between:
- what is evidence of what values were at the valuation date, which might include information from later (as, for example, a sale not completed until a week later but which was being negotiated at the same time), and
- changes in values after the valuation date.

21.6.18 Events happening after the valuation date that change the value of the land are not relevant. The surrender of a tenancy after the event was considered in the development option case, *Gaze v Holden*, where the tenancy of a farm was surrendered after the valuation date. It was held that date was critical so that the subsequent surrender could not be reflected in the valuation. Similarly but this time in the claimant's favour, in *ADP&E Farmers v Department of Transport* events after the valuation date meant that hope value on some retained land impacted by the scheme fell away, but the Tribunal held that the loss should be assessed on the position as at the valuation date.

21.6.19 Special purchasers – On occasion, property may be of interest to a special purchaser, something which ordinarily falls outside the definition of market value (and so disregarded under the Red Book). In *Solarin v Wandsworth Borough Council*, Mr Solarin had paid what was agreed to be a very high price for his house which the Council then acquired by CPO. While the Council argued that he had paid that price as special purchaser who should be ignored, the Tribunal, in a decision reminiscent of tax cases, decided that his existence would be taken into account by the market and thus compensation was awarded at that higher level.

21.6.20 Hierarchy of evidence – Acquiring authorities will often put forward evidence of other settlements on the scheme to substantiate their offer. They may also argue that twenty or thirty settlements reached at comparable levels is a substantial weight of evidence, particularly in a thin market. However, the decision in *Zarraga v City of Newcastle* was very clear on this point. Mr Zarraga produced one open market comparable sale which was at odds with a number of settlements on the same scheme produced as evidence by the local authority. The Tribunal Member found in favour of Mr Zarraga saying

> I am in no doubt myself that this open market sale, although a solitary transaction, must be given greater weight than all the corporation's settlements. The sale blows like fresh air through the whole debate.

That follows the established understanding of the hierarchy of evidence, considered at several points in this book, with open market, arm's length unconstrained transactions at its peak.

21.6.21 Development value – The question of hope value and development value is a vexed area, though less often met in a straightforward rural case, if there is such a thing. However, the increased opportunity for building and other development (including that under Class Q in England) may make it a more routine element. Where an alternative use is in prospect, the valuer should have regard to:
- extant planning permissions
- development zoning in any relevant local plan policy
- the potential for a Certificate of Appropriate Alternative Development under s.17 of the Land Compensation Act 1961 and s.65 of the Planning and Compensation Act 1991 or s.232 of the Localism Act 2011

21.6.22 Where planning permission exists, the position is relatively straightforward in finding and adjusting comparables. That will differ between the site for a few houses on an old yard next to the village through to the major urban extension site. In the latter case, detailed development forecasts will be required (see Chapter 15) and consideration will need to be given to the period over which the land would come forward for development and to what extent that might be influenced by the scheme, as explored in *Viscount Camrose v Basingstoke* and *Myers v Milton Keynes DC*.

21.6.23 Where there is not a current planning permission, the confidence with which the valuer can approach the question of hope value will vary with the position of the land in respect of current planning policy, the extent to which that policy overall might meet statutory requirements, particularly where the five-year land supply question is relevant, and the extent to which that development may be contingent on other issues, whether those are related to the scheme or simply the order in which the Local Planning Authority wishes to see development coming forward.

21.6.24 Where there is sufficient confidence that planning permission for development would be forthcoming were it not for the scheme, then the claimant may well apply to the planning authority for a Certificate of Appropriate Alternative Development under s.17 of the Land Compensation Act 1961 as recast by s.232 of the Localism Act 2011, following a number of Tribunal and Court of Appeal decisions including *Spirerose v Transport for London*.

2.6.25 This provides a route to establish what planning permission might have been forthcoming. If a Certificate is granted, then that permission can be assumed for the purposes of assessing value, even though the land will not be developed for that use. It only applies to the land being acquired and cannot be applied to retained land to assist with the assessment of severance or injurious affection claims for example.

21.6.26 *Spirerose* also explored the assumptions to be made when considering the potential for planning permission. The Tribunal had based its assessment of compensation on the assumption that planning permission would be forthcoming and the Court of Appeal upheld that approach which would have allowed the full development value to be claimed. However, the House of Lords took the contrary view and decided that the assessment should be based on the assumption that planning permission would probably have been granted. Given the element of risk inherent in that assumption and adopting the Tribunal's deliberations on value, the compensation was reduced by approximately one third.

21.6.27 Marriage value and development – The prospect of development may also give rise to questions of marriage value which was tested in *Stokes v Cambridge Corporation*, one of those cases beloved of rural valuers because it can be applied in a wide variety of circumstances, including compulsory purchase.

21.6.28 Mr Stokes owned some land which was compulsorily acquired by the Council for industrial development, subject to a condition that satisfactory ownership could be provided. The only access to the property was over land which the acquiring authority owned – seen as holding Mr Stokes's prospects to ransom. Whilst the fact of its ownership had to be ignored, it was reasonable to assume that a hypothetical owner would sell the land, not least as it would enable development of their land. Thus, the question was: when assessing the value of Mr Stokes's land what deduction should be made for the cost of acquiring the access? The decision was that it would be one third of the uplift in value of Mr Stokes's land. The market could be expected to exact that ransom with sufficient left for Mr Stokes to proceed and sufficient taken for the neighbour to be satisfied that value had been achieved and both were better off than if no deal had been done.

21.6.29 Whilst the underlying 'ransom' principle has been applied to a number of cases, the proportion of uplift has varied with the circumstances of the case. In *Batchelor v Kent CC*, where there were alternative accesses, 15% was applied, but in *Ozanne v Hertfordshire CC*, the Tribunal awarded 50% of the uplift on the basis that there was only one practical access and that figure was used in *Foster v HMRC*. Other decisions have ranged between these figures, commonly choosing 15%, 30% or 50% with a slight departure of 45% in *Persimmon Homes (Wales) Ltd v Rhondda Cynon Taff County Borough Council*.

21.7 Claims based on value

21.7.1 Land taken – While it might be assumed that this will be the most fundamental and probably the largest part of the claim, very often with agricultural property the claims for severance and injurious affection may often amount to a significantly higher figure, especially where the property taken is confined to agricultural land and the injurious affection claim includes the impact on the farmhouse.

21.7.2 The basis of the claim is market value, with no additional compensation due to the element of compulsion. This market value may be confined solely to agricultural use unless development value, whether confirmed by planning permission or simply hope value, is in play.

21.7.3 In considering market value, the valuer may consider how best the farm could be lotted for sale. That could also apply to the assessment of injurious affection and severance, although arguing that the land might be more valuable offered for sale in lots might run against the severance claim.

21.7.4 As with other valuations, the starting element will be an assessment of comparable sales, whether for the property as a whole, which after all is the basis of the equivalence principle, or its constituent elements. Whilst the actual amount of the claim for the land taken may be relatively small in the overall scheme of things, these comparables will also inform the assessment of the potentially far larger claims for severance and injurious affection. In the latter context, it will be important to identify separate comparables for any houses which may be affected by the scheme.

21.7.5 Severance and injurious affection – Whilst the claim for the land taken deals with just that, the value of the land expropriated, severance and injurious affection are the two heads of claim addressing the diminution in value of land that the claimant retains after the compulsory purchase. That land has to be in the same ownership as the land taken. In *Cooper v Northern*

Ireland Housing Executive, a house with joint owners was not "held with" land owned by one of them that was taken for a road, so limiting the claim that could be made for the house (see 21.10 later on).

21.7.6 While there is no market in severance or injurious affection, comparables are still key to the assessment as ultimately these heads of claim represent the difference in the value of the retained land before and after the acquisition of the land taken and the completion of the works. That is not to say, however, that reference to previous settlements cannot be used as comparables for the impact on affected property. That means that valuers practising in this area will wish to give some thought to key elements of comparability for these settlements which are normally expressed as a percentage of the original value of the affected property. Timing and, to a lesser extent, the nature of the land may not be as critical as they are reflected in the market comparables. However, comparable issues might include:
- the proportion of the land taken as part of the whole,
- the degrees of separation involved in any severance,
- the proximity of the house to the scheme, and the nature and scale of the house for reasons explored further later on,
- the nature and scale of the holding and
- the benefit and impact of any accommodation works.

21.7.7 The provisions for these two elements of any claim are set out in:
- section 7 of the Compulsory Purchase Act 1965 for England and Wales
- section 48 of the Land Clauses Consolidation (Scotland) Act 1845 for Scotland
- article 8 of the Land Compensation (Northern Ireland) Order 1982.

S.7 provides that compensation is to be due for:

> . . . the damage, if any, to be sustained by the owner of the land by reason of the severing of the land purchased from the other land of the owner, or otherwise injuriously affecting that other land by the exercise of the powers conferred.

21.7.8 Severance – This claim assesses the diminution in the value of the property as a consequence of the loss of the part taken for the value of the remainder because of any parts of that remaining property being severed from each other. It is naturally more likely to arise with a linear transport scheme than a site acquisition for say a school development which will take a block of land in a ring fence. It can result in adding several miles (and so cost) to the travelling between parts of the farm. This may be more marked still where, as with some new major roads, agricultural vehicles are barred from using them.

21.7.9 The amount of the claim will turn on the area of land, the nature of the holding, the nature, topography and orientation of the severed part and the extent to which any accommodation works may mitigate that effect.

The prospect of accommodation works seems less likely now than was the case in the past. There are a number of bridges over the M5, for example, that were provided to allow access from one side of a farm to the other, but this seems a less common solution nowadays. Where a bridge or underpass is offered, then a bridge is generally rather more useful; underpasses which were tall enough and wide enough for tractors and mowers in the 1980s will not necessarily accommodate a self-propelled silager now.

21.7.10 The most damaging types of severance might include:
- isolation of a dairy unit from the bulk of the grazing land
- isolation of grain or potato storage from the bulk of the growing land
- isolation of the house from the land or buildings in terms of utility for commercial farms and amenity for lifestyle holdings
- severance of fields into smaller, less economic parcels
- severance compromising the exercise of sporting rights, particularly in high value areas.

21.7.11 The basis of compensation is the diminution in the value of the retained land and that can give rise to some difficulty where the bulk of the impact is from additional travel costs. That said, the before and after valuation is likely to reflect that at least in part. One particularly frustrating element of negotiation can arise for claimants when good-sized arable fields are severed and the acquirer counters a claim form severance with a Betterment argument on the basis that the severed land will now suit the pony paddock market.

21.7.12 Injurious affection – This claim covers those elements of diminution in value of the retained property which are not attributable to severance; on occasion, it can be difficult to distinguish between these two claims. The claim might encompass:
- the continuing impacts of the scheme, noise or light pollution from a road or runoff or dust on neighbouring land
- the loss of marriage value between the various parts of the holding in being part of a larger whole
- the impact on the amenity of the land from major infrastructure development, particularly relevant in lifestyle areas
- the impact on the buildings, particularly where the scheme is close enough to disturb the more nervous types of livestock, horses being the obvious example
- disrupted drainage
- the impact on any diversified activities which rely on a peaceful environment
- the impact on sporting activity
- the impact on the dwelling or dwellings.

21.7.13 Again, the impact on the value of the retained land and property is the issue and here the nature of that land and property and the market that the property would attract is important.

21.7.14 This seems particularly relevant when considering the dwelling house. The compensation is likely to be expressed as a percentage reduction from the original value. The conclusion might be that for a pair of semi-detached farm cottages, with a value of £300,000 each, the impact of a nearby road scheme reduced the value to £255,000 or by 15%.

21.7.15 On the same farm, the main farmhouse is exactly the same distance away from the new road as the cottages, from which it is about 100 metres removed. The value of the house before the road scheme was £1,000,000. Simple logic might suggest that the diminution in value was 15% or £150,000, but is that right? The purchaser with a £1 million budget can be far more discerning than one with a £300,000 budget. At that price point, depending on the area, he is likely to be looking for something completely unbothered by main roads or noise and with a high level of privacy. Thus, it may be that the value of the farmhouse is reduced by a significantly greater value until it gets to a bracket where the merits of the house outweigh the impact of the road. Depending on the circumstances that may be £800,000 or even £700,000, but in any event likely to be more than the proportionate diminution in value of the cottages. Consequently, in considering the injurious affection on a house, there is more involved than simply copying the last settlement.

21.7.16 Nationally Significant Infrastructure Projects and ancillary development – As explored earlier, the Housing and Planning Act 2016 gives developers promoting NSIPs the opportunity to include related land for residential development in the Development Consent Order. That will give rise to similar difficulties as in the example set out earlier. Whilst it will be of limited consolation to a landowner who sees the development uplift being seized by the developer using the no-scheme world principle to purchase the land at existing use value, it will be important to ensure that the full range of compensation arguments are brought to bear on the part of the owner.

21.7.17 Thus, extending the previous example to a trunk road scheme which is of a scale to qualify as an NSIP and assuming:
- the affected property is a farm with a substantial farmhouse and reasonable buildings and 200 hectares of land
- the road would take 8 hectares, but the developer has added a further 12 hectares of contiguous land for residential development
- that land is in full view of the principal rooms of the house
- there remains a further 20 hectares of land around the proposed development land which is not to be acquired and is likely to be zoned as a "Green Wedge" by the Local Planning Authority.

21.7.18 The presence of the roadway would already give rise to a claim for land taken, severance and injurious affection, both to the house and land. However, how much extra should that compensation be as a

consequence of the proposed housing development? The following may be considered:
- the additional injurious affection to the house resulting from the housing development
- the additional injurious affection in respect of the 20 hectares of retained land which cannot now be developed but in particular will now become vulnerable to trespass, stock bothering, fly-tipping and probably complaints about farming activity if the land has to be accessed close to the new development
- the extent to which the farm is now over-capitalised
- the loss of economic efficiency of the holding which, taking the impact on the land both proposed for housing and adjoining it, has been reduced by 15% in size after allowing for the road.

21.7.19 The first point is perhaps worth exploring a little further. There will be injurious affection to the house and any cottages on the farm as a consequence of the road and its intrusion into the environment of the farm and the house. The owner will receive compensation for the impact of the road on the farmhouse, but in this case the development land is in full view of the house and this will add further to the impact. A farmhouse with a view of a road with countryside beyond is one thing in value terms, one with a view of a road and a housing estate beyond might be very different. As set out earlier, the more valuable the dwelling the more marked that impact is likely to be.

21.7.20 Temporary possession – For England, the Neighbourhood Planning Act 2017 has enacted arrangements, although these are not yet implemented, to allow acquiring authorities to take temporary possession of land that may be required, for example, to access a scheme during the work but which will not be required by the scheme itself and so to be returned on completion of the scheme. It helps to avoid the difficulty which used to face acquiring authorities when they had to acquire land outside of the legal Order, sometimes because the need for it was not recognised by designers when the Order was drawn. Such acquisitions would have to be negotiated on commercial terms, often under considerable time pressure and negotiating with an owner's agent well aware of that constraint. It does, however, also mean that they can now be rather less accurate in defining the land that they acquire than was previously needed. As long as the Order takes a corridor wide enough to take, for example, a railway, then any surplus land can be returned once the railway has been designed and built.

21.7.21 The logic seems reasonable, although not to the client who, even if not hoping for a generous licence fee for a site compound, will be concerned when or in what condition the temporary land will be returned. What if the land in question, previously part of the arable rotation, or one of the better drives in a shoot, is returned with trees and other landscaping planted on it?

21.7.22 However, the temporary regime has a degree of permanence about it which could cause significant difficulty for claimants; providing the preliminary requirements are met, a question of getting the CPO or DCO right:
- the acquiring authority may take temporary possession of the land specified
- the temporary period may be very considerable, unless it is limited by the claimant serving a counter notice, within 28 days of notice of entry, in which case it will be for:
  - 12 months where a dwelling is involved; or
  - 6 years where there is no dwelling
- the authority may use the land as if it had acquired full rights in the property in accordance with the use specified in the notice
- a claimant is entitled to compensation for loss or damage; however, where the claimant is carrying on a trade and the compensation includes disturbance for damage to that trade, that will only be payable if the claimant has to quit the land.

21.7.23 Where bare land is taken for 18 months, for example, that might have limited consequences, but elsewhere the impact could be significant, long lasting and difficult to recognise. This introduces a new set of challenges for valuers.

27.7.24 The claimant is entitled to receive compensation from the acquiring authority for "any loss or injury" they sustain as a result of the temporary possession (s.23(2)). Where the disturbance is to a trade or business, the compensation

> includes compensation for any loss which the claimant sustains by reason of the disturbance of the trade or business consequent upon the claimant having to quit the land for the period of the temporary possession.
> (s.23(4))

That seems to set the agenda for the claimant who has an obvious motivation to quit the land, particularly where the works involved will have a marked impact on the business activity. This might be more relevant for business premises such as a garage or pub, but it might equally apply to a farm with holiday accommodation or other diversified income.

21.7.25 However, there is a need for caution here, particularly until such matters have been tested in the courts or by use. What if the acquirer subsequently argues that the claimant was too precipitous in moving? Or indeed need not have moved at all and thus failed to mitigate their loss? Or seeks to challenge what is meant by ". . . consequent upon the claimant having to quit the land . . ."?

21.7.26 It is clearly important for the agents to discuss the approach to compensation in advance and at least reach some agreement on the principles, although the quantum is only likely to be fully assessable when the temporary occupation is complete. Even a discussion of the principles will be difficult. Is a key access compromised? What happens if the temporarily acquired property includes buildings? Will they remain or are they to be demolished or adapted and will the latter be reversible? And the fundamental question: how long is temporary? The acquirer's agent may not know, particularly where the temporary possession is more precautionary than premeditated.

21.7.27 This all suggests a challenge. As usual, prior discussion and clarity are the best protection for both parties. Thus, even where the precise nature of the ensuing works may not be clear, it seems sensible for the agents and parties to agree:

- the anticipated nature of the temporary period
- the works likely to take place and the impact on the temporary land
- any actions that the claimant may be required to take based on those assumptions, including where possible an explicit agreement that those are acceptable.

21.7.28 That will at least set some parameters for the subsequent negotiations. That agreement may need to be revisited over time if the acquirer's plans change.

21.7.29 In many cases, it may be that the property involved is bare land and the occupation is relatively short, perhaps as a compound or materials dump. In such a case, valuers will be well versed in the remedial works required and the approach to compensation, albeit the claimant will no longer have a commercial licence fee.

21.7.30 However, where the occupation is for a longer period (and the term can be for six years if there is no house involved even if the owner serves a counter notice to limit the term), other more significant and more knotty problems may come into play, for example:

- if the claimant is contracted to supply feedstock for an AD or meet production targets for a processor and the loss of land means that the contract cannot be met, or premium payments based on targets are lost
- if the claimant has invested in new fixed equipment on the farm which will now not be fully utilised due to the temporary loss of land
- if the claimant is running a diversified business where being able to offer continuity of service is fundamental to repeat business, where the temporary land is part of a clay pigeon shoot or cross country course or a glamping site for example

21.7.31 All these have the potential for very substantial disruption to the business, but there are two other potential areas of claim for issues which could be particularly damaging to the claimant, one more tangible, though no less intractable, than the other: where the temporary land taken includes or prevents the full use of the farm buildings and strategic land.

21.7.32 A farmer whose access to farm buildings is denied or compromised for even part of a season, such as valuers might be used to with utilities work limiting access, will have difficulty managing the farm. Picture the situation if those buildings are lost for three or four seasons. Hopefully, acquirers will recognise the problem and seek to avoid it, but if that should occur there are relatively few sets of spare farm buildings around the country which the claimant might occupy. Co-operative grain storage might assist on a pure combinable farm but there will still be other facilities required. Livestock and dairy farms will find it most likely impossible to find sufficient spare accommodation and whilst the former may in some parts of the country be able to change enterprises to summer fattening, seasonal dairy farming does not seem a viable prospect.

21.7.33 For some, the most viable option may be to mothball the current enterprise and, where there is some flexibility, contract farm or even let out the farm. This brings a number of sizeable potential claims. Most equipment works whilst it is being used and breaks down if left unused; a parlour unused for three years will need wholesale repairs if not renewal before operating again. Once having ceased the current enterprises, the claimant may find it hard to start up again if there is a gap of a number of years and, in the meantime, family members will need to find other employment and staff will be made redundant.

21.7.34 A mothballed farm, particularly one let on an FBT in the interim brings all sorts of financial and fiscal challenges. What about the Inheritance Tax implications, particularly if a death were to occur during the temporary cessation of business? How will this work in terms of Capital Gains Tax on any sales which might have been contemplated and the potential or otherwise for Rollover or Entrepreneurs' Relief, both with their associated time limits? If nothing else the acquirer should be forewarned of the sizeable professional costs involved from the team of advisors who will be required.

21.7.35 That scenario is full of practical challenges and likely to be expensive for all involved. The difficulties with temporary possession of strategic land may be less tangible and somewhat speculative, in turn not helping to assess the impact, but could be even more marked in financial terms.

21.7.36 Returning to the 200-hectare farm considered in relation to NSIPs and housing development, rather than include the additional 8 hectares of land in the DCO the acquirer has served notice that part of the land, including the access is to be acquired temporarily for the duration of the scheme, possibly up to five years. The claimant serves a counter notice on a precautionary basis fixing the maximum time at six years. The Local Planning Authority puts out a call for housing sites, particularly seeking some for early release. The land in question would be well suited for development and the owner promotes it except:

- he cannot say the land is under his control
- he cannot be certain when it will be available, other than sometime six years hence

- he cannot be sure what the nature of the land will be in terms of its readiness for development or indeed potential contamination (another possible head of claim) after that time
- he may not be able to get access to undertake the various preliminary surveys the LPA will wish to see in a more detailed submission, although in fairness ecological benefits may be rather more diminished than they would otherwise have been.

21.7.37 The Local Planning Authority, in considering the sites, simply cannot accept the land in question because it does not satisfy the criteria for selection because of the problems described earlier. It shifts the focus of development to other areas for urban extension (perhaps even if the scheme behind the temporary possession makes the land more suitable for development).

21.7.38 The owner has missed the potential for a development sale on the 8 hectares in this round. At least the owner still has the site and the position may not be as bad as in the previous example; the development may only be delayed, but if the focus for development has changed to another area that delay may be extended. That represents a very expensive temporary loss of occupation. Considerable professional time and cost will need to be expended in marshalling the arguments and substantiating a claim. Once again, if the scenario can be foreseen it should be raised at the outset, although one can imagine some acquirer's reaction.

21.7.39 Betterment – The other side of the development coin is the question of betterment, where the value of the claimant's retained land is improved by the scheme. The classic example is a bypass which leaves some of the claimant's land between the town and the new bypass, which as the new development boundary offers the claimant development opportunities. The acquiring authority is entitled to reduce the amount of compensation by this betterment, but only so far as to reduce the compensation to zero. The aggrieved claimant at least does not end up owing the authority money for having his land taken.

21.7.40 The long applied and much argued principle was codified by the Neighbourhood Planning Act 2017 which added a new section, 6B, into the Land Compensation Act 1961. A further new section, 6C, entitles the claimant to additional compensation in the event that adjoining land is reduced in value as a consequence of the scheme.

21.7.41 In considering arguments for betterment, the valuers should have regard to:
- the extent to which there is a direct causal link between the scheme and the benefit
- the extent to which there is a specific benefit, so the greater prospect of development, which could reasonably be expected to be implemented – of which more later on

- the extent of the betterment
- the challenge of actually achieving that betterment
- consequential damage of loss resulting from the betterment.

21.7.42 Thus, the mere improvement in values of property in an area generally is not something that the acquiring authority should seek to recover in respect of the claimant's dwelling for example. The benefit needs to be more specific; the potential to sell land for development, for example.

21.7.43 Where enhanced development opportunity is the argument, then the question of the value of that opportunity is subject to all the debates that have been explored earlier and in Chapter 15 in respect of the difference between hope value and value with a planning permission granted. Except for the smallest sites, where that begs the question whether or not any scheme would be viable, it is unlikely that the owners of the land would be able to promote this themselves. They will need a promoter partner to take on the task and the ultimate share to the landowner would be significantly reduced from the headline sales figures which will be quoted as comparables. Any such figures need to be adjusted for the uncertainty of securing permission; the residual value might be much less, perhaps even 20% of the gross development value.

21.7.44 That begs a further question: even with a larger scale scheme, does the current value warrant the work required to secure the betterment and, in particular, the further damage that may occur? As an example, the owner of a farm finding it severed by a bypass with some 4 hectares now with enhanced development potential may be interested in the opportunity if the land is remote from the house and buildings. However, as in the previous example, what if the proposed development land is in full view of, or indeed close to the house? The owner will receive compensation for the impact of the road on the farmhouse, but a housing development of 4 hectares close to the house will have a further adverse impact on the value of the house which will not be compensated for as pursuing the development would be of the owner's own volition. In many circumstances, the owner will not wish to pursue that opportunity, particularly if there are no other houses from which to farm the land, and it would be unreasonable in those circumstances for a betterment claim to reduce their compensation.

21.7.45 Blight – Arguably at the opposite end of the spectrum to betterment is blight. The prospect, possibly for years, of a potential compulsory purchase scheme can cast a blight of an owner's use of a property and the market's view of it. The law defines more limited circumstances in which an owner with property affected by plans for infrastructure or regeneration schemes can serve notice compelling the acquiring authority to acquire the land. This entitlement to Statutory Blight varies between the parts of the United Kingdom.

21.7.46 England and Wales – The provisions are set out under s.149 of the Town and Country Planning Act 1990 with eligibility limited to:

- owner-occupiers of hereditaments with an annual rateable value of no more than £36,000 in England though £44,200 in Greater London (both set in 2017) and £36,000 in Wales (set in 2019)
- owner-occupiers of residential property
- owner-occupiers of agricultural property

21.7.47 For these purposes, leaseholders with an unexpired term of more than three years will count as owner-occupiers. Personal representatives of the previous and mortgagees who are entitled to exercise their rights of sale will also be eligible to serve a blight notice if they are unable to sell their interest except at a substantially lower price than would have been the case before the proposed works (s.160). However, the rules exclude higher value business and other non-residential properties, landlords and shorter term or annual tenants.

21.7.48 The classes of blighted land to which the provisions apply are set out in Schedule 13 to the 1990 Act (as amended by the Planning and Compulsory Purchase Act 2004) and include land impacted by compulsory purchase. Where the owner-occupier and the land both qualify, and the proposals for the scheme are sufficiently well advanced that they are more than a rumour, the owner-occupier can serve a notice on the acquiring authority compelling it to purchase the land in advance of it being required. Subject to certain exceptions, the applicant must have attempted to sell the land before serving the notice and failed to be able to do so other than at a substantially reduced price.

21.7.49 The acquiring authority has the opportunity to serve a counter notice, within two months, refusing to purchase on any of the grounds that:
- the land will not be acquired or affected by the proposals
- the notice server does not have a qualifying interest
- there is no blight and that inadequate efforts have been made to sell
- the land will not be required for at least 15 years.

21.7.50 Where the counter notice is served, the owner then has one month to refer the matter to the Upper Tribunal. As with all such issues there is much to be gained by discussions in the shadow of the notices.

21.7.51 There will be occasions where proposed works will impact on properties, but Statutory Blight will not apply, typically because the property is not required for the scheme. In those cases, there is the opportunity for the acquirer to use discretionary powers to purchase those properties by agreement where hardship would otherwise arise. This principle of discretionary blight is often captured in specific schemes such as those offered by HS2.

21.7.52 Elsewhere in the United Kingdom – Similar provisions for statutory blight apply in Scotland, although with a maximum rateable value limited of £30,000 and Northern Ireland with the alternatives of the rateable limit of a maximum annual value of £19,685 or a maximum capital value of £2,100,000 (which, of no interest at all other than that this is a book on valuation, is observed to be a rate of return of 0.9%).

21.8 Disturbance and other claims

21.8.1 Disturbance – The disturbance claim covers those elements of loss which are not to do with the value of the land; that distinction is recognised in the rules in section 5 of the Land Compensation Act 1961 and explored in further detail in the next section. It can encompass a very wide range of often temporary impacts including:
- damage to retained land during the construction process
- loss of or damage to crops
- future crop loss
- additional costs in moving stock
- losses on early sale of stock
- losses and costs involved in straying or escaping stock
- damage to drains and flooding
- impact on shoot profits
- interruption to diversification income: farm shops for example
- additional fencing costs
- claimant's time and expenses
- damages such as interruption of services
- damage to buildings
- costs of cleaning dwellings
- expenses including professional fees
- removal costs if dwellings are vacated
- compensation for lost timber.

21.8.2 The nature and extent of the disturbance claim will vary with each farm and scheme. Whilst some items will be common to most claims, such as loss of crops on working and access areas, others will be particular to the holding, lost revenue to a wedding venue for example.

21.8.3 Such is their nature that many items cannot be foreseen when the initial model form of claim is lodged in response to the notice to treat. The valuer should reserve their client's position to enable a claim to be made at a later date. At the same time, as with utility cases, the claimant should be advised to keep a note of all the time he is engaged with the scheme, whether that is additional time dealing with stock, liaison with the contractor or meetings with engineers and agents.

21.8.4 Basic and occupier's loss payments – These additional payments are provided for England and Wales by the Land Compensation Act 1973. The potential for these payments should also be reflected in any settlement reached privately in the shadow of CPO. They are available in all four UK administrations in slightly different forms and at different levels.

21.8.5 There are three payments:
- Home Loss Payment – where the claimant has to leave their dwelling provided that has been their residence for more than one year and claimed as an additional 10% on the compensation payable on the property subject to maximums (in July 2020) of:
    - in England – £65,000 with a minimum of £6,500
    - in Wales – £62,000 with a minimum of £6,200
    - in Scotland – £15,000
    - in Northern Ireland – £45,000 with a minimum of £4,500
- Basic Loss Payment – where the Home Loss Payment is not available. In England, Wales and Northern Ireland, this is 7.5% of the value of the claim subject to a maximum claim of £75,000. In Scotland, this is only available as a Farm Loss Payment based on three times the annual profit in the three years ending with the date of displacement and payable if the claimant takes up farming again within three years.
- Occupier's Loss Payment – claimable by the occupier of the property taken as:
    - an additional 2.5% of the value of the claim
    - plus the land amount of which is the greater of £300 or £100 per hectares for the first 100 hectares of land taken and £50 per hectare for the next 300 hectares or part thereof
    - plus £25 per square metre for any building lost
    - subject to a maximum of £25,000 in England and Wales

In Northern Ireland the allowances are the same, but the claimant is entitled to the greater of 2.5% of the claim or the buildings amount or the land amount again subject to the maximum of £25,000; this is not available to someone occupying only under conacre. There is no equivalent in Scotland, but tenants can claim the Farm Loss Payment, noted earlier.

21.8.6 Accommodation works – These are works which may mitigate the impact of the works on the claimant's property and may range from triple glazing dwelling windows to new field accesses or a bridge or underpass.

21.8.7 Accommodation works thus serve to reduce the claim for injurious affection and, where alternative access routes are provided, severance. There is no specific entitlement to claim individual accommodation works, but they are routinely agreed between the parties as a practical solution to problems which compensation might not fully recompense and which can mitigate the compensation payment. If nothing else, the cost of the acquirer's contractor providing those works with the economies of scale available to them and access to the design skills required where complex engineering works are involved should offer a saving.

21.8.8 The ideal time to discuss accommodation works is at the start of the scheme, not the end. It is always worth the claimant and valuer reviewing what might be useful as the acquiring authority will not have considered site specific accommodation works in that much detail. It will routinely consider issues such as the provision of double or triple glazing as noise mitigation on a road scheme, for example, but there is no reason that it should be aware of the works that may be required on an individual farm, whether those are visible (such as replacement of field accesses) which they might capture from inspection or hidden (such as diverting the irrigation network). It is all too easy for the claimant, having been used to something for years, simply to assume that it will be appropriately replaced. The agents need to capture that information as soon as possible so that accommodation works can be planned and the requirement for them or their absence does not come as a surprise when the contractor is packing up the site.

21.8.9 Whilst, hopefully, a detailed schedule of accommodation works will be agreed pre-entry, there will be some matters which are simply not foreseeable until contractors are on site. Routine contact between the agents will help to get agreed works added to the programme as soon as possible. Similarly, whilst there are some accommodation works which are common to nearly all schemes, such as fencing, other works required will vary depending on the individual circumstances. Precedents from past schemes are helpful, but it would be wrong to assume that because the last local road scheme, a bypass to the village, automatically included provision for triple glazing that the next proposed one, involving relatively minor junction improvements and a cycle path, will do the same.

21.8.10 Fees, interest and advance payments – Fees and advance payments have been covered in section 21.4 earlier but are addressed in greater detail in the CAAV numbered publications Good Practice in Statutory Compensation Claims and particularly Professional Fees in Compensation Claims.

21.8.11 Whilst the advance payment will be based on the acquirer's estimated claim at the outset, this figure may increase over time and as the works develop and where meaningful changes occur the claimant's valuer should seek a further advance payment.

21.8.12 The claimant is entitled to interest on the outstanding balance of the claim when paid. The good news ends there. Interest is paid at 0.5% below the Standard Rate, essentially the Bank of England Base Rate, now subject to a minimum of 0%, and so has been at that minimum since the financial crisis. This is hardly an incentive for acquiring authorities to expedite payment. Following consultation on that point the government has proposed to introduce a new Statutory Instrument with a rate of 8% (above the standard rate consistent with other statutory late payment provisions). However, it almost immediately withdrew this in the face of concerns that this might lead claimants to delay, taking advantage of the premium rate, an intriguing concept for those acting for businesses facing a cashflow gap of tens or hundreds of thousands of pounds. A further consultation was held in summer 2018, but as yet there is no change.

21.9 Compulsory purchase and let holdings

21.9.1 Introduction – In the case of a let farm where part or all of the premises is taken under compulsory purchase, the separate interests of the landlord and tenant are entitled to be compensated under the same headings as an owner-occupier. The claims do not fall separately between the two, with the tenant claiming disturbance and the landlord the rest for example; each will have potential claims for their interest in the land which has been taken and both may suffer from injurious affection and severance. Having mentioned disturbance, it is unlikely that the landlord will have a disturbance claim beyond professional fees.

21.9.2 The landlord's interest – The value of the landlord's reversion in the land taken is subject to the same basic principles as apply to owner occupied land with issues such as alternative development equally in play. The approach to the valuation should adopt the approach set out in Chapter 11 with the nature of the tenancy, the security of the tenant, the prospect of vacant possession and income and expenditure all coming into consideration.

21.9.3 However, there are also specific considerations to be borne in mind as a consequence of the valuation being for compulsory purchase. Principal amongst these is the fact that the Notice to Treat fixes the interests in the property (the alternative GVD effectively doing the same) at the date of the notice. Thus, where something unforeseen happens after the event, most notably if the scheme which is the subject of the CPO acts as a catalyst for a negotiated surrender by the tenant, that cannot be taken into account.

21.9.4 That does not mean that the potential for vacant possession, whether of whole or part, should be ignored. Where the parties are already in negotiation over the future of the holding, and that would still be taking place in a no-scheme world then the impact on the interests should be taken into account. Similarly, where the grant of planning permission would give rise to a Case B Notice to Quit, supported by a part resumption clause if necessary, that should be reflected. The closer the prospect, the greater the influence that will have on value; the more remote, the less. The exception to that approach is that the landlord's agent cannot make use of the proposed scheme to justify the argument that a Case B Notice to Quit could be served for that purpose and hence seek vacant possession value for the land to be taken in the scheme.

21.9.5 The landlord may also be entitled to compensation for severance and injurious affection in the same way as an owner-occupier, save that the impact will be on the value of the holding subject to tenancy. In these circumstances, it is likely that in most instances the valuer will wish to use a Term and Reversion approach to valuing the investment interest, capturing the impact on the reversionary value where these factors are likely to be most marked.

21.9.6 The landlord will also be entitled to Basic Loss Payment at a rate of 7.5% on the claim subject to a maximum of £75,000. While, at the time of writing, there have been proposals for this payment to be reduced to 2.5% up to a maximum of £25,000, the government has not brought this forward.

21.9.7 The landlord may also be interested in potential accommodation works, particularly the extent to which they may mitigate the impact on the reversionary value of the holding. The tenant will also be interested in these matters, in terms of their impact on the utility and future profitability of the holding. There is much to be said for landlord and tenant's agents working together here as, for much of this, their interests are aligned. Where the landlord owns multiple holdings which will be affected, it may be that some pragmatic rationalisation of those tenancies could help mitigate the impact on the future management of the holdings.

21.9.8 That is in the control of the landlord and the affected tenants working together, however, and not something which should reduce the compensation payable for severance or injurious affection to the holding. It would be difficult to stretch the responsibility to mitigate losses this far when a solution can only be achieved by all of the parties working together, so that it is beyond anyone of them to mitigate their loss in isolation. It might also be that some financial investment would be required to achieve this, in which case that could form part of any claim, providing that this is not claimed alongside full compensation for the diminution in value of the holding as a result of the damage which the rationalisation will remedy. As often in a rural valuer's work there is a technical role here, assessing liabilities for compensation, and the equally important diplomatic one of working with the parties, who may not always see eye to eye, to achieve a practical and beneficial outcome.

21.9.9 The tenant's interest – The Tenant will be entitled to compensation for the loss of his interest in the land taken. This could include:
- the value of his tenancy (subject to the election explained later on)
- severance and injurious affection on his interest in the remainder of the tenancy and in certain circumstances on other land which they farm
- the end of tenancy compensation due at the end of the tenancy
- disturbance
- the Occupier's Loss Payment

21.9.10 The value of an agricultural tenancy is considered in Chapter 32. The same principles apply in this context, subject to the complication that the most likely scenario is that only part of the land will be affected and thus the valuation will be in respect of the tenanted interest in that part. That said, given the principle of equivalence, and considering the diminution of the value of the farm and thus the tenancy overall this is better seen as a proportion of the whole rather than the specific part.

21.9.11 In England and Wales, the acquirer has two routes to extinguishing an agricultural tenancy:
- by a conventional notice of entry under the Compulsory Purchase Act 1965
- by taking the landlord's interest and then as landlord to serve a Case B notice under the Agricultural Holdings Act 1986 or any relevant break clause in an FBT.

21.9.12 Where, in the case of a 1986 Act tenancy, the acquiring authority acquires the tenant's interest by first acquiring the landlord's interest and then serving a Case B Notice to Quit, the tenant will be entitled to compensation under s.60 of that act (see Chapter 29), with basic compensation of either one or two years times the rent payable for the land taken together with an additional four years rent payable for reorganisation. If the Tenant does not opt for the alternative route, then the compensation for the tenancy will effectively be capped at six times the rent payable, though the tenant could have a claim under s.62 for any benefit lost from if the notice to quit is shorter than to a term date at least twelve months hence.

21.9.13 In the alternative, the tenant's interest could be acquired by notice of entry under the Compulsory Purchase Act 1965 in which case the basis of compensation would be the value of the tenancy.

21.9.14 Whichever approach is used, s.59 of the Land Compensation Act 1973 allows the 1986 Act tenant to opt for compensation to be payable under this basis instead of the s.60 disturbance compensation where:
- their interest is not greater than year to year
- the Case B notice to quit follows the notice to treat served on the landlord
- the tenant cannot serve a counter notice.

21.9.15 In those circumstances, the tenant can serve notice that he wishes to be compensated under s.20 of the Compulsory Purchase Act 1965 where compensation will then be based on the unexpired term of the tenancy. As once opted for this the tenant cannot revert to a s.60 basis, the Tenant's agent will need to do some arithmetic and, given the challenge of valuing the tenancy, some speculation before his client opts for this alternative.

21.9.16 Severance and injurious affection may apply to the Tenant's interest as well as the landlord's interest in the reversion. Typically, items of claim may include the additional costs of any new fencing (such as against a road), the cost of future hedge laying when any replacement hedge is sufficiently well grown and the additional costs in farming the smaller fields that may be left. Where the tenant has invested in fixed equipment for which he would receive compensation at the end of the farm tenancy and that compensation would be reduced as a consequence of the investment being superfluous in whole or in part, then the tenant could be entitled to compensation for that superfluous element. When the issue of costs or diminution in value was tested, the Lands Tribunal decided in *Gooderam v Department of Transport* that diminution in value was the appropriate approach.

21.9.17 The prospect of the tenant securing a rent reduction as a consequence of severance and injurious affection may be argued by the acquiring authority to reduce that head of claim. Whilst acknowledging the possibility in *Minister of Transport v Pettit*, the Lands Tribunal felt there was uncertainty about this being achieved at the next review and so adopted a period of five years as appropriate to value the impact on the business. That assumes that the diminution in value will be fully recoverable through the rent review, which might not always be the case. Once the value of the tenancy has been assessed, then this is reduced by four times the rent, being the reorganisation payment under s.60 described earlier (s.48 Land Compensation Act 1973 and again not applying to FBT tenants).

21.9.18 The approach to the compensation that might be claimed on the end of a tenancy is addressed in Chapter 29. Depending on the nature of the holding and the land being acquired, and here the approach would be specific to the land rather than proportionate, the tenant might be entitled to payment for:

- Unexhausted and Residual Manurial Values
- beneficial labour to manure
- tenant's pastures
- growing crops, whether on a seeds and tillages basis or discounted harvest value depending on the state of growth
- crops in store, where there are any buildings on the ground
- tenant's improvements and fixtures, the latter more likely with bare land in terms of fencing, new gates or water troughs.

21.9.19 The tenant will also be entitled to claim for any disturbance due to the works under the various headings and others set out in respect to owner-occupiers in an earlier paragraph. Any attempt to reduce disturbance claims to a definitive list ends in frustration; some Heads of Claim, crop loss and claimant's time for example, are long established. However, when the last edition of this book was written, it was probably not at the forefront of Gwyn Williams's mind that solar panels do not work as well if covered with dust from nearby road construction work, so that cleaning the panels becomes a viable claim!

21.9.20 The tenant will also be entitled to Occupier's Loss Payment at a rate of 2.5% on the claim, subject to a maximum of £25,000. Whilst there have been proposals for this payment to be increased to 7.5% up to a maximum of £75,000 the government has, not at the time of writing, brought this forward.

21.10 Claims where no land is taken

21.10.1 Most of this chapter has dealt with the position where the affected party has land taken by a scheme with the opportunity for the full range of compensation payments under the various headings set out previously. There are also provisions for compensation for those affected by a scheme who do not have land taken, but these are either harder to access or less generous than the terms for those who do lose land.

21.10.2 Rural valuers acting for claimants are rather less likely to encounter these claims but those acting for the acquirer may encounter such "Part 1 claims" in their hundreds. The claims arise under two provisions:
- Injurious affection to other land under:
  - S.10 Compulsory Purchase Act 1965 in England and Wales
  - S.1 of the Second Schedule of the Acquisition of Land (Authorisation Procedure) (Scotland) Act 1947 in Scotland, extending provisions of the 19th century railway legislation to other schemes
  - Article 18 of the of the Land Compensation (Northern Ireland) Order 1982
- Depreciation to the value of property due the use of the scheme under:
  - Part 1 Land Compensation Act 1973 in England and Wales
  - Part 1 Land Compensation (Scotland) Act 1973 in Scotland
  - Part II Land Acquisition and Compensation (Northern Ireland) Order 1973

21.10.3 The s.10 procedure, otherwise known as the *McCarthy Rules*, applies to the execution of the works and extends the right to compensation to those who have not had land taken. Essentially, the claimant has to prove that they have suffered injury or damage for which, were the works not taking place under statute, they would be able to pursue a claim for nuisance.

21.10.4 The rules operate in a complex way so that the prospect of compensation can be remote; the injurious affection must be as a consequence of the works and not of the use of the scheme, the damages must be to land and property interests, not business interests, and the damage should be related to the works. The most obvious damage is obstruction to a right of access, but there might be a right to a claim for loss of profits for example under the separate precedent of *Argylle Motors (Birkenhead) Lt v Birkenhead Corporation*.

21.10.5 Set against that background, Part 1 claims are far more accessible. These provide that where someone has had no land taken by a scheme but there is a diminution in value to their property as a consequence of the use of the scheme, they are entitled to compensation for that loss in value.

21.10.6 Claims cannot be submitted until 12 months after the scheme has come into use, sometimes a little complex with a phased road scheme, for example, and they are limited by the Statute of Limitations. The claimant is entitled to compensation as a consequence of the physical factors arising from the use of the scheme; the physical factors are defined as:
- noise
- vibration
- smell
- fumes

- smoke
- artificial lighting
- the discharge onto land of any solid or liquid substance.

21.10.7 Whenever the claim is made, the valuation date is the *first claim day*, which is 12 months after the works came into use.

21.10.8 In terms of the impact on the enjoyment and value of land and property, particularly houses, the key factor missing from the list is the impact on the view. This is often the most significant factor for value but it has to be ignored because it is not defined as a physical factor. The assessment is sometimes characterised as the "switched-off" value. the difference is not that between "road" and "no road", as is the case with injurious affection, but between a road with traffic and a road without traffic.

21.10.9 That, in turn, has a significant impact on the level of compensation payable for injurious affection to a house where land has been taken, where the view and unsightliness will be reflected in the compensation, and a Part 1 claim where it will not. The difference will vary for every case, but it is marked. As we shall see in a later example, an ordinary injurious affection claim might be 20%, but a Part 1 claim is more likely to be in the order of 5%, and markedly less where the property is some distance from the road.

21.10.10 Compensation is limited to the existing use value of the property; there is no compensation for the effect on development value.

21.10.11 There are a number of specialist firms who pursue Part 1 work anywhere in the country, a phenomenon also seen with oversail claims in the electricity sector. They are very effective in monitoring infrastructure works, transport schemes in particular, so that they can then market their services to affected householders.

21.10.12 Their ideal scenario is a housing estate sitting against a new road where there are various roads of houses progressively further from the works. If a claim can be agreed for the most distant house from the road on the estate, even at 0.25% of pre-scheme value, then that automatically brings every other house on the estate into play at ever increasing rates as they are nearer to the road. For the valuer acting for the acquiring authority, the opposite approach is to be favoured, settling compensation for properties nearest the road if possible and hoping to negotiate any claims away within the first couple of roads back from that point. Proximity is not the only variable and topography and the extent of any mitigating works will be important considerations.

21.10.13 For the valuer advising the acquiring authority, the critical issues are to assemble good data in advance of the first claim date. That will include comparable sales data and other physical data including noise readings. At the risk of repetition, the author being sufficiently scarred by the experience to mention it twice, there is no fun in negotiating Part 1 claims where your instructing authority, coming to you after the event, has no pre-scheme noise data.

Example

Notes

1) *Previous editions have shown an example claim as it would be submitted; many firms will have their own templates for such claims and consequently this example considers the various issues and potential calculations involved but not in the format of a formal claim.*
2) *Notes are provided in italics to explain elements where necessary.*
3) *The circumstances of each claim will be different which may in turn change the approach adopted. This example is assessed from the claimant's viewpoint, hopefully realistically.*
4) *The assumption is that possession was taken on 1 September 2017 so that values are being assessed either at that time, for identifiable elements of the claim, land take, injurious affection or severance, or in 2019 for the events which took place during the works.*

The holding

West Farm (seen first in Chapter 10) is a 100-hectare (250-acre) mixed farm in the south-west of England with a four-bedroom period farmhouse in fair condition, if a little outdated, adequate farm buildings mainly 25 years old and approximately 100 metres removed from the farmhouse, a mix of predominantly Grade III pasture and arable land all in a ring fence. There are no realistic opportunities for alternative development.

The farm is owner occupied. It is run as a mixed arable and beef holding with 69 hectares (170 acres) arable and 30 hectares (75 acres) pasture, the balance of the land being buildings, woodland and waste.

Proposed works

The farm is to be bisected by a single carriageway bypass around the village. All legal formalities were completed with entry taken on 1 September 2017. At the time the initial claim was constructed the contractors had started on the scheme and at the commencement were anticipated to be on site, intermittently for between 18 months and two years.

The design showed that the bypass would:
- take 3 hectares of arable land for the road
- take 3 hectares temporarily for access and as a working area, the likely temporary period being no more than two years from entry
- pass within 100 metres of the farmhouse and in full view of the principal rooms of the house
- sever approximately 30 hectares (75 acres) of which 25 hectares is arable and 5 hectares pasture from the remainder of the holding
- and that access to the farmstead may be affected temporarily by increased traffic during the works, but the existing public highway to the farmhouse and buildings will not be affected permanently.

The completed scheme complied with the design and met those targets. The contractors left the site in the main in May 2019 but had to return for some final works to the road surface and some accommodation works and quit the site fully in July 2019.

The Disturbance valuation is being considered in September 2019.

Prior negotiations

Fortunately, the owner of West Farm instructed his valuer in good time so that she was able to discuss the potential impact on the holding with him and serve a model claim form on her client's behalf in response to the Notice to Treat.

There have been a series of meetings between the scheme engineers, both from the Unitary Authority representing the Highways Agency and the retained contractors who are constructing the road on a design and build contract, and the District Valuer, acting for the authority having tendered successfully for the work.

As a consequence of those meetings the following had been agreed:

Accommodation works

a) The current water supply to the severed land will be re-laid under the bypass in a protective sleeve.
b) One field water trough currently in the field boundary between NG 1234 and 5678 on the line of the road will be replaced on completion of the work.
c) The design shows that the road will be in a shallow cutting for approximately half of its length through the farm including opposite the farmhouse. There will be further landscape planting in that area within the line of the permanent land take and a specification will be provided.
d) Double glazed sash windows, to match the existing window design, will be provided for the farmhouse to reduce noise from the road.
e) There will be a new surface water drain constructed either side of the road and existing field drainage will be connected into those drains:
   - the owner of West Farm has provided a plan of the drainage scheme from the drainage engineer's original plan submitted to MAFF for grant aid, though he cannot now confirm that it is correct or that there are no other drains in the land
   - the contractor is to:
   - connect any drains exposed into the new land drain unless agreed otherwise with the owner through the agent; and
   - to provide a map of the drains as laid.
f) Gateways will be provided between NG 1234 and 5678 and 4231 and 8765 to enable access to fields severed by the new road and with no direct access to the existing public lane to the south of the farm.

g) The road will bounded by a new hedge as set out in the scheme specification (*mainly quickthorn indigenous species mix*) to be fenced on the field side by a stock fence as set out in the scheme specification (*sheep netting and two strands of barbed wire above on tanalised posts*).
h) The temporary working area will be carefully restored to its pre-works condition by the contractor with topsoil stripped and reserved for reinstatement before working commences. The authority will reimburse the claimant for subsequent restoration works required to return to effective arable production including stone picking, weeding and restoration of fertility as required.

Other

i) **To be noted.** Despite representations made on behalf of the client the acquiring authority are unwilling to construct a bridge over the bypass to enable direct access to the severed land. (*Fortunately, the positioning of a roundabout on the bypass means that the additional travel distance to and from the land is only approximately half a mile.*)
j) The claimant's agent's proposed basis of charging (time and expenses) and proposed fee and travel rates were acceptable. Given the nature of the scheme, the acquiring authority accepted that routine liaison would be required between claimant and agent, although this was not expected to require more than one site visit per quarter over the life of the scheme unless unforeseen emergencies arose in which case the acquirer should be notified as soon as possible over any excess time.
k) The claimant's solicitor's estimate of fees to be incurred was agreed.
l) The acquirer would reimburse the claimant for the agent's time in meetings and liaison between client and authority prior to the authority taking entry. Whilst noting that the total time to date was 20 hours, which included time advising the client prior to engagement with the authority, the agreed time to include the forthcoming site takeover meeting was 15 hours plus VAT and expenses.

There was a **site takeover meeting** when the contractor took entry where the owner was able to meet the contractor's Agricultural Liaison Officer and protocols were agreed over communication as well as confirming:

a) The access and egress routes to the site for contractors' vehicles
b) That the final scheme design was as the last iteration discussed at the previous meeting between claimant, engineers and agents the previous month
c) That the project timetable was as previously indicated in information provided at the last meeting.

d) Further confirmation, these being matters of particular concern to the client that:

- There was no intention to use the temporary land to site any structures; and
- That it would be returned as restored arable land

Potential Claim

Head of Claim		Amount of Claim
1	**Land Taken**	
1.1	Permanent Land Take	
	3 hectares (7.5 acres) land taken	
	Market Value – £20,000 per hectare (£8,100/acre) for Freehold Interest	60,000
1.2	Temporary Land Take	
	3 hectares (7.5 acres) lost for two seasons	
	Licence fee on the basis of income foregone	
	3 hectares @ £800 per hectare per annum	4,800
1.3	Basic Payment Entitlements	
	Assuming 6 hectares of entitlements lost	
	3 hectares on land permanently taken and 3 hectares through lack of use and to be replaced	900
	6 hectares @ £150 per hectare	
	Total for land taken	**65,700**
2	**Severance**	
2.1	Area severed from the rest of the farm, including the farmstead and comprising:	
	25 hectares arable land	
	5 hectares pasture	
	Issues:	
	a) The claimant has identified the difficulties of the loss of access but as agreed prior to the commencement of works no direct access has been provided and consequently extra travel time is required for tractors and machinery for arable working and otherwise for visiting stock.	
	b) The severed fields will be less economic to work than was previously the case with what had been 5 fields extending to 45 hectares now split into 9 fields extending to 42 hectares (3 hectares lost to the road) including 6 arable fields smaller than 5 hectares	
	c) The farm was previously in a ring fence with access to all the fields either directly internally, or via quiet public highways around the perimeter. The added value of the severed land being occupied with the remainder and thus potentially attracting a premium from the lifestyle market is lost.	

Head of Claim		Amount of Claim
	d) Production area impacted by traffic from farm machinery tracking across land from highway where previously there had been direct field access	
	Overall Impact on Holding	
	30 hectares south of road – diminution in value from comparables of land sales £2,500 per hectare (issues a to d)	75,000
2.2	Area severed from original fields and remaining north of road	6,000
	12 hectares arable land (balance of 15 hectares originally in fields now severed less 3 hectares taken for road)	
	Item b) only from 2.1 earlier diminution in value 12 hectares at £500 per hectare	
	Total for severance	**81,000**
3	**Injurious Affection**	
3.1	The farmhouse will be severely affected by traffic noise, car lights at night (and in the day now with an increasing number of cars with automatic lights) and the view will be significantly impaired, particularly where the road is not in a cutting.	
	Value of the farmhouse with adjoining 4 hectares (10 acres) (as would be lotted if the property were for sale)	
	£700,000	175,000
	Diminution in value *(only marginally reduced by double glazing which is only of benefit in the house)*	
	25% of value	
3.2	Approximately 7 hectares of land in total adjoining the road to the north and south will be affected by proximity to the road	
	Issues	
	a) Land affected by noise, fumes, dust and run off particularly where not in the cutting	
	b) Drainage likely to be impacted by interruption of land drains	
	Preliminary allowance made but position reserved until impact is clear	
	c) Risk of stock getting onto bypass from pasture fields adjoining the road	
	Diminution in value of 7 hectares at £1,500 per hectare	10,500
3.3	Impact on entire holding – 97 hectares	
	(avoiding double counting on land included in severance claim – sometimes difficult to distinguish the impacts – to illustrate that in this instance claim limited to 93 hectares as 4 hectares included with the farmhouse – acquirer might argue that because of severance claim this should be on 63 hectares but the heads of claim are different)	

Head of Claim	Amount of Claim

    a) 1,800 metres of hedge and fence to be maintained against the road including annual maintenance of the fence, cutting and laying of hedge and annual hedge maintenance once established, replacement of fence with similar guard fence when redundant *(note cost estimate would be prepared to inform assessment of impact on value)*

    b) Diminution in value of land due to reduced area and impact of road particularly on potential lifestyle buyers *(in addition to impact on farmhouse)*

Overall diminution in the value of the retained land impacting 93 hectares (100 ha less land taken 3 ha and land with farmhouse (3.1 earlier) 4 ha) at £1,000 per hectare   93,000

Total for Injurious Affection   **278,500**

4 Loss of Timber & Amenity

(Arguably the amenity element has already been allowed in the Injurious Affection claim earlier and it could be included in that heading although there is an actual loss of timber – which might never have been realised if these are amenity trees)

4.1 Loss of 5 mature oak trees
15 cubic metres oak at £100 per cubic metre   1,500

4.2 Amenity value at £250 per tree   1,250

Total for Timber and Amenity   **2,750**

Summary of Claim

1) Land Taken   65,700
2) Severance   81,000
3) Injurious Affection   278,500
4) Timber and Amenity   2,750

Total Claim to Date   **427,950**

5 Disturbance and Crop Losses

This will be the subject of a future claim when the contract is completed

6 Basic and Occupier's Loss Payment

To be calculated on agreement of claim.

7 Accommodation Works

The acquiring authority will execute at their expense and to the claimant's complete satisfaction the works set out in the attached Schedule, making good any damage arising from the execution of the works

Head of Claim	Amount of Claim

   For completeness, a list of accommodation works could be attached within the claim and that was the traditional approach.

   There is some convenience in attaching these as a schedule so that the schedule can be updated over time:

   a) As the works are completed and signed off

   b) If amendments are agreed – perhaps the need for specific drainage points which need to be noted or restoration of a damaged gateway

   c) If as part of the negotiation the claimant agrees to undertake or procure works himself and reclaim the cost (which should include his management time)

8 Plans

8.1 The acquiring authority will provide a new farm plan, calculating the revised areas of the affected fields and provide a digital copy, on a software platform to be agreed, and 5 paper copies to the claimant

8.2 The acquiring authority will provide a plan showing the new drainage system as connected as a digital copy, on a software platform to be agreed, and 5 paper copies to the claimant

9 Fees

The acquiring authority will pay the claimant's reasonable agent's and solicitor's fees on the terms as agreed

10 Advance Payment

An advance payment of 90% of the claim to include 90% of the claimant's valuer's and professional fees to date

Note – as set out in the text the Advance Payment will be based on the acquiring authority's estimate – in this case set by the District Valuer

11 Interest

The acquiring authority will pay interest at the statutory rate on the remaining amount of this and subsequent claims (after payment of the advance) from the date of entry 1 September 2017 to the date of payment

Worth including even though 0% at present – things might change

Schedule of accommodation works

As per agreement made prior to entry dated 15 August 2017

a) The current water supply to the severed land will be re-laid under the bypass in a protective sleeve.

b) One field water trough currently in the field boundary between NG 1234 and 8765 on the line of the road will be replaced on completion of the work.

c) The design shows that the road will be in a shallow cutting for approximately half of its length through the farm including opposite the farmhouse. There will be further landscape planting in that area within the line of the permanent land take and a specification will be provided.
d) Double glazed sash windows, to match the existing window design, will be provided for the farmhouse to reduce noise from the road.
e) There will be a new surface water drain constructed either side of the road and existing field drainage will be connected into those drains:
    - the contractor is to:
        - connect any drains exposed into the new land drain unless agreed otherwise with the owner through the agent; and
        - to provide a map of the drains as laid
f) Gateways will be provided between NG 1234 and 5678 and 4231 and 8765 to enable access to fields severed by the new road and with no direct access to the existing public lane to the south of the farm.
g) The road will be bounded by a new hedge as set out in the scheme specification *(attached)* to be fenced on the field side by a stock fence as set out in the scheme specification *(attached)*.
h) The temporary working area will be carefully restored to its pre-works condition by the contractor with topsoil stripped and reserved for reinstatement before working commences. The authority will reimburse the claimant for subsequent restoration works required to return to effective arable production including stone picking, weeding and restoration of fertility as required.

During the works

Whilst the works generally go quite well and liaison between claimant and contractor works well through the Agricultural Liaison Officer (ALO) in particular there are, as always, some unforeseen problems.

These would form part of the disturbance claim and a limited number of examples are set out for illustration later on. Experienced practitioners will appreciate that there an almost inexhaustible supply of events which can beset a farm and indeed a project once the contractors are on site and connoisseurs of the subject may wish to revisit earlier editions for some other examples. However unusual or esoteric the issues may be the basis of claim will most commonly be either lost revenue, for crops unable to be harvested or stock fallen or needing to be sold at an inopportune time, or the cost of remedying damage, replacing damaged gates for example. That said regard must be had to the extent of any consequential loss; were this a dairy holding for example a fallen cow would also mean a reduction in milk sales.

The other elements of the disturbance claim will fall into similar categories to those explored in the utility claim in the next chapter, including future loss in productivity on land which has been used as part of the working area and claimant's time.

Issues during the contract

a) Delays in access to the farmstead were anticipated, but in practice they were worse than expected with five occasions when access was delayed for more than half a day, including one preventing a domestic heating engineer getting through for a routine service to the central heating.
b) Whilst the works were some distance from the farmhouse dust from the works and mud on the road increased the need for routine cleaning as well as contract cleaning of carpets and curtains as agreed by the ALO.
c) The guard fence against the works was poor and cattle grazing the pasture field got out on three occasions and trampled the adjoining field of winter wheat, which was reseeded after the first occasion but abandoned after that. Crop inspected by claimant's agent and ALO and extent of area damaged agreed.
d) A number of ditches were blocked during the course of the work, particularly north of the road and had to be cleared. That caused flooding to NG 2743 causing loss of grazing over part.

All the various incidents were noted by the claimant and reported to his Agent and the ALO as appropriate with emergency incidents reported straightaway.

Potential claim for disturbance

Head of Claim			Amount of Claim	
1	**Farmhouse damages and losses**			
1.1	Access to farmstead interrupted on five occasions			
	Payment for inconvenience on 5 occasions @ £150		750.00	
	Domestic heating engineer abortive call out charge (receipt attached)		69.00	
				819.00
1.2	Additional Cleaning			
	Cost of additional cleaning of farmhouse due to dust and mud during main working period 40 hours @ £12 per hour		480.00	
	Contract cleaning of carpets and curtains (receipt attached)		432.50	
				912.50
1.3	Farmhouse requires internal and external redecoration 2 years earlier than would have been the case.			4,000.00
	Estimated cost £10,000 (5-year cycle) 40% cost			
	Total Farmhouse Claim			5,731.50
2	**Farm Damages and Losses**			

Head of Claim		Amount of Claim	
2.1	Dust contamination to arable fields adjoining working area Total area 6 hectares winter wheat grown as milling wheat, but sample rejected and only fit as feed wheat 8 tonnes per ha @ £20 per tonne for 6 ha	960.00	
2.2	Damage to winter wheat in NG 1234 after collapse of Temporary guard fence against works – agreed area of damage 2 hectares		
	a) First incident 20 October 2017 Crop reseeded as agreed 2 ha seeds, fertilisers, sprays and cultivations @ £250 per ha	500.00	
	b) Second incident 25 May Agreed that affected area would be abandoned Loss of output 2 ha @ 8 tonnes per ha @ £165 per tonne 2 ha straw @ £35 per ha	2,640.00 70.00	
	c) Additional time in rounding up stock – 3 occasions 6 hours @ £15 per hour	90.00	3,300.00
2.3	Works to drainage		
	a) Clearing drains north of area taken for road 250 metres cleared by contractor (receipt attached)	937.50	
	b) Consequential flooding to NG 2743 meant grazing was lost over 1.5 ha from September 2018 to March 2019 – 1.5 ha @ £60 per ha	90.00	1,027.50
2.4	Restoration of Working Area Temporary area restored by contractor but with claimant completing restoration as agreed in the accommodation works Area 3 hectares		
	a) Additional expenses in cultivation and re-establishing seed bed for arable production @ £100 per ha		300.00
	b) of compound fertilizer, lime and farmyard manure. The land has been tested and the pH value is 4.8; the agronomists' recommendation for treatment per ha is:		
	12 tonnes lime @ £28/tonne 30 tonnes FYM @ £20/tonne 600 kg 21:8:11 compound @ £320/tonne	336.00 600.00 192.00	
	Cost per ha 3 ha @ £1,128 per ha	1,128.00	3,384.00
	c) Clearing surface debris Stone picking and other debris 2 men 2 days plus tractor and trailer @ £373.95 per day		1,495.80
	e) Fence Removal Removal of temporary fencing alongside pipeline 2 men 1 day plus tractor and trailer @ £373.95 per day		373.95

Head of Claim		Amount of Claim	
	f) Weed control Spraying affected area with herbicide 3 ha @ £70 per ha		210.00
2.5	Loss of future profits Reduction in output for 4 years following return of land – estimated output average £1,100 per ha *Average of rotation output over 4-year period*		
	2019/20 60% reduction	660.00	
	2020/21 35% reduction	385.00	
	2021/22 20% reduction	220.00	
	2022/23 10% reduction	110.00	
	Total	1,375.00	
	3 hectares @ £1,375.00 per ha		4,125.00
2.6	Loss of BPS income on temporary land 2 seasons @ £234 per ha per annum on 3 ha		1,404.00
	Total Farm Claim		**11,292.75**
3	**Claimant's Time and Costs** Additional time incurred by the claimant in liaising with contractor and meetings with acquiring authority, claimant's agent and District Valuer As per diary kept throughout project period (excluding time in 2.2 earlier)		
	37 hours @ £30 per hour		1,110.00
	Cost of telephone calls, emails etc 40 @ £2		80.00
	Total Claimant's Time		1,190.00
	Summary of Claim		
	1) Farmhouse Claim		5,731.50
	2) Farm Claim		11,292.75
	3) Claimant's Time		1,190.00
	Total Claim		**18,214.25**
4	**Accommodation Works** The acquiring authority will execute at their expense and to the claimant's complete satisfaction the works set out in the attached Schedule, making good any damage arising from the execution of the works *This may not have changed from the previous claim, but hopefully it will have done to the extent that some of the works will be completed.* *Thus, it may be appropriate to carry forward the Schedule perhaps with a note of the works completed and outstanding.*		
5	**Fees** The acquiring authority will pay the claimant's reasonable agent's and solicitors fees on the terms as agreed		

Head of Claim	Amount of Claim
6 **Interest**	
The acquiring authority will pay interest at the statutory rate on the remaining amount of this and subsequent claims (after payment of the advance) from the date of entry 1 September 2017 to the date of payment	
Worth including even though 0% at present – things might change	
7 **Reservations**	
Whilst the intention should be to settle as much as possible there may be particular issues where the claimant will want to reserve his position – in this case drainage and the future loss of output on the temporary land may issues to revisit after things have settled down from the work	

Example 2 – let farm – the tenant's claim

Notes

1) *For these purposes assume that all the circumstances are as in Example 1 except that the farm is let as set out later on.*
2) *Again, this example serves only to illustrate some of the ?s explored in the main text. Thus, the example is limited to the Tenant's equivalent of the initial claim for the owner-occupier. The issues explored in the owner-occupier's disturbance claim would fall to be claimed by the Tenant, but that has not been repli˙˙˙˙˙ here.*
3) *Notes are provided in italics to explain elements where necessary.*
4) *Again, the assumption is that possession was taken on 1 September 2017 so that values are being assessed either at that time, for identifiable elements of the claim, land take, injurious affection or severance, or in 2019 for the events which took place during the works.*

The holding

West Farm (seen first in Chapter 10) is a 100-hectare (250-acre, ...xed farm in the south-west of England with a four-bedroom period farmhouse in fair condition, if a little outdated, adequate farm buildings mainly 25 years old and approximately 100 metres removed from the farmhouse. a mix of predominantly Grade III pasture and arable land all in a ring fer There are no realistic opportunities for alternative development.

The farm is let on an Agricultural Holdings Act tenancy which commenced in 1970. The current Tenant succeeded to his Father in 2005 and is 55; the Tenant has two children, neither of whom are working on the farm at the moment and both have started careers elsewhere.

The farm is run as a mixed arable and beef holding with 69 hectares (170 acres) arable and 30 hectares (75 acres) pasture, the balance of the land being buildings, woodland and waste.

The letting is on a relatively standard model clauses agreement and the current passing is £17,500 per annum, last reviewed in 2017. The Tenant has invested in a Dutch Barn and a covered yard, neither of which will be taken in the scheme. The entitlements are all the Tenant's. The profit rent is estimated to be approximately £20 per hectare, although that may change as 2025 and 2027 approach.

The vacant possession value of the farm is estimated to be £2.5 million as set out in Chapter 10. If a remaining term of 20 years is assumed, factoring in a degree of uncertainty over the likelihood of succession, then the investment value is conveniently close to half that figure or £1.25 million.

A summary of the potential claim is set out overleaf, but before turning to that it is perhaps worth reviewing the key arguments concerning the value of the tenancy in this example.

Value of the remaining term of the tenancy – The uncertainty around the potential for succession creates considerable room for discussion. The Tenant's agent will wish to argue that there is still plenty of time for one or other of his children to return to the farm and indeed their chosen careers, accountancy and GP, could easily be pursued from the family farm. The stronger the case for succession the stronger the argument is for a greater proportion of the vacant possession premium to go to the Tenant. The District Valuer will likely wish to argue the contrary position suggesting that there is every prospect that there will be no succession and the appropriate consideration, if this method is adopted, is to consider the likely remaining time that the Tenant will be on the holding.

Just before doing so, however, they may wish to pause for some reflection and some arithmetic. Settlement needs to be reached with the Landlord as well; and the value of the Landlord's investment rises or falls with the degree of uncertainty of succession. Thus the DV may wish to pause to consider how strongly to pursue this argument.

The Landlord's agent, as we shall see later on, has no such need to pause for thought; it will be clear to them, at least in their negotiating position if nowhere else, that it is only a matter of time before the incumbent Tenant with no prospect of succession and the prospect, at the time of writing, of a radically changed regime of agricultural support, will decide to retire early.

For the purposes of illustration in the example the "middle line" has been adopted. When considering the value of the tenancy the approach adopted in this example has been to consider the value of land taken as part of the whole tenancy, covering the entire holding, rather than basing this on the value of the individual area of land taken.

Potential claim

Head of Claim		Amount of Claim	
1	**Value of the Tenant's interest in the land taken**		
1.1	Permanent Land Take		
	3 hectares (7.5 acres) land taken from the whole		
	Market Value of the farm per hectare		
	With vacant possession	25,000	
	Subject to the existing tenancy	12,500	
	Vacant possession premium	12,500	
	Tenant's share of VPP @ 50%	6,250	
	3 hectares @ £6,250	18,750	
	Less 4 year's rent – 3 ha @ £175 per ha		
	(the Tenant having opted to claim for the value of the lost term rather than under s.60)	2,100	
		16,650	
1.2	Temporary Land Take		
	3 hectares (7.5 acres) lost for two seasons		
	Licence fee on the basis of income foregone		
	3 hectares @ £800 per hectare per annum	4,800	
1.3	Basic Payment Entitlements		
	Assuming 6 hectares of entitlements lost		
	3 hectares on land permanently taken and 3 hectares through lack of use and to be replaced	900	
	6 hectares @ £150 per hectare		
	Total for interest taken		**22,350**
2	**Outgoing Tenant's Payment**		
2.1	Unexhausted Manurial Values 3 ha @ £45/ha	135	
2.2	Residual value of feedstuffs 3 ha @ £25/ha	75	
2.3	Unexhausted value of lime applied 3ha @ £18/ha	54	
2.4	Loss of crops on land taken		
	(Slightly moot point given the timing of possession – arguably there is no loss here)		
2.5	Loss of 400 metres stock fencing		
	Tenant's Fixture, 4 years old @ £3 per metre run	1,200	
	Total for Outgoing Payment		**1,464**
3	**Injurious Affection to Tenancy**		
3.1	Fencing liability		
	1,800 metres of hedge and fence to be maintained against the road including annual maintenance of the fence, cutting and laying of hedge and annual hedge maintenance once established, replacement of fence with similar guard fence when redundant		
	a) Annual maintenance 40 hours @ £15 per hour	600	
	YP @ 4% in perpetuity		
	(The DV might argue for the remaining term)	25	15,000
	b) Cut and lay hedge in 10 years @ £18 per metre	32,400	
	PV of £1 in 10 years @ 6%	0.5584	18,092

Head of Claim		Amount of Claim	
	c) Replacement of fence in 20 years @ £4 per metre	7,200	
	PV of £1 in 20 years @ 6%	0.3118	2,245
	There is an argument here about maintenance and the need for replacement, but no matter how well maintained ultimately the fence will rot		
3.2	**Redundant Fixed Equipment**		
	In this case the nature of the holding and the location of land taken means that none of the Tenant's fixed equipment is redundant, but this may often be an argument elsewhere		
	Total for Injurious Affection		**35,337**
	Summary of Claim		
	1) Interest Taken	22,350	
	2) Outgoing Tenant's Payment	1,464	
	3) Injurious Affection	35.337	
	Total Claim to Date		**59,151**
4	**Basic and Occupiers Loss Payment**		
	To be calculated on agreement of claim		
5	**Disturbance and Crop Losses**		
	This will be the subject of a future claim when the contract is completed		
6	**Accommodation Works**		
	As Item and of the Owner's Claim		
	The agents for Landlord and Tenant will have to liaise over this issue		
7	**Plans**		
7.1	The acquiring authority will provide a new farm plan, calculating the revised areas of the affected fields and provide a digital copy, on a software platform to be agreed, and 5 paper copies to the claimant		
7.2	The acquiring authority will provide a plan showing the new drainage system as connected as a digital copy, on a software platform to be agreed, and 5 paper copies to the claimant		
8	**Fees**		
	The acquiring authority will pay the claimant's reasonable agent's and solicitor's fees if applicable		
9	**Advance Payment**		
	An advance payment of 90% of the claim to include 90% of the claimant's valuer's and professional fees to date		
	Note – as set out in the text the Advance Payment will be based on the acquiring authority's estimate – in this case set by the District Valuer		
10	**Interest**		
	The acquiring authority will pay interest at the statutory rate on the remaining amount of this and subsequent claims (after payment of the advance) from the date of entry 1 September 2017 to the date of payment		
	Worth including even though 0% at present – things might change		

Example 2 let farm – the landlord's claim issues for consideration

The Landlord's claim will echo the owner-occupier's initial claim in terms of the Heads of Claim. However, the level of compensation will be based on the Market Value of the Landlord's interest, so the investment value. The approach to interpreting that may vary on the particular circumstances.

Land taken – the Landlord's claim will be essentially as the owner-occupier's but subject to the Tenant's interest. Here the Landlord's Valuer will wish to adopt a different argument to the Tenant's Valuer. As we have seen earlier the Tenant's argument will be based on the firm prospect that there is every prospect for succession the Landlord will wish to argue the contrary position; neither of the Tenant's children are interested in the farm and if they wanted to return they would be showing signs of that now. Hence, for the Landlord's valuer the Landlord's investment value is higher than the Tenant's argument would suggest. Intriguingly, they are not arguing with each other but separately with the District Valuer.

Severance – clearly this will be a function of the investment value of the holding and subject to all the same arguments as that for the value of the land taken. Given the paucity of evidence of sales of investment farms it is unlikely that the Valuer will be able to find any direct market evidence of investment sales which provide comparables for the impact of Severance, although comparing the market for whole farm and bare land investment sales may help, once the valuer has been able to unpack all the special purchaser issues which often impact on investment sales.

In the alternative the Landlord's valuer may look to the owner-occupier market and, assuming that they adopt a Term and Reversion approach to the valuation, they will be able to reflect those in the value of the Reversion with or without the scheme. As a third alternative the valuer may decide the simplest approach, in a world shorn of direct comparables and where a good deal of mental gymnastics is required to convert what evidence may be available into a justifiable argument, is to apply Severance settlements achieved elsewhere in an owner-occupier context on a proportionate basis to the investment value of the holding.

Injurious affection – the same arguments run in respect of Injurious Affection with the added complication that the largest element of claim here will be in respect of the farmhouse and that would otherwise be valued as part of a smaller lot, most likely sold with 4 hectares (10 acres and in the earlier example and Chapter 10).

On that basis and assuming that the most appropriate approach is to consider this on a Term and Reversion basis and adjust the Reversionary value of the farmhouse to reflect the impact of the road, the valuer will then have the additional interest of deciding how best to approach the investment value of the farmhouse and 4 hectares. This is an exercise the author has indulged in routinely, albeit more often as part of discussions of the rationalisation of a tenanted holding. A number of approaches are available,

amongst which the much-vaunted valuer's gut feeling quite often seems to hold sway. However, to substantiate or test that, and indeed to offer some rational support for the conclusion when it comes to negotiation two methods are immediately available:
- a simple proportionate approach, applying the proportion of the vacant possession value which the farmhouse lot represents to the overall investment value
- undertaking a separate Term and Reversion valuation of the farmhouse with the added frisson of excitement of deciding what the rent should be and whether one should adjust yields

Ultimately the same test applies as with so much of the rural valuer's work – does this look like a sensible estimate of market rection and hence value?

Further reading

Compulsory Purchase and Compensation, Barry Denyer Green (Routledge, 2018)
Compulsory Purchase and Compensation: The Law in Scotland, Jeremy Rowan Robinson (Sweet and Maxwell, 2009)
Good Practice in Statutory Compensation Claims (CAAV, 2019)
Professional Fees in Compensation Claims (CAAV, 2nd Edition, 2020)

22 Cables and pipes for electricity, gas, water and sewerage

22.1 Introduction

22.1.1 With great lengths of pipes and cables inevitably crossing rural land to serve both town and country, dealing with claims arising from the exercise of rights by utility companies forms a significant element of many valuers' work, whether representing one of the utility companies, who mainly now use external consultants to deliver valuation and other professional services, or the person whose land is affected.

22.1.2 This is also an area of practice which gives rise to considerable complaint from valuers on both sides about the approach to the negotiation and settlement of claims. Some of this is beyond the scope of valuers, particularly the delay in settlement of agreed claims by utilities whether by omission or systemic design, but elsewhere poorly or lazily constructed claims are often blamed for failure to agree and the consequent delays. As with all areas of work, there is no substitute for the careful and comprehensive gathering of background information from affected parties and supporting evidence which will assist both in the assessment and settlement of claims and in minimising as far as possible the practical impacts of work on a rural business.

22.1.3 Underground pipes and cables and overhead cables can all limit what might be done with farmland or other property. The line itself has to be protected from development (so limiting where farm fixed equipment might be sited) and damage and can limit planting (as for an orchard) or other work. Overhead cables have similar effects and limit irrigation or sterilise possible development while poles and pylons inhibit modern machinery. Gas mains are further protected under the Planning Advice for Developments near Hazardous Installations, produced by the Health and Safety Executive, sterilising a greater width than for water mains. Underground electricity cables also need a wider protected area than overhead cables.

22.2 Pipelines

22.2.1 Utilities may lay pipes using powers for statutory undertakers under the Water Industry Act 1991, for water mains and sewers, the Gas Act

1986 as amended by the Gas Act 1995 for gas mains, and the Pipe-lines Act 1962 for commercial pipelines (as for oil).

22.2.2 There are essentially two elements to a claim:
(a) the capital payment, whether recognition payments for sewer and water mains or claims for gas and commercial pipelines, and
(b) claims for damages arising following the laying of any pipelines and associated works

22.2.3 Capital payments – Capital payments following the laying of sewer and water pipes are a recognition payment for the effect of the presence of the pipeline on the whole of the claimant's property to reflect the fact that no easement is taken. Following the decision in the *St John's College* case in 1990, the recognition payment, based on 50% of the capital value of the land, conventionally extends to the full working width, in that case 20 yards, and not as had previously been common practice, the notional easement width.

22.2.4 As well as the oft-quoted financial settlement, a point often overlooked from that case but, arguably of far greater significance, is that the damage of injurious affection caused by the pipe is to be considered on the basis of the property as a whole, rather than just the line of the pipe.

22.2.5 While the approach using half of the value of the working width as a way to settle the effect on the property has been accepted practice since that case was decided in the common absence of direct evidence, latterly there have been some challenges with water companies arguing that the marked increase in the price paid for agricultural land should not be seen to increase the loss to the claimant. The argument then sometimes develops that rising land values are due to scarcity with the consequent interest of special purchasers and not really a true reflection of some underlying "market value" which hence should not be reflected in the capital payment. This argument has been pursued with varying degrees of vigour and success, depending on the circumstances and the skill and robustness of the claimant's valuer and the strength of comparable evidence.

22.2.6 Where, unusually, there is good evidence to hand as to the actual effects of a pipeline on the value of property that may get directly to the matter. That was at issue with the prospect of a fresh judgement in this area offered by the Northern Ireland case, *Chivers*, referred to briefly in the previous chapter. Mr and Mrs Chivers owned some land with the benefit of planning permission for residential development, having sold two other blocks for development. Northern Ireland Water Ltd (NIWL) served notice of their intention to lay a sewer pipe across the land to a new pumping station on part of the sold land which would service the development. The Chivers claimed a recognition payment based on diminution in the value of the land because a strip was sterilised by the pipe and incapable of residential development and so losing some house plots.

22.2.7 The case came to the Northern Ireland Lands Tribunal where NIWL argued that, under the Northern Irish legislation, it was only obliged to pay for physical damage. It was not liable to pay for any other depreciation in value because it had not taken any interest in the land. The Chivers' counter-argument was that not paying them compensation for the loss would be "unusual and unfair", arguing that the relevant Water and Sewages Service Order 1973 provided for compensation for loss "in consequence" of the works, which meant more than simply physical damage from the work. The Chivers' argument was accepted by the Tribunal, which gave rise to the question of what the right level of compensation should be but that was held over for a future hearing.

22.2.8 NIWL took the case to the Northern Ireland Court of Appeal which considered the two distinct questions:
- what was the legal nature of what is taken to lay a pipe?
- what damage is compensatable?

22.2.9 On the first question the Court referred to *St John's College* and its conclusion that the law conferred power to do work rather than acquire any interest in the land. It reaffirmed that no wayleave or easement was created and felt that the phrase "recognition payment" was not an appropriate expression in the circumstances.

22.2.10 Once again reviewing *St John's College*, the Court decided that there were two elements which might satisfy the limitation that compensation was only payable for damage to the property caused by or as a consequence of the works:
- any physical damage which had not been fully and adequately reinstated and any loss of use during the period of the works; and
- any diminution in value of the land as a whole, based on the before and after value of the land

22.2.11 Thus, whilst this fell under the slightly different Northern Irish legislation, the Court broadly confirmed the decision in *St John's College* that diminution in the value of the affected person's whole property was to be compensated. The missing piece in the puzzle is what this actually means for the assessment of compensation. With the different issue at the original hearing, the Tribunal did not hear any evidence on value and the Court of Appeal could not address the matter fully but, in upholding the Tribunal decision, it referred the matter of the amount of compensation back to the Tribunal. However, it is understood that the parties settled the matter before that hearing took place and the sum remains undisclosed.

22.2.12 Aside from the payment for the line of the pipe, there may be additional payments if there are any further structures or apparatus on the land including marker posts, chambers, valves and manholes. The level of payment will turn very much on the location; those in the middle of a field are much more intrusive and thus attract more compensation than those in a boundary hedge, although that sentiment may not always be shared by a hedging contractor who finds that he has bent his hedge cutter on an invisible marker post. Utilities will sometimes bury hardware below cultivation depth, which significantly reduces routine interference but can lead to them making more of a mess when they need to dig to find the manhole on later visit. It is one of those strange natural phenomena that valve chambers and similar constructions which are shown well below the depth of a subsoiler on the engineering drawings seem, in practice, to struggle back nearer the surface; it is far from unknown for cultivators to be broken on such "buried" structures.

22.2.13 Easement payments – In contrast to the position with water or sewer pipes, commercial pipelines and gas mains require an easement. These enterprises do not benefit from the same compulsory powers as are afforded by the Water Acts and consequently National Grid (for gas) and commercial pipeline installations normally pay an easement payment of 75% of the vacant possession value of the land affected over the relevant width. Consequently, as with the land market generally, payments will vary not just with the quality of land but with the size of the field, small paddocks generally being more expensive per acre, and the location, particularly the proximity to a dwelling.

22.2.14 In addition, payment is normally made to the occupier (whether the owner or a tenant) essentially to encourage cooperation. In the case of larger schemes this payment has often been agreed between the promoters and the CLA and NFU. Increasingly as consultation becomes more relevant to the delivery of infrastructure schemes, these payments, along with other recognition payments (for example for access for surveys) may be discussed with relevant stakeholder groups which may include a diverse range of national and local groups sometimes including valuers' organisations. The extent to which this may change with the introduction of wider powers in respect of access for surveys is explored in the chapter on compulsory purchase. Suffice it to say, attitudes vary very considerably between utilities and, particularly, their engineering contractors and land acquisition consultants sometimes according to the basis on which they have been procured.

22.2.15 Again, payments are made for manholes, chambers and other structures.

22.2.16 Project management – The best managed projects will have Field Officers or Agricultural Liaison Officers appointed to work with landowners, occupiers and their agents to mitigate as far as possible the impact of the work and the subsequent claim for damages. Some Field Officers are very accomplished in fulfilling this difficult role where success is often measured by achieving balanced levels of dissatisfaction amongst the parties. However, even the best may be let down by what appear to be the increasingly complex and fractured lines of communication within and between utilities and their various different levels of contractors.

22.2.17 In some cases, where notice of access is given early enough, it may be possible to mitigate some of the losses by planning grazing or cropping or agreeing entry dates with the contractor to limit damage. Unfortunately, the contrary appears increasingly to be the case with either limited notice of proposed works or general notice of the works being given relatively early but then very short notice for particular works. If anything, this is more difficult to manage and it is far from unusual for entry to a field to be requested with less than 12 hours' notice.

22.2.18 Whether this is due to pressure on contractors or lack of appreciation of the impact on owners and occupiers is to an extent academic: the impact is the same. The role of the agents, acting on either side, should be to ensure that there is an awareness of those impacts, particularly where they may give rise to substantial items of claim, short-notice access interfering with planned vegetable lifting for example. Where such difficulties are likely to arise, or even where they have arisen, it is important for the valuers to recognise that this will need careful management. As in other areas, there is much to be gained by reducing as far as possible the scope for surprises, for both clients and other agents, when the claim is submitted.

22.2.18 As with compulsory purchase, the utility should be taking a record of condition on entry for their own benefit which should be agreed with the claimant's valuer. A site meeting should be held pre-entry to check the programme of work, areas affected, access routes, welfare and storage compounds and the like. Ideally, there should also be a meeting at the end of the works to ensure that the site is handed back in a clean condition free of any of the detritus of a major infrastructure scheme and any specific areas of damage are noted. The claimant's valuer should ensure that the utility will meet the costs of attendance at these meetings, which if properly conducted will save time in negotiating the claim later.

22.2.19 An increasing problem is one of biosecurity, particularly with vulnerable crops or livestock, for example closed intensive pig and poultry units. Such farms will now have biosecurity plans. Those selling direct to consumers or to retailers or dairy companies with stringent standards will be particularly cautious about standards of cleanliness for contractor and other vehicles. Settling biosecurity protocols is an increasingly important part of managing a project of this type and the cost of, say, the temporary loss of a dairy contract or downgrading of a farm's quality rating with a particular retailer could have extremely adverse consequences and lead to a substantial and contentious claim.

22.2.20 Damage claims – The claim should address all the heads of claim raised by the infrastructure scheme. As with other elements of compensation practice, the principle is one of equivalence; the affected party should be left in the same position after the activity of the water company as they would otherwise have been and financial compensation is used to remedy any shortfall.

22.2.21 The individual Heads of Claim will vary with the nature of the work and the property affected but is likely to include some or occasionally all of the following:

a) *actual crop losses* – whether in the ground or occasionally in store, and including where crops grown for feed are superfluous due to works reducing stocking capacities and stock carried.

b) *future crop losses* – over a number of years which might range from two to ten years, dependent on damage caused and the nature of the soil but generally is for three or four years. Generally, these are related to the crops being grown when the pipeline goes through, but additional claims might arise where there is an interruption in the rotation, particularly where higher value crops are grown.

c) *cost of restoration of ground* – including stone picking and clearing any residual rubbish, waste and spare pipe fittings etc which can be a problem on some sites.

d) *cost of replacing lost fertility and improving soil condition* – where works have led to compaction or clagging of soil.

e) *weed control* – weeds grow liberally on pipe tracks and there is also the risk of them being introduced to clean land with imported topsoil or other material with Japanese knotweed and the cost of its removal a growing concern.

f) *the additional cost of maintaining hedges and fences* – where hedges have been re-planted, this will particularly include the cost of laying hedges in the future, likely to be around ten years after planting, along with the cost of maintaining guard fences and beating up to replace failed new plants in the interim.

g) *cost of replacing topsoil* – both where any is lost, or more frustratingly where it is not properly set aside and is reinstated below the subsoil! The right should also be reserved to claim where subsidence occurs after a period of time. The tell-tale sign for a problem is a perfectly level reinstatement over the pipe; reinstatement where the line of pipe is, at first, clearly visible under a slight mound of replaced soil is more likely to deliver a level surface long-term.

h) *cost of repairs or reinstatement of drainage* – pipeline schemes may often cut across drains and it is important that they should be properly reinstated after the pipe is laid. Failed reinstatement may not become apparent for a number of years, arising sometimes from delayed subsidence, and thus it is helpful to collect drainage maps from the farmer before entry so that contractors can look out for the drains but also so that the scope for disagreement is reduced in the future.

i) *claimant's time* – the claimant should be paid for all the time he spends on the matter, whether with the utility's agents, contractors, or his own valuers and solicitors as well as rounding up stray stock, telephone and

j) *livestock losses* – whether arising from stock injury or fatality from straying on to the line, provided that is not arising from negligence on the part of the farmer. There might also be a claim for lost production, although that is one of the hardest heads of claim to establish and test where, for example, long term intrusive works interfere with milking regimes giving rise to lost production.

k) *lost Basic Payment Scheme (and potential successors) and environmental payments* – arising where the infrastructure scheme interferes sufficiently with farm management and lasts long enough that areas are not eligible for payment. There will be a claim both for the lost payment, including any entitlements that might have been lost due to lack of use and for the costs of mapping the affected areas (currently as Temporarily Ineligible Features). Similar claims may arise with environmental payments whether through lost revenue or costs in reinstating to a particular specification.

l) *all professional fees and associated expenses* – which used to be based on Ryde's Scale, last revised in 1996. Some utility companies still try to pay on the basis of say Ryde's, usually plus a percentage. However, the basis of the fee should be quantum meruit, as affirmed by the Upper Tribunal in *Poole v South West Water* (following the earlier Lands Tribunal decision in *Environment Agency v Matthews*). Time records should be kept as the utility company will seem some justification for the fee.

m) *interest on the agreed claim*. However, it is usually difficult to recover this from the pipe-laying authority, except on the easement payment.

22.2.22 Calculating the easement recognition payment – Every claim will be different but an indicative claim, based on a new water pipe, is set out in the following example. Professional fees and interest should be claimed in addition. Claimant's time will be included in the damage claim, an example of which is set out in Example 2.

Example 1

A new 150mm diameter water main is laid through a 200-hectare (500-acre) owner-occupied mixed farm. The total length is 280 metres and the working width is 15m. There are four marker points and three valve chambers (two in the fields and one in a hedgerow); the land is Grade III strong arable capable of producing 10 tonnes per hectare of winter wheat in a reasonable season. The suggested claim in these circumstances (every case is different) is:

a)	**Pipeline** Area Affected 280m run x 15m working width = 4,200m² = 0.420 ha Market Value £20,000/ha (£8,100/acre) X 50% = £10,000 / ha: 0.420 ha x £10,000			4,200.00
b)	**Marker Posts** 4 no x £25			100.00
c)	**Valve Chambers** 2 no in field @ £500 1 no in hedgerow		1,000 100	1,100.00
d)	**Claimant's Professional Fees** Valuers' and Solicitors' Fees and costs			
e)	**Interest** On total of claim from date of entry to date of payments			

Note: as set out earlier the rate of 50% of Market Value applies for claims under the Water Act. The rates of easement payment may be higher for other utilities which do not have the benefit of a robust statutory code.

22.2.23 Example pipeline claim – Details of the scheme, including the line and length of the pipeline and the number and location of any apparatus will be provided with the Water Industry Act notice so that the agents should be able to agree that element of the claim prior to possession being taken, although, in practical terms, sometimes timescales may not allow that. However, in the balance of the claim, dealing with damages can only be assessed after work has been completed and it may be necessary for the claimant to reserve rights to return to the claim if problems are anticipated, as where land drainage is impacted or subsidence appears likely.

22.2.24 Example 2 includes the capital and damages claim for work not covered by the Water Industry Act. Whilst the two elements of the claim are taken together in this example, again the capital payment, for an easement in this case, will be assessable before the damages claim.

Example 2

An oil pipeline is crossing the same 200-hectare (500-acre) owner-occupied mixed farm. The total length is 680 metres and the working width is 20m. There are four marker posts and three valve chambers (two in field margins and one in the hedge). The pipeline crosses four fields and work took place over the autumn of 2018 and the spring and summer of 2019. Cropping during the works is as follows:

NG 1841	Winter Wheat	0.24 ha working area
NG 2538	Winter Wheat	0.30 ha working area

Cables and pipes for electricity, gas, etc 321

NG 5451 Sugar Beet followed by Spring Barley 0.40 ha working area
NG 2200 Pasture Ley 0.42 ha working area

There are established hedges between each of the fields and access is via a metalled farm track. The farm has the capacity to store all grain produced.

The claim might be as follows:

Head of Claim		Amount of Claim	
1	Easement Payment		
1.1	Pipeline Easement		
	Area affected 680 m run x 20 m working width = 13,600 m² = 1.36 ha		
	Market Value £20,000 per ha (£8,100/acre)		
	X 75% = £15,000 per ha: 1.36 ha x £15,000		20,400.00
1.2	**Marker Posts**		
	4 no x £25		100.00
1.3	**Valve Chambers**		
	2 no in field margins @ £400	800.00	
	1 no in hedgerow	100.00	900.00
	Total to Summary		21,400.00
2	Loss of Crop		
2.1	NG 1841 – Working Area 0.24 ha		
	a) 2019 – Loss of milling wheat yield 8.5 tonnes per ha		336.60
	2.04 tonnes @ £165 per tonne		
	b) 2019 – Loss of straw 4 tonnes per ha		
	0.96 tonnes @ £35 per tonne		33.60
2.2	NG 2538 – Working Area 0.30 ha		
	a) 2019 – Loss of Feed Wheat yield 10 tonnes per ha 3.00 tonnes @ £140 per tonne		420.00
	b) 2019 – Loss of straw 4 tonnes per ha		
	1.20 tonnes @ £35 per tonne		42.00
2.3	NG 5451 – Working Area 0.40 ha		
	a) 2018 – Loss of sugar beet yield 66 tonnes per ha Site works prevented lifting sugar beet in the area south of the pipeline – Total area lost 1.58 ha @ 66 tonnes per ha 104.28 tonnes @ £25 per tonne		2,607.00
	b) 2018 – Loss of sugar beet tops 1.58 ha @ £60 per ha		94.80
	c) 2019 – Loss of Malting Barley yield 5 tonnes per ha 0.4 ha only 2 tonnes @ £155 per tonne		310.00
	d) 2019 Loss of straw 4 tonnes per ha 1.6 tonnes @ £40 per tonne		64.00
2.4	NG 2200 – Working Area 0.42 ha		
	a) 2018 – Loss of winter grazing @ £75 per ha 0.42 ha @ £75		31.50
	b) 2019 – Loss of spring summer and autumn grazing @ £300 per ha 0.42 ha @ £300		126.00
2.5	Basic Payment Scheme		
	a) 2019 – Loss of claim 1.36 ha @ £232.84 per ha		316.66

Head of Claim		Amount of Claim	
	b) Costs for remapping to exclude working areas on RPA Digital Mapping	200.00	
2.6	Future Loss of Profits		
	Ongoing reduction in yield on arable crops over working area for 3 years post 2019 based on a proportion of 2019 losses		
	2.1 2019 claim	370.20	
	2.2 2019 claim	462.00	
	2.3 Adjusted for working area and sugar beet crop only	660.00	
	2.4 66 t/ha on 0.4 ha = 26.4t @ £25/t	24.00	
	Tops 0.4 ha @ £60/ha 2018/9	157.50	
	Total	1,673.70	
	Estimated losses as a proportion of 2019 claims		
	Year 1 50% loss x £1,673.70	836.85	1,422.65
	Year 2 25% loss x £16,73.70	418.43	
	Year 3 10% loss x £1,673.70	167.37	
	Total to Summary		**6,004.81**
3	**Reinstatement of Hedges**		
3.1	Cost of first cut and laying of replanted hedges in 10 years		
	NG 1841/2538 – 18m		
	NG 2538/5451 – 15m		
	NG 5451/2200 – 30m		
	NG 2200/track – <u>16m</u>		
	Total 79m		
	79m run @ £20/metre	1580.00	
	PV in 10 years @ 5%	0.6139	969.96
3.2	Annual maintenance cost		
	Cost of fence maintenance, beating up and trimming until established		
	3 hours per annum at £30/hour	90.00	
	YP 10 years @ 5%	<u>7.7217</u>	694.95
	Total to Summary		**1,664.92**
4	**Reinstatement of Working Area**		
4.1	Arable land NG 1841, 2538, 5451 = 0.94 ha		
	Cost of additional expenses in cultivating working area and re-establishing seed bed @ £100 per ha		94
4.2	Pasture ley NG 2400 = 0.42 ha		
	Cost of cultivating to establish seed bed and reseeding @ £300 per ha		126.00
4.3	Restoration of Fertility		
	Application of compound fertilizer, lime and farmyard manure. The pipeline has been tested and the pH value is 4.8; their agronomists' recommendation for treatment per ha is:		
	12 tonnes lime @ £28/tonne	336.00	
	25 tonnes FYM @ £20/tonne	500.00	
	600 kg 21:8:11 compound @ £320/tonne Cost per ha 1.36 ha @ £1,098 per ha	<u>192.00</u> 1,098.00	1,398.08

Head of Claim		Amount of Claim	
4.4	Clearing surface debris Stone picking and other debris 2 men 1.5 days plus tractor and trailer @ £373.95 per day (based on CAAV costings)		1,121.85
4.5	Fence Removal Removal of temporary fencing alongside pipeline 2 men 1 day plus tractor and trailer @ £373.95 per day *NB sometimes (often) settled by the farmer retaining the fencing*		747.90
4.6	Weed Control Spraying the affected area with herbicide to control weed growth knapsack or ATV sprayer for small area 1.36 ha @ £100 per ha		136.00
	Total to Summary		**3,623.83**
5	**Reinstatement of Trackway**		
	Repair metalled farm track where it has been damaged by the passage of contractor vehicles; scrape out and level ruts and fill with clean stone, reinstate drainage ditch cut across track opposite gate to NG 2200 2 men 2 days plus tractor and trailer @ £373.95 per day Materials – clean stone	747.90 150.00	**897.90**
	Total to Summary		
6	**Consequential Losses**		
6.1	Livestock damage		
	a) Sheep escaped from NG 2200 due to poor guard fence against pipeline and grazed emerging sugar beet plants in NG 5451 Losses on 4 ha sugar beet affected 20% of crop thus say 53 tonnes @ £25 per tonne		1,320.00
	b) Sheep trespass in NG 5451 also disturbed the pre-emergent spray. Weeding required through two passes with steerage hoe. 8 ha affected @ £39.40 per ha		315.20
	c) Loss of two breeding ewes (in lamb) which fell in to trench through failed guard fencing 2 ewes @ £80 Loss of lambs 2 ewes @ 175% lambing percentage 3.5 lambs @ £60/lamb	160.00 210.00	370.00
6.2	Disturbance to Rotation		
	a) Farm rotation disturbed in NG 2538 and adjoining field NG 2387 with delayed access meaning Spring Barley being sown instead of Winter Wheat over 8 hectares Loss in grain sales per ha: 10 tonnes winter wheat/ha @ £140/t 1,400.00 Less 5 tonnes spring barley/ha @ £125/t <u>625.00</u> Loss per ha 775.00 8 ha @ £775 per ha		6,200.00

Head of Claim	Amount of Claim	
b) Soil heap set aside from trench interfered with drainage causing flooding from NG 2200 to adjoining arable field 1.6 ha Winter Wheat failed to germinate and area resown with Spring Barley 1.6 ha loss of grain sales as described earlier @ £775/ha Cost of cultivation and resowing @ £120.75/ha	1,240.00 193.20	1,433.20
Total to Summary		**9,638.40**
7 **Claimant's Time and Costs**		
a) To claimant's time:		900.00
a) liaising with company's representatives, contractors and own agent		
b) additional time involved in stock management whilst works undertaken in NG 2200 as per attached time schedule 30 hours @ £30 per hour		
b) To machinery and labour to tow contractor's vehicle out when stuck in NG 2538 Tractor and Driver 2 hours @ £43.59		87.18
Total to Summary		**987.18**

9 **Additional Items (subject to future appraisal)**

9.1 Drainage

The company to make good any damage caused to the drainage of the land as a consequence of these works to include paying for any loss of crop or other losses arising as a result and the cost of the restoration works on production of an invoice by the claimant

9.2 Subsidence

The company to make good any subsidence resulting from these works, removing and restoring topsoil or providing a minimum of 0.3m depth of topsoil of a quality and type equivalent to that lost where necessary

9.3 Subsidies and other payments

The company to meet any fine or shortfall in BPS or other payment suffered by the claimant as a consequence of penalties imposed by the RPA for any failure to comply with statutory requirements under BPS or any other relevant scheme resulting from these works

9.3 Fees

The company to pay the claimant's valuers fees and expenses on a time basis in accordance with the timesheets provided together with any solicitor's costs.

9.4 Interest

The company to pay interest on the agreed claim from the date of entry at either the statutory rate or 3% over the Bank of England Minimum Lending Rate whichever shall be the higher

Head of Claim	Amount of Claim
Summary	
1 Easement Payment	21,400.00
2 Loss of Crop	6,004.81
3 Reinstatement of Hedges	1,664.92
4 Restoration of land	3,623.83
5 Reinstatement of Track	897.90
6 Consequential Losses	9,638.40
7 Claimant's Time	987.18
Total	44,217.04

22.2.25 Every claim will be different and there will be matters of discussion and dispute with between valuers. In some cases, these will be fundamental, particularly where high value crops or livestock are involved. The example claim is constructed on behalf of the claimant as it is for the claimant's valuer to make the claim. There is no onus on the utility's valuer to draw matters to his opposite number's attention. That said, difficulties can be minimised where both valuers are:
- realistic in their approach – drawing on practical solutions to what in the main are practical problems, so recognising the extra unit cost of dealing with small areas of affected land for example
- sensible in their ambition – the aim is, or at least it should be, to reach agreement on a reasonable and adequate settlement for the loss, damage and inconvenience caused to the client
- collaborative where possible – much difficulty can be avoided if the client is notified where there are changes to working programmes and the company's valuer is forewarned of any significant or surprising claims for consequential loss

22.2.26 The age-old doctrine has been that privity of contract lies between the utility company and the client. Whilst in practice the client will have far more contact with the contractors, the practice amongst some acquirers of seeking to lay liability off on the contractor should be resisted firmly. If the contractor has done something which has added to the damage, then the starting point should be that it is up to the utility company to pursue matters with them, not the affected occupier. The contractor is there at the express purpose of the utility company which is liable for his actions and they are liable to the claimant for compensation. That position can be more challenging when the contractor goes outside of the contract, the basic principle being that the employer or client of a contractor is not liable for negligence or other torts the contractor may commit. However, the Lands Tribunal case *Donovan v Dwr Cymru* considered the situation where the contractor went outside of the area shown on the plan with the Water Industry Act notice. The Tribunal decided that the contractor had entered the land solely as agents for Dwr Cymru and consequently the utility company could not avoid its statutory liability to the landowner.

22.2.27 That is a good starting point for the claimant's agent, albeit the decision does not bind higher courts. However, it is worth the claimant's valuer just being a little circumspect before phoning up their opposite number to complain about a pipe dump outside of the working area. A quick call to the client might reveal that they were there because he had agreed they could be and has already had the fee from the contractor! More generally, there may be agreements to settled for value with the contractor for a compounds or other facilities, which would be entirely separate from the compensation claim against the utility.

22.2.28 Payment – The timing of the payment is a matter of negotiation between the parties. However, where there is a substantial claim delays in payment can cause cashflow issues for claimants which will not be wholly remedied by interest on the claim. This is especially so for a claimant operating close to their overdraft limit, more likely in these days of tighter overdraft management by both banks and businesses. As a general rule, the easement payment can be agreed before entry and there is then no reason that it should not be paid soon after, rather than waiting on the finalisation of the claim which could be one or two years later.

22.2.29 Again some management of client expectations is required here. Utility companies have become more corporate, initially when they moved from the public sector to the private sector and subsequently both as they have been traded and new owners have imported their own systems which they continue to adapt. This can all have an impact on payment times both for clients and for agents. Some utility companies refuse to pay agents direct, or even sometimes to notify them that payment has been made the claimant. The problem can be further exacerbated by arcane practices within corporate accounting teams which, designed to ensure accuracy and resist fraud, add to delays. The client will not be too worried about the relevant companies' QA procedures but will be frustrated by the delay in receiving payment of a settlement agreed between agents months ago.

22.2.30 Somewhat surprisingly, unless you are of a cynical bent or have reached the state of grace of a senior practitioner for whom little is surprising any longer, the combined technological might of utility accounts departments and modern electronic money transfer has combined to make payment markedly slower than used to be the case when cheques were written and posted by hand. While an interesting philosophical discussion, it is not one to be had with a frustrated arable farmer using a wet morning to catch up with the agent to check "where his cheque from the water company has got to".

22.3 Electricity line wayleaves and claims

22.3.1 General – In common with other utilities, there has been a significant increase in construction activity in the electricity sector and this is forecast to continue, repairing and replacing existing infrastructure and providing connections to new generating sites. In parts of the country, the absence of transmission capacity has been a major constraint on the development of new renewable schemes and there are major schemes proposed or underway to connect renewables, both on and offshore, and new nuclear sites to the National Grid.

22.3.2 The most significant and intrusive activity and the most substantial compensation claims usually arise on the construction of 400 kv and 132 kv overhead transmission lines. The former are usually erected by the National Grid and the latter, which would previously have been maintained by area electricity boards, are now the remit of local distribution companies. As elsewhere in infrastructure provision the position has been complicated by:

- the demands of transparency and consultation, particularly where a scheme is classed as Nationally Significant Infrastructure which means that many more routes and variations will be published and consulted on than was previously the case. That leads to a greater number of landowners facing the anxiety of being potentially affected and greater demands for access for survey and feasibility work.
- the increasing use of third-party contractors to deliver schemes under hybrid arrangements which create a commercial incentive for them to reduce compensation payable to affected landowners as far as possible.

22.3.3 As explored in the previous chapter, this latter issue has been noted as a particular challenge by claimants' valuers who complain that these contracting companies are rather too careless over their activities with the consequent impact on private property rights, lacking the general appreciation of how to work with landowners and the principles of compensation now seen to have been previously demonstrated by utilities.

22.3.4 This is perhaps indicative of a wider problem where the delivery of infrastructure for broader society benefit is seen to trump individual property rights with concerned landowners perhaps too easily perceived as luddite obstacles to progress. There are, however, two sides to every story and this characterisation of the utility side as less virtuous would be regarded with some scepticism by those agents acting for utility companies who have been physically threatened by objectors to new powerlines for example.

22.3.5 Notwithstanding this changing and, unfortunately, generally less consensual environment, the underlying principles for those involved in electricity infrastructure work have not changed from previous editions of this book. Claims may arise from temporary losses during the construction work and long-term impacts on the freehold and, in some cases, leasehold interest in the property affected.

22.3.6 Losses arising as a result of constructional work – The claim is likely to include crop losses, both current and future, damage to soil structure through the traffic of heavy plant and occasionally the storage of heavy materials and damage to farm infrastructure, including water supplies and drainage. In addition, there may be occasions where there are losses in animal production or increased costs due to enforced changes in grazing or management.

22.3.7 The claim will also often include timber cut down and trees lopped, whether amenity trees or commercial forestry; one of the trickier claims to substantiate may be subsequent loss in timber due to windblow as explored later on.

22.3.8 In addition to the physical losses there may well be associated losses in support payments and grant income. Any lengthy interruption in occupation, and some electricity companies and their contractors now leave scaffolding in place for far longer than the actual required period of working to enable easy return access, is likely to impact on entitlement to Basic Payment or a subsequent post-CAP replacement. Less common, but perhaps more likely in the future, is the risk that losses in habitat as a result of construction works may adversely affect the occupier's ability to qualify for higher tier environmental support in the future.

22.3.9 The construction of the claim is likely to follow that in respect of the laying of a water main, again electricity schemes may well require site compounds or storage dumps to facilitate the scheme which are likely to fall outside of the rights provided either by the relevant compulsory code for a new scheme or the wayleaves allowing access for repair or replacement works. As such the licence fee for the facility will be a matter of commercial negotiation between the parties. One recent development is the increasing use of mobile WCs, commonly referred to as "welfare facilities", on sites. Whether or not a licence fee may be chargeable for the facility may turn on whether it is sited in the wayleave area or outside.

22.3.10 Damage to the freehold – As previous editions have rightly noted, electricity lines are unsightly, mar the landscape, interfere with views from farmhouses and cottages, interfere with irrigation systems and sporting rights and generally depreciate the value of a freehold. In addition, the apparatus is getting bigger, particularly when connecting major new generation sources into the grid. For example, the proposed transmission line towers linking Hinkley C Nuclear Power Station in Bridgwater Bay to the National Grid at Avonmouth at 150 feet are more than 60 feet taller than those of the previous generation.

22.3.11 Whilst lines cause all these problems, compensation can be claimed where the claimant enters into a deed of grant, granting a perpetual easement for the line in return for compensation.

22.3.12 In those circumstances the claim should recognise, as appropriate:
- the easement itself
- the impact on the farmhouse and other dwellings
- any impact on the utility of the farm, including day to day issues such as irrigation and cultivations and longer-term challenges such as the siting of buildings
- any impact on woodland or sporting rights

22.3.13 Basis of claim – the approach to such claims was explored in two cases in the early 1980s: *George* and *Clouds Estate Trustees*.

22.3.14 The Lands Tribunal decision in the first case, *George v South Western Electricity Board*, established that a claim could be constructed either:
- on an elemental approach (as with the pipeline examples mentioned earlier); or
- by taking an overall view of the diminution in the value of the property as a consequence of the line, reflected by a percentage depreciation of the agreed value of the holding. In that case, the Tribunal awarded 5% of the value of the holding as compensation.

22.3.15 Which approach is more appropriate will turn on the individual circumstances, but the potential should be assessed using both methods in the first instance.

22.3.16 The second Lands Tribunal case, *Clouds Estate Trustees v Southern Electricity Board*, addressed the question of the annual "rental" payments agreed between the electricity industry and the NFU and CLA. The decision established that these did not make any allowance for such issues as loss of visual amenity, nuisance value and damage to shooting which, possibly with other relevant matters, could be the subject of an additional claim. It is important to take care to include all these additional claims when a permanent easement is being negotiated. One fears for those clients who are not advised, taking the imprimatur of an industry-wide agreement as indicating all to which they are entitled. In any event the NFU and CLA rates can only be a guide with individual claims turning on the specific circumstances of the case.

22.3.17 Potential heads of claim – With those two cases in mind, there are a range of additional items which might form part of a claim, depending on the nature of the property and the scale and location of the apparatus.

22.3.18 The easement payment, routinely agreed as an annual payment between the electricity industry and the NFU and CLA, is capitalised by multiplying it by 20 years' purchase which is increased by 50% where there is a second line and it is a small field. This is normally at the arable rate of payment except where the land concerned is not arable land and cannot physically be ploughed.

22.3.19 Where dwellings are involved, the impact on a dwelling is reflected by assessing the adverse impact on value, usually expressed as a percentage of the Market Value of the property. The difficulty is in assessing the percentage deduction in what is, by definition, a subjective exercise. As with all other valuations, comparable settlements will be important evidence, but the critical issue is the nature of the property and its outlook rather than the scale or proximity of the apparatus. Thus, an overhead line erected in a busy urban fringe environment with roads, railways or other infrastructure in view may have less of an impact on the attractiveness of the farmhouse to a purchaser than the same line erected in a remote location with limited other constructions in the landscape.

22.3.20 The view from the house is regarded as the key impact, but there may also be a diminution in value from the simple physical proximity of apparatus. Pylons located close to the house or the yard may have a limited impact on the view from inside looking out simply because of their location and the position of the windows, but the apparatus may have a very marked impact on the view of the house from the drive which, in turn, could have a very significant impact on the marketability and value.

22.3.21 Pylons and lines may have an impact on farming activity. One of the more obvious limitations is that irrigation cannot be practised under overhead transmission lines for fairly obvious reasons. A relative minority of farmers irrigate or indeed have the need to; however, where they do, that is normally only for high value crops and consequently the interference and loss would be considerable. Mindful of the impact of climate change there is every prospect that more irrigation will be required in those areas where it is practised.

22.3.22 Where such a claim is involved, this is likely to involve an estimate of the average annual net loss over the area affected capitalised by such a yield as would be used for valuing the farm. This is often a contentious area of claim with alternative arguments being that irrigation may not be required every year or that the claim should be based on the capitalised loss in rental value. However, the latter approach does not fully reflect the investment in both reservoir and irrigation equipment which, in the worst affected cases, may now be superfluous and hence wasted. The difficulty may be exacerbated where the farmer is already committed to various supply contracts which can no longer be met.

22.3.23 In some circumstances the orientation of pylons and transmission lines may inhibit access to some parts of the land, particularly the case where a pylon is sited close to a field corner with the scale of cultivation and harvesting equipment on broadacre arable farms now such that considerable room is required to enable efficient access. That may ultimately see some land lost to production.

22.3.24 Sporting rights may be seriously impacted, whether an overhead line goes through drives or coverts on a shoot or over pools on a fishing river or coarse ponds. Such claims are probably best assessed as a reduction in rental value then capitalised at the relatively low yields which apply to sporting, perhaps at 3% to 4% yield. In the case of shooting, new lines may change the overall approach to the shoot and in the most affected cases a claim might also include for the relocation of breeding and release pens to suit the new configuration of the shoot. There may be very significant differences in the level of claim between lines impacting rights exercised by serious commercial shoots and those exercised by the local farmers' syndicate, although in practice the value of the right should be based on more than just the identity of the occupier. Where coarse fishing ponds have been developed as a diversification and let on day-tickets, a loss of profits approach may be appropriate.

22.3.25 A line passing through a wood will permanently sterilise that area as trees cannot be grown on the route of the line or on close proximity thereto. This might make for a corridor of 20 metres plus width. It may also increase operational costs in the future, particularly where the line goes across, rather than up and down, a sloping site and is therefore directly in the fall line when it comes to harvesting. A claim should include total losses for the future, essentially the loss in production from those areas given that harvest costs will still be incurred on the rest of the compartment, plus additional future costs consequent on the line. There is also likely to be increased risk of windblow to the surrounding area from disturbance of the canopy.

22.3.26 Aside from the woodland, and associated sporting loss, the construction of overhead lines may have an impact on the increasingly common use of woodland for leisure activity, whether formal or informal. As the extreme case, an unfortunately routed line could require radical changes to an overhead "treetop adventure" course or a mountain bike course with very substantial financial losses for both operator and landowner. Less directly intrusive lines might still see a downturn in visitor numbers to a woodland, or indeed a farm visitor attraction, with consequential reductions in entry fees, where charged, and adverse effects on receipts in associated cafes or farm shops. Glamping and high voltage lines probably do not mix.

22.3.27 A new line may also impose future limits on the siting of buildings which cannot be constructed under overhead lines or constrain future development. Assessing and settling these claims can be problematic given that there is likely to be a degree of speculation over the nature and prospect of any development.

22.3.28 Where there is a substantive claim, this may be assessed either using a residual method, applying the appropriate discount for uncertainty and timing, or by assessing the hope value. In some cases, the prospect of impact on future development may be eased by a "lift and shift" clause in any consent where smaller lines are involved. However, this is not always the easy solution it may seem, particularly where undergrounding may be the most practical solution for the relocated cable. The added conductivity of the soil means that the area of land sterilised when the cable is buried is significantly greater than the airspace previously occupied by the line and the consequent impact, in terms of lost plots on a residential development site for example, buried may be far greater than might be imagined. Landowners and advisors should seek detailed design advice from the relevant distribution company before triggering any clause.

22.3.29 However the claim is constructed, solicitors' and valuers' fees and interest at the statutory rate (which sadly at the time of writing is 0%) should be claimed. The damage claim should also include payment for the claimant's time, which is usually considerable, spent dealing with contractors and managing the farm and business around them. As a practical point, as with water schemes, some valuers make a point of giving their clients a diary to record their time involved in dealing with the claim.

22.3.30 Necessary wayleaves – The issue of oversailing power lines and depreciation in the value of property arises both for new construction and in necessary wayleave cases under the Electricity Act 1989, subsequently amended by the twin Necessary Wayleave regulations introduced in October 2013. Necessary Wayleave cases arise where there is no permanent easement and a landowner serves notice on the electricity company to remove apparatus. This this may be a consequence of proposed development or simply where a wayleave lapses due to the passage of time. Where a landowner serves such a Notice, the electricity company can apply for a necessary wayleave. If that is granted, matters of compensation, if not agreed, are, for England and Wales, now referred to the Upper Tribunal (Lands Chamber). An earlier case referred to the Lands Tribunal, *Naylor v Southern Electricity Board*, awarded compensation of £4,000 for the depreciation in the value of a house due to oversailing power lines.

22.3.31 A more significant sum was awarded in the Court of Appeal case of *Arnold White Estates Ltd v National Grid Electricity Transmission Plc*. In this case, an appeal from an award of compensation by the Upper Tribunal (Lands Chamber), the Respondent had entered into a contract to sell some land for residential development which was conditional on the removal of an overhead power line. The Respondent served notice of removal on National Grid which applied for and secured a Statutory Wayleave so that the line did not need to be removed and the matter of compensation was referred to the Tribunal. It awarded compensation of £5,829,476 based on the difference in value of the land with and without the line. National Grid appealed arguing that the loss incurred was purely a personal contractual loss outside of the provisions of statute. The Court of Appeal, however, considering the specific provisions of Schedule 4 of the Electricity Act 1989, concluded that the Tribunal had interpreted the position appropriately and upheld the Tribunal's award.

Further reading

The Powers of Utility Companies in Relation to Land (CAAV, 2009)
Professional Fees in Compensation Claims (CAAV, 2nd Edition, 2020)

23 Masts and cables for communications

Note – This chapter reviews the new Electronic Communications Code, a relatively complex area of law for which decided cases are still clarifying the law and practice is still developing. It also considers the previous Code where it applies to agreements made before 28 December 2017.

23.1 Statutory position

23.1.1 Telecommunications apparatus such as phone masts, fibre optic cables, copper telephone cables and other apparatus being used for public electronic communications networks have long had their own distinctive code of law and approach, now called the Electronic Communications Code. This was overhauled by the Digital Economy Act 2017, with the new Code applying from 28 December 2017, having been inserted into the Communications Act 2003. Some issues for agreements earlier than that, such as valuation and some operators' freedoms, remain under the previous Code under the Telecommunications Act 1984.

23.1.2 This Code, applying across the whole United Kingdom, provides its own security regime for operators and is very distinct from the law for the historic utilities of water, electricity and gas, being founded on the principles of agreement and consideration (rent) based on price, not compulsory purchase requiring compensation. Indeed, it makes ancillary provision for compensation alongside its main provision for the consideration to be paid for agreement.

23.1.3 The Communications Act 2003 then has entirely separate provisions for the compulsory purchase of property.

23.2 The basics of the Electronic Communications Code

23.2.1 The Code provides rules for the relationship between site providers and operators under which operators can seek Code rights to install, operate and keep their apparatus. The agreement providing for that may then often take effect as a lease. Operators may be either those running communications networks or infrastructure providers (those in business to produce apparatus for network operators to use). These must be Code operators, approved as such by Ofcom.

23.2.2 The role of the Code is to provide a framework for resolving the possible tensions between a site provider with suitable land or a building and the public's interest in having the assured use of good communications networks, including those provided by mobile and fixed-line operators. This does not only involve the lease of a site or the agreement for a cable to cross property but also wider Code rights and supporting needs such as for access on terms to be agreed, interaction with other uses of the site and other land, the impediments it may pose for the repair or redevelopment of a building, and the limitations that may be imposed under the rules regarding non-ionising radiation.

23.2.3 The Code makes the presumption that all apparatus is on land by agreement with the site provider. In this respect, it is fundamentally different from the approach of say s.159 of the Water Industry Act in England and Wales whereby a water company can simply take rights on short notice. The justification for statutory intervention is the public's interest in access to fixed or mobile communications services. The Code's purpose is to provide operators with power to seek access to land and then defend its retention there where the apparatus achieves that end, while recognising some of the issues that arise for the site provider affected by it.

23.2.4 An agreement must be a "Code agreement" if the operator is to have the benefit of the Code rights that protect the operator. That requires it to be in writing, signed by the parties, state the period for which the agreement is granted and state what, if any, period of notice is needed for termination.

23.2.5 The Code very largely leaves it to the parties to settle their own agreements, negotiating over the terms that are important to them, including payment. The Code has required Ofcom to publish a Code of Practice for negotiation and behaviour as well as standard terms for agreements and forms of notice.

23.2.6 If the parties do not agree, the operator can refer the issue to the tribunal (the Upper Tribunal (Land Chamber) in England and Wales, the Lands Tribunal for Scotland, or – currently – the county court in Northern Ireland) to determine if it can impose Code rights and, if so, what are the terms for the agreement for that.

23.2.7 The access test requires both of two tests to be satisfied:
- that the impact on the site provider can be compensated in monetary terms
- the benefit of the additional connectivity to the public outweighs the prejudice to the site provider.

23.2.8 If that test is met, then it is for the tribunal to determine the terms of the agreement, including the consideration to be paid for it. That assessment will turn on the terms that have been settled.

23.2.9 Where a case is referred to the tribunal, it is to set such terms as it thinks appropriate, but these will always include provisions for:

Masts and cables for communications 335

- payment of consideration
- the least possible loss or damage
- how long the right is exercisable

and then must consider:

- providing for termination
- enabling the owner to require repositioning and temporary removal. This might be needed for building repairs, proposed works that could conflict with the apparatus as located, or redevelopment.

As well as these issues, parties are likely to want to see provisions over access, the operators often wanting rights to immediate access for repairs and site providers often wanting to be able to regulate it to protect their business, property and personal interests including relations with third parties such as other tenants.

23.3 Consideration (rent)

23.3.1 The Upper Tribunal has made it clear that:

> Where terms for the grant of Code rights are agreed the parties may obviously provide for the payment of consideration at whatever rate they choose, but if the matter falls to be determined by the Tribunal it must do so by applying the valuation criteria and assumptions prescribed by paragraph 24 (1)–(3).
>
> (*EE Ltd v London Borough of Islington*)

23.3.2 While a site provider and operator can agree any terms and rent they please, where the tribunal is ordering a *new agreement* for Code rights it is, in setting the consideration for the agreement, to find the market value of the agreement with a definition essentially close to the standard definitions of market value, but it is to do so on the four special assumptions made by paragraph 24 of the Code that:

- the apparatus is not being used for the purposes of an electronic communications network (now known from Tribunal decisions as the "no network" assumption and to be distinguished from the "no scheme" assumption of compulsory purchase compensation law)
- the operators' new rights to assign and qualified freedoms to upgrade or share apparatus are disregarded so far as they are not provided under the agreement
- but in all other respects, the Code rights granted are taken into account so that the apparatus can be considered to be there, so reminding parties of the real world
- there is always at least one other site available. While market value would anyway disregard ransom value, this is there to ensure that is excluded but does not require assuming more than one other site or its full suitability.

23.3.3 In summary, the new Code:
- takes a market value approach to finding the price for the specific agreement
- does not adopt a compensation for loss basis as seen for the historic utilities.

This consideration is separate from any claim for compensation under paragraphs 25 or 84 to 86 of the Code.

23.3.4 In practice, positions have become sharply polarised between operators and site providers as operators used the advent of the new Code as the occasion to demand what might only be nominal rents on the renewal of mast agreements, typically proposing the near elimination of the preceding rents. While there were few references to the courts under the old Code which had generated very limited case law, the new Code has already seen much greater reference, controversy and litigation. At the time of writing, a growing body of Tribunal decisions is beginning to develop greater understanding of what the Code means in practice for valuation and other issues, but this process is not complete.

23.3.5 The Upper Tribunal offered observations on valuation in its decision in *Cornerstone Telecommunications Infrastructure Limited v Compton Beauchamp Estates Ltd* in which the Tribunal found that it could not give the applicant operator the Code rights it sought. It noted first that:

> There is a limit to what we can usefully say without first determining the terms of the agreement. . . . It is not necessary for us to reach any conclusion on the appropriate consideration in this case, nor would it be possible to do so without resolving the disputed terms of the agreement, a number of which may influence the valuation.

23.3.6 The Tribunal then discussed the approaches that had been canvassed in the case to the statutory valuation provisions for a new agreement, observing that:
- it was not right for valuing small sites to apply values proportioned from sales of large blocks of land: "it is not consistent with the evidence we received . . . of open market agreements relating to small rural sites let for non-telecommunications purposes". Further, that might not reflect the reservations that a landowner might have about having a Code operator with its needs as a tenant.
- there were then difficulties in applying an investment yield derived from agricultural lettings to this situation, especially without equivalent rent review provisions
- if working from evidence of previous lettings, the need to justify how any adjustment is made to their values for the new statutory assumptions is made
- while evidence from agreed values for new Code lettings might be "persuasive" as market evidence, this had so far lacked detail such as

"whether there was any particular pressure on the operator in a particular location, nor whether the parties took into account the fact that the alternative to a negotiated agreement might be a Tribunal determination involving the 'no network' assumption . . . or if a coherent basis for adjustment could be suggested"
- evidence from "transactions in respect of similar rights for non-telecommunications purposes – weather stations, air-traffic control stations and the like . . . has the advantage that it does not require adjustment to reflect the no-network assumption. It might therefore also be useful. Its value is likely to increase if it can be shown that the reference land may realistically be of interest to those types of user. The prospect of planning permission being forthcoming may also be a relevant consideration."

23.3.7 The Tribunal concluded with the perennial advice that:

As in any valuation exercise, it is necessary to stand back and look at the product of the valuation method which has been adopted; if it seems improbably high or low, as [the operator's expert's] figures seem to us, it is necessary to ask critical questions.

23.3.8 The approach to pre-existing agreements is considered later on.

23.4 Code rights

23.4.1 Assignment, upgrading and sharing – Once the new Code agreement is in place, the operator will:
- be free to assign the agreement to another Code operator. Any provision limiting that or requiring payment is made void.
- have the qualified freedom to upgrade or share apparatus so long as this does not:
  - impose an additional burden on the site provider
  - have a more than minimal adverse visual impact.

These rights do not apply to agreements made before the new Code took effect.

23.4.2 Renewal and termination – For England and Wales, the Upper Tribunal has held in *CTIL v Ashloch Ltd* that where an operator seeks renewal of a Code agreement from before 28 December 2017 and that agreement has the protection of Part 2 of the Landlord and Tenant Act 1954 (and, by extension, the equivalent Business Tenancies Order in Northern Ireland), then renewal should be pursued under that legislation. The interaction of the new Code's valuation provisions for consideration with the rent provisions on renewal under business tenancy legislation has been first tested in *Vodafone v Hanover Capital* which distinguished between them with the renewed rent being set on the basis of operators competing for the site in the market place.

23.4.3 In Scotland, a pre-existing agreement will continue under tacit relocation, appearing to require the operator to terminate that agreement if seeking a renewed one on different terms.

23.4.4 Otherwise and where a new Code agreement's primary purpose is to grant Code rights, it will fall outside the right to renew provisions in England and Wales of the Landlord and Tenant Act 1954 and the similar provisions in Northern Ireland. Instead, its renewal or termination will be under the rules of the new Code.

23.4.5 The core mechanism allows the site provider to serve 18-months' notice of termination to expire on or after the end of the present agreement. The operator then has three months in which to serve a counter notice and then three months in which the operator is able to refer the issue to the tribunal. These formal procedures provide a framework, if required, for continuing negotiations.

23.4.6 If the matter does reach the tribunal, it is to consider whether the access test (see 23.2.,7) is failed. The site provider's notice for termination must be founded on a breach of the agreement, non-payment of rent or redevelopment for the tribunal to consider it.

23.4.7 There are further provisions for requiring the removal of apparatus, but at each point the operator can invoke Code rights if it wants them.

23.5 Assessment of compensation

23.5.1 Independently of consideration (rent), compensation can be claimed under the new Code:
- by a site provider, provided that the agreement has been imposed by the Tribunal, under paragraph 25 for loss or damage from the exercise of the Code rights granted, whether at the time or later. That payment can be set by the Tribunal or referred to arbitration.
- by a site provider where apparatus is removed under paragraph 44 of the Code.

23.5.2 Under paragraph 84, this can include but is not limited to payment for:
- expenses (including reasonable legal and valuation expenses)
- diminution in the value of the land (assessed using rules 2 to 4 of the six rules for compulsory purchase compensation)
- costs of reinstatement.

23.5.3 Neighbours can make claims for injurious affection under paragraph 85.

23.6 Further developments?

23.6.1 Further cases are coming forward at the time of writing while appeals to higher courts are awaited in both *Compton Beauchamp* and *Ashloch*.

23.6.2 With the economic imperatives seen by government for high quality communications and changing needs of new technologies (such as the delivery of 5G), further legislation is likely in this area over both:
- development control regarding Code infrastructure, already benefiting from substantial permitted development rights which vary between parts of the United Kingdom
- the installation of apparatus, with England looking to legislate to ease the legal processes for the installation of cables within blocks of let flats but with possible wider implications where property is let.

There may even, in time, be further provisions for the operation of the Code or amendments to it while Tribunal and court decisions may continue to develop understanding of the Code.

Part 4
Agricultural tenancies

24 Introduction to agricultural tenancies

24.1 Throughout Great Britain, but not in Northern Ireland, tenancies of agricultural units ("holdings") that are used for trade or business are governed by specific legislation, reflecting the distinctive issues of farming and its history. In particular, most farming sees the tenant necessarily investing in the landlord's land to produce output, whether crops or grass and so livestock. That has been seen to require a framework to support that investment and land management and so the legislation started with laws protecting the tenant's rights in fixtures and rights to compensation for improved fertility. The legislative framework has then evolved in response to changing pressures and concerns, generally becoming more prescriptive until the 1980s and much less so since with corresponding consequences for landlords' willingness to let new land.

24.2 The main legislation is:
- in England and Wales:
  - the Agricultural Holdings Act 1986, consolidating earlier law and essentially for tenancies with origins before September 1995, with greater security of tenure and more prescriptive rules
  - the Agricultural Tenancies Act 1995, with Farm Business Tenancies, essentially governing tenancies granted from September 1995.

Enacted for England and Wales, devolution now means that Wales can and has varied such legislation, not always in step with England, creating issues especially for practitioners in the Welsh marches likely to be working with both regimes.

- in Scotland:
  - the Agricultural Holdings (Scotland) Act 1991, consolidating earlier law and providing long-term security of tenure but rarely used in recent decades for new lettings
  - the Agricultural Holdings (Scotland) Act 2003 creating the shorter Limited Duration Tenancies (LDTs) and Short Limited Duration Tenancies (SLDTs) as well as amending the 1991 Act
  - the Land Reform (Scotland) Act 2016, amending both the earlier Acts, replacing Limited Durations Tenancies with Modern Limited Duration Tenancies (MLDTs), and other measures not all of which were yet implemented at the time of writing.

24.3 In conjunction with the tenancy agreement, such legislation defines the asset that is the tenancy and so important for valuation issues not only of the tenancy itself and so for the landlord's reversion but also for transactions between landlord and tenant, whether:

- voluntary, as for a negotiated surrender or the tenant buying out the landlord or
- statutory, as in Scotland under the 2003 Act's right for a tenant to pre-empt a landlord's sale of the reversion and the relinquishment and assignation provisions proposed under the 2016 Act

with interactions then for compulsory purchase, divorce and business partnership valuations, tax and other matters (see Chapter 32).

24.4 More particular points for the valuer lie in:

- the rules for rent reviews – essentially rental valuations with much associated procedure (see Chapters 26 to 28)
- compensation between the parties for improvements and dilapidations to the holding by the tenant (see Chapters 29 and 31)
- compensation where the tenancy is terminated or otherwise intruded on through no fault of the tenant (see Chapter 30).

Other matters may arise during the tenancy (see Chapter 25) or under the tenancy agreement itself.

Further reading

Schedule of Time Limits (CAAV, 2006)
Surrender and Regrant of Agricultural Tenancies: A Review of Issues (CAAV, 2010)

25 Issues during a tenancy requiring valuation

25.1 Introduction

While the tenant has the full use of the let property subject only to the matters reserved by the landlord and obligations imposed in the tenancy agreement, both often considerable, and the prime landlord's obligation is to allow the tenant quiet enjoyment of the holding, a number of issues requiring valuation can arise during a tenancy.

25.2 Exercise of landlord's reservations and covenants

Tenancy agreements will typically require the landlord to compensate the tenant for damage caused in the exercise of reservations under the agreement, as where felled trees are extracted over farmland or intrusive surveys ahead of potential development are conducted causing loss of crop or other disruption.

25.3 Landlord's improvements to the holding

25.3.1 A landlord may agree to make an improvement, such as building or drainage, to improve the holding or meet a legal requirement, such as over slurry storage. The 1986 Act tenant has the distinctive power under s.11 to seek to require the landlord to make an improvement needed to meet statutory requirements for a reasonable farming activity.

25.3.2 Such a work is likely to improve the value of the holding and, in all parts of the United Kingdom, that may be reflected in an agreement between the landlord and tenant under which the tenant gives the landlord some benefit for that, often as a specific charge or an extra rent. That is a commercial agreement that needs to suit both parties for it to proceed.

25.3.3 For England and Wales, s.13 of the 1986 Act then provides that in the absence of an agreement, the landlord can serve a notice on the tenant under which the rent is increased by the rental benefit of the improvement after allowing for any element that was grant funded.

25.4 Game damage

25.4.1 A tenancy agreement will typically reserve game (pheasants, partridge, grouse and ptarmigan) and sporting more generally to the landlord who may then exercise the sporting rights directly or sometimes let them to a third party sporting tenant. In some situations, a third party may hold the sporting rights independently of the landlord.

25.4.2 The tenant nonetheless has some statutory rights in respect of rabbits and hares under the Ground Game Act 1880 applying throughout the United Kingdom, with further but limited rights on moorlands in England and Wales under the Ground Game (Amendment) Act 1906.

25.4.3 As the preservation and management of game can have a substantial and detrimental impact on both arable crops and grass, s.20 of the Agricultural Holdings Act 1986 provides a basis for claims by tenants in England and Wales for damage by any wild animals or birds that the tenant is not able to control under the tenancy agreement.

25.4.4 Where the tenant does not have the power to take and kill game, a claim may be made against the landlord for damage by that game. S.20(5) then entitles the landlord to be indemnified for that payment by a third party holding sporting rights.

25.4.5 The tenant must give written notice to the landlord within a month of becoming aware of the damage and give the landlord a reasonable opportunity to inspect it before the crop is harvested or, if harvested, removed from the land.

25.4.6 To make a claim, the tenant must then serve a further notice on the landlord specifying the claim within one month after the end of the tenancy year in question; by default that will be between 29 September and 28 October unless a different date has been agreed. That will usually build in a delay of up to a year and so requires good records, photographic and written, to provide good evidence.

25.4.7 The claim will ordinarily be for loss of crop requiring an assessment of the lost yield and its value so far as it can be shown to arise from game damage and not other causes. There might be further losses where feed or other produce has had to be bought in instead. This is all a matter for proof.

25.4.8 If the compensation is not agreed, it is to be determined by arbitration or by a third party jointly appointed by landed and tenant.

25.5 Damages and dilapidations

As shown by the decision in *Crewe v Silk* concerning damages for disrepair, it is possible for a landlord to pursue a claim for dilapidations or damages during the course of a tenancy. For consideration of the valuation issues, see Chapter 31.

25.6 Changes to repairing obligations

25.6.1 Where the repairing obligations of the parties are adjusted, this may be reflected by compensation or a change in the rent.

25.6.2 Under the 1986 Act for England and Wales, this is statutorily provided for when the Model Clauses are changed with compensation where liability is transferred for an item of fixed equipment that has not been properly maintained but the month allowed as the time window for this is tight.

25.6.3 Where there is a determination under s.8 revising repairing clauses that differ from the model clauses, that determination can also vary the rent. Compensation can also be due where an item transferred has not been properly maintained.

25.6.4 The Scottish 1991 Act makes provision under s.46 for compensation to be payable where liability for an item of fixed equipment for any failure to maintain it by the party from which it is being transferred. Again, this requires the service of a notice within a month of the change.

25.6.5 Compensation in such cases would typically be for cost of putting the item into good order.

Further reading

The Model Clauses for Agricultural Tenancies in England and Wales (CAAV, 2019)

26 Rent reviews under the Agricultural Holdings Act 1986

26.1 Introduction

26.1.1 When the law in 1947 imposed extended, potentially lifetime, security of tenure for both existing and new tenancies under what became the Agricultural Holdings Act 1986, it made provision for rent reviews to allow for changing economic circumstances.

26.1.2 While the landlord and tenant of what is now a 1986 Act tenancy can mutually agree at any time to change the rent, the present legislation for this is set out in s.12 and Schedule 2 of the Act and provides for:
- a minimum three-year period between enforceable reviews
- a mechanism using an initial notice to trigger a potentially enforceable rent review under s.12 (the "s.12 notice")
- a dispute resolution mechanism, generally using on arbitration but with alternative means for the final and binding use of a third party, such as an expert, added in 2015
- a basis for the assessment of the rent at dispute resolution (but thereby offering potential guidance to the parties), last significantly revised in 1984 and set out in Schedule 2.

The procedure

26.2 Serving a s.12 notice: what and when?

26.2.1 What? – This notice is the trigger without which no rent review can be unilaterally taken to arbitration to achieve an answer. Without that recourse, the rent could only be changed by agreement. Either party may serve the notice but, once served, the other can rely on it. It cannot be unilaterally withdrawn. However, it does not require that a review actually happen.

26.2.2 The s.12 notice must identify the parties to the tenancy, identify the holding and demand that the rent properly payable be referred to arbitration as at the next termination date. That formal wording can make it sensible to write a covering letter explaining matters.

26.2.3 With what period of notice? – The s.12 notice has effect for the "next termination date" after it is served. This is calculated on the same basis as when a notice to quit would ordinarily take effect if served on the same day. That means the review date will between 12 and 24 months after the notice is served.

26.2.4 When? – The s.12 notice will only be valid if:
- the review it triggers would not be less than three years from the last rent change (as defined)
- is effective for the "next termination date"
- is served correctly.

26.2.5 Schedule 2 paragraph 4(1) requires a minimum of three years between rent reviews. That period starts from the most recent of:
- the date when the tenancy commenced
- the last increase or reduction in the rent for the whole (an agreed standstill does not count)
- any previous direction by an arbitrator, including one that the rent remain unchanged.

Once that minimum period has passed, a review can be considered for any subsequent year until a new change re-triggers the cycle.

26.2.6 As the rent of a holding may have changed for a variety of reasons as well as by a review, Schedule 2 provides seven exceptions where a variation in rent does not re-trigger the rent review cycle. These are where:
- the rent has been varied by an arbitrator when either recording the terms of the tenancy in writing or reviewing the repairing and other provisions against the "Model Clauses"
- the rent has been varied to reflect a new landlord's improvement and that variation was agreed before or within six months of the completion of the improvement.
- a resumption of part of the holding by notice to quit is accompanied by a proportional reduction in rent (however, an agreed reduction on a surrender of land will still re-trigger the cycle)
- a change in the amount of rent because of changes in VAT or the implementation or revocation of an option to tax
- the landlord's interest in the holding has been divided and the rent then appropriately apportioned between the new landlords
- a boundary adjustment or variation in the terms of the tenancy see a change in the rent
- a surrender and regrant of the tenancy occurs and the rent is unchanged (so that this event is not treated here as a commencement).

The Act's Schedule 2 and case law provide more detail on these occasions.

26.2.7 How? – S.93 of the Act sets out the rules for proper service with the notice to be delivered to the recipient, left at their address or posted to their address using a registered letter or recorded delivery service. It can be served on the recipient's agent if they are responsible for the management or farming. Where there are joint landlords or tenants it is best practice to serve notice on both. It is sensible to ask for acknowledgement of receipt of the notice or have other proof, lest this be argued later.

26.3 Negotiating the rent

26.3.1 To take a rent review forward, the party wanting a change in the rent should open discussions with the other, usually with one or both parties being represented by a valuer. Those discussions might commonly revolve around the current fortunes of the relevant sectors of agriculture, issues concerning the holding, and knowledge of the rental market. These discussions can provide a good opportunity to consider any other issues about the tenancy at the same time, such as investment or succession. Especially if there are significant issues between the parties, it is sensible to bear in mind the statutory basis on which the rent would be settled by an arbitrator or expert.

26.3.2 Sufficient time should be allowed for facts to be gathered, views assembled, discussions and negotiations to be had and considered with advice and reflection and then, as may be necessary, adjusted for a settlement to be reached. During that process, one or both parties may need to accept a new understanding of the relevant issues and current circumstances: that can take time.

26.3.3 Save in the simplest of cases, those negotiations are likely, according to the case and circumstances, to involve meetings, an inspection of the holding (or key parts of it) and correspondence. Open letters may record the writer's then position on the rent (and possibly other matters) which may then be available as evidence to the arbitrator. Where a letter offering or exploring an offer for possible settlement is expressed to be "without prejudice" then it is privileged until any final settlement is reached. That means that it cannot be used later as evidence unless both parties agree.

26.3.4 **Recording the result** – The overwhelming majority of rent reviews are settled by agreement. The result should be recorded with any other points arising in a memorandum between the parties, signed by both and kept with all other documentation for the tenancy.

26.4 Recourse to dispute resolution (see also Chapter 35)

26.4.1 As the s.12 notice will only be effective for a review as at a particular date, typically one of the conventional term dates, any party wanting a review should consider their position ahead of that date if the negotiations have not been concluded. The options are:
- to consider if negotiations may yet be settled
- to let the matter lapse
- to serve a further s.12 notice for review 12 months later

- to agree with the other party on the appointment of an agreed person as either:
  - an arbitrator whose decision would be final and binding for the intended review date
  - a third party to act, say, as an expert but whose decision would be final and binding

If either of these are done, the person must be appointed before the review date.

- to apply before the review date to an appointing professional authority (any one of the President of the CAAV, the Chair of the ALA or the President of the RICS) for the appointment of an arbitrator. This can be done by just one party, whether or not they served the original s.12 notice. Provided the application is made in time, it does not matter if the appointment is made later.

26.4.2 Even where an arbitrator or expert has been appointed or an application made, the parties can continue their negotiations and settle the matter by agreement without the cost of dispute resolution. Very often, an appointment is requested as a means to preserve the power to negotiate since without that appointment the original s.12 notice would have no force after the review date.

26.4.3 The arbitrator or expert can only decide the rent and not any other issue that may have been involved in the negotiation.

26.4.4 Especially where there are such other issues, mediation (which can be seen as a means of enhanced negotiation) may offer a way to an answer that is agreed between the parties, possibly resolving deeper issues between them than just the rent.

26.4.5 The parties to a rent negotiation might also find assistance in using early neutral evaluation, asking a respected third party for a view on the rent or a key issue between them to give both parties a sense of perspective on the issue and the outcome before they become seriously committed to a dispute process.

26.5 The valuation of the rent

26.5.1 Schedule 2 prescribes the basis on which the rent is to be assessed, what must be considered and what is to be disregarded when the rent is to be determined by an arbitrator or other third party, such as an expert.

26.5.2 The rent properly payable for the holding is defined by paragraph 1 of Schedule 2 as

> the rent at which the holding might reasonably be expected to be let by a prudent and willing landlord to a prudent and willing tenant, taking into account (subject to sub-paragraph (3) and paragraphs 2 and 3 below) all relevant factors, including (in every case) the terms of the

tenancy (including those relating to rent), the character and situation of the holding (including the locality in which it is situated), the productive capacity of the holding and its related earning capacity, and the current level of rents for comparable lettings, as determined in accordance with sub-paragraph (3) below.

26.5.3 Described by the Court of Appeal in *Childers v Anker* as a self-contained provision ("a complete statutory code... [to] be applied without addition, or, I add, subtraction"), this definition has been the subject of close consideration and a number of judicial decisions.

26.5.4 While attention is often directed as the later, specific parts of that definition, its key but sometimes overlooked core is

> the rent at which the holding might reasonably be expected to be let by a prudent and willing landlord to a prudent and willing tenant, taking into account (subject to sub-paragraph (3) and paragraphs 2 and 3 below) all relevant factors.

That is now analysed taking each part in turn.

26.5.5 "the holding" – It is the holding as a whole that is being assessed with its physical nature and the legal definition given by its tenancy agreement. It is a single unit for valuation. While approaches, especially when working from comparables, based on analysing its parts can offer assistance, they are not let as separate freestanding parts but are parts of the whole that is to be valued. That whole also offers the latent value of the opportunities that a tenant can unlock within the terms of the tenancy agreement; this is discussed later on.

26.5.6 "at which the holding might reasonably be expected to be let" – This is the reasonable expectation of the person making the decision, the arbitrator or expert. What would someone at the time with knowledge of the holding and the circumstances reasonably expect it to be let for? It is not what one or other actual party might think is a reasonable figure for them. It implies a view of the future.

26.5.7 "by a prudent and willing landlord to a prudent and willing tenant" – This assumes a hypothetical landlord willing to offer the tenancy and a hypothetical tenant to take it when forming that reasonable expectation. This does not consider the actual parties with their characteristics, characters, histories, motives, resources and limitations. A conventional market rent definition would refer to such parties as willing, meaning that they are willing to enter into the letting but, with the tension between them, not keen beyond reason. This definition also requires them both to be prudent, with the tenant perhaps not venturing more than is necessary to achieve the letting on a sustainable basis and the landlord perhaps seeking a sustainable return in that role.

26.5.8 "taking into account . . . all relevant factors" – This phrase allows consideration of *any* factor that is relevant save where it is excluded or qualified by the paragraphs of the Schedule mentioned. This is important not only because of limitations in the points that are then required by Schedule 2 but because so much has changed in the farming world since Schedule 2 was enacted in 1984, with the evolution of subsidies, the rise of cottage lettings and other business activities, agri-environment agreements and other factors. Save as excluded by Schedule 2, anything relevant can be considered, not just the four statutory factors usually discussed and reviewed later on.

26.5.9 Exclusion – "subject to paragraph . . . 3" – This requires the review to disregard two key factors about the tenant:

- that the tenant is in occupation of the holding. While the actual tenant might be among the potential bidders, the hypothetical tenant is not in occupation and so personal points that might flow from that such as a contract are ignored while it cannot be taken that the tenant will pay more to stay in situ. As an example, it would be usual to disregard access to the Young Farmer Top-Up element of the Basic Payment regime as that turns on the age of the individual tenant. However, if it becomes clear this was a general attribute of tenants who might bid for the holding, then it might come to be considered for such effect as it might have. The tenant's Basic Payment entitlements would be disregarded as they are not part of the land, but entitlements let with the holding would be considered.
- any dilapidation that the tenant has caused. The rent is to be assessed as though the tenant has fulfilled his maintenance obligations under the tenancy agreement with no discount for poor maintenance. However, if the landlord has not fulfilled his responsibilities, then those deficiencies can be reflected in a lower rent.

26.5.10 Exclusion – "subject to paragraph 2" – improvements and grant aid – This requires the disregard of the benefit of tenant's improvements. The tenant is not ordinarily to pay rent to the landlord on his own investment – nor, to the extent that they were grant aided, landlord's improvements. These provisions can be important and are considered later on.

26.5.11 Exclusion – "subject to sub-paragraph (3)"- influences on comparables – This exclusion is considered later on in the discussion of the treatment of evidence from comparable holdings.

26.5.12 That opening part of the core definition of the basis for rent review establishes the property that the Act says is to be valued and requires that rent is to be what could be expected to be agreed for at that time between two prudent and willing parties. Schedule 2 then directs attention to four statutory factors which are now discussed, followed by a return to "all relevant factors".

26.6 The four statutory factors

26.6.1 Within the overall duty to consider all relevant factors, the arbitrator or expert is required by Schedule 2 to have regard in every case to four particular aspects:
- the terms of the tenancy (including those relating to rent)
- the character and situation of the holding (including the locality in which it is situated)
- the productive capacity of the holding and its related earning capacity
- the current level of rents for comparable lettings.

While regard has to be given to each of these, they do not exclude anything else that may be a relevant factor.

26.6.2 The weight given to each of the statutory factors in an assessment is to be determined in the circumstances on the arguments and evidence put. The Act does not give any of these factors any especial or overriding importance in determining the rent, simply requiring that regard has to be had to them. The weight that may or may not then be given to each will be a matter of the circumstances, the arguments and the quality of the evidence available for each.

26.6.3 The statutory factors can be seen as falling into two classes:
- the terms of the tenancy and the character and situation of the holding are really the legal and physical nature of the asset that is the holding, defining what is to be valued with its location, opportunities, liabilities and limitations.
- the other two factors concern economic matters and so what might be earned from farming the holding and what others are paying for such holdings but does not exclude other points.

26.7 "The terms of the tenancy (including those relating to rent)"

26.7.1 The terms of the tenancy can usually be determined from the tenancy documents. The tenancy agreement should be examined, along with any other relevant documents or memoranda which vary the terms. Where there is no written tenancy, the behaviour of the parties should be identified and considered.

26.7.2 Some points in agreements may be overridden by statute, either in the operation of the agreement or under the legislation for rent review process.

26.7.3 The landlord may have given permission for activities outside the tenancy agreement. This may have been done expressly in writing or he may be found to have waived a breach of the agreement by acquiescing in it over time if it can be shown that he knew of it.

26.7.4 The effect of individual terms will vary with the nature of the holding; for a bare land holding, repairing liabilities may be less significant than for a holding with dwellings and buildings.

26.7.5 Terms that can be significant in assessing rent include:
- the obligations on the tenant to repair and maintain the holding – this may be governed by the model clauses or see the tenant carry full repairing and insuring liabilities or apply some other division
- the user clause – this may limit the tenant to a specific agricultural enterprise, such as dairying, or to agricultural uses only or allow other uses. It may impose limitations on how the business may adapt and potentially limit the rent.
- a clause requiring the tenant to reside in the farmhouse which may in turn limit the opportunity for the landlord to argue that the holding has marriage value
- prohibitions or limitations on subletting
- any obligations to make improvements
- on the very rare occasions when it is the case, the absence of a bar on assignment
- payments under associated agreements for diversification.

These will be matters for individual appraisal.

26.8 "The character and situation of the holding (including the locality in which it is situated)"

This factor concerns the physical nature of the holding with its opportunities and problems. An initial list of possible points could include:
- the size of the holding
- layout and compactness as a unit
- location (particularly in terms of access to local and if appropriate specialist markets but also restrictions on use and access to subsidies, such as the Basic Payment region),
- landscape character and attractiveness
- farm access (including local infrastructure and public access)
- the nature of buildings and other fixed equipment
- soil types
- topography and aspect
- rainfall
- drainage
- regulatory controls.

26.9 "The productive capacity of the holding and its related earning capacity"

26.9.1 This seeks first to establish what the holding can physically produce when competently farmed and then to assess the financial outcome of that production. The determination of the rent is to have regard to that, but the weight given to it will vary between cases.

26.9.2 While that financial outcome has to be considered, the Act does not say how it is to be assessed or how it is then to be taken into account. If the analysis uses gross margins and overheads, that outcome may be positive or negative but could anyway be outweighed by other factors, especially for a holding where there is a greater opportunity for income from, for example, environmentally based income.

26.9.3 While this factor is commonly approached as a process of preparing a budget, most budgets for farm businesses also include other items in addition to what is in the related earing capacity, such as subsidies, cottage rents and agri-environment agreements. It is conventional to arrive at a divisible surplus with the next step in the discussion concerning how that might be divided between landlord (rent) and tenant (return on his labour, management and finance).

26.9.4 "Productive Capacity" – The Schedule defines "productive capacity" as

> the productive capacity of the holding (taking into account fixed equipment and any other available facilities on the holding) on the assumption that it is in the occupation of a competent tenant practising a system of farming suitable to the holding.

26.9.5 This assessment will require consideration of the soil, the prevailing climate, the capacity of the buildings and other fixed equipment, drainage and other relevant matters. The disregard of tenant's improvements and grant aid to landlord's improvements considered later in this chapter is relevant.

26.9.6 With the assumption of a hypothetical competent tenant, the first task is to find the farming system suitable to the holding. This may often be close to the one followed by the actual tenant, but this should not be influenced where:
- the actual tenant is pursuing a specialist enterprise with skills, contracts or investment available to him which others might not be expected to undertake
- the actual tenant is not farming the holding to its potential.

26.9.7 This assessment may be more complicated now than used to be the case with many tenants farming more complex units, including more than the holding which is the subject of the rent review. In some circumstances, there might only be a marginal difference as where a dairy tenant rents or owns some off-lying land for youngstock rearing. In such cases, only minor adjustment, whether reducing the capacity or more likely allowing a charge for the land in the budget, may be required from the actual system, provided that it is reasonable. In other cases, the tenant may be farming a substantial area in addition to the holding and may have distinct economies of scale or marketing opportunities which would not be available to the hypothetical tenant of the holding being reviewed.

26.9.8 The output of the productive capacity assessment should be expressed as physical output, for example, in tonnes of wheat, litres of milk or numbers of lambs. This exercise can be supported by knowledge of local farming circumstances as well as information from the available costings handbooks, adjusted as appropriate. The farmer's own figures may be relevant as also may be those of the landlord if also farming in the vicinity. For local or actual figures, it would be conventional to use rolling yields, averaged over, say, three or five years to take account of seasonal distortions, since the assessment is to be relevant to the coming three years or so.

26.9.9 "Related Earning Capacity" – This is defined to mean

> the extent to which, in the light of that productive capacity, a competent tenant practising such a system of farming could reasonably be expected to profit from farming the holding.

In other words, how much money might the hypothetical competent tenant be expected to make from the physical output that was determined in the assessment of productive capacity? It is a judgment of the expectation of profit in the band of competent tenant.

26.9.10 The classic approach to this is to:
- identify the output for each enterprise under the chosen farming system
- produce a gross margin for each enterprise
- total the gross margins
- there may be enterprise specific overheads to deduct (some labour may be in this) these costs not concerning other enterprises on the holding
- deduct the fixed costs/overheads for the whole holding (that is, the costs that are not attributable to any one enterprise). As the object is to find the rent, the rent is not included in this
- arriving at a net figure.

26.9.11 The values for inputs and produce sold should be those that would reasonably be expected to be in the minds of the hypothetical tenant and landlord when settling a rent for the coming period. That may often be conditioned by past figures, though the most recent figures and trends may predominate. An average value over the last three years may not be relevant to people judging the next three.

26.9.12 In considering costs, there may be a judgment of the extent to which the tenant would use contractors for some operations.

26.9.13 The budget – While that covers what the Act requires, this is customarily developed as a larger budget for the holding, including subsidies, non-farming and agri-environment agreement income with their associated costs.

26.9.14 The outcome is the net pre-rent surplus.

26.9.15 The Act requires the arbitrator or expert to have regard to the related earning capacity in forming the reasonable expectation of the rent but gives no further direction. That is a matter for the facts of the case.

26.9.16 The practicality of that can be illustrated by the simple point of a holding with a few acres but a useful house. The outcome of the productive capacity and related earning capacity may be limited but the holding may nonetheless offer a significant residential attraction to a farming tenant, albeit within the terms of the tenancy agreement (and so see the discussion of dwellings at 26.14 later on). Such other factors may be found to outweigh the results of the budget. In other circumstances, the budget can be found to have a greater importance.

26.9.17 Customary practice is that the pre-rent surplus is divided between landlord and tenant. Illustrations often use an equal split, but the actual division is also a matter of knowledge of the marketplace and could be specific to the circumstances. In principle, the tenant might expect a higher return for systems that entail higher risk or greater working capital than for other farming systems. It may be possible to derive the appropriate split from budgetary analysis of holdings offered as comparables. If other factors such as a hypothetical tenant's finance costs are brought into account, that may only lead to a corresponding adjustment of this split as the final determination of the rent should not be an artifice of the calculation.

26.9.18 Where tenant's improvements (see later on) have not been disregarded ("black patched") in establishing the suitable farming system for this exercise, they might now commonly be valued out as a deduction from the landlord's rental share of the pre-rent surplus.

26.10 "The current level of rents for comparable lettings"

26.10.1 As well as considering what profit might be achieved by farming the holding, the arbitrator or expert is also required to have regard to the rents agreed or likely to be agreed in respect of "comparable agricultural holdings on terms, (other than terms fixing the rent payable) similar to those of the tenancy under consideration". In short, that is looking at what other people are agreeing and making allowances or adjustments where there are differences between the comparables and the subject holding, the most basic and traditional means of valuing property.

26.10.2 With its focus on comparable holdings let on similar terms, the emphasis has often been taken to be on rents for other holdings let under the 1986 Act. That would be limited to evidence from reviews or succession negotiations as no entirely new 1986 Act tenancies will have been let since 1995 – and few in the years immediately prior to that. Following the comments of the Scottish Court of Session in *Morrison-Low v Paterson*, under the relatively similar s.13 of the Agricultural Holdings (Scotland) Act 1991, that does not necessarily exclude the possibility that some FBTs may qualify as comparable holdings and may be let on similar terms with the need to make appropriate adjustments for differences. That decision with its context and reasoning is more directly applicable to the 1986 Act than the analogous reasoning in *Spatholme v Greater Manchester Rent Assessment Committee*, a residential tenancy case under the Rent

Act, in which it was held that rents for assured tenancies could be taken into account when assessing the rents for protected tenancies, providing that an adjustment was made for any identified scarcity (see later on).

26.10.3 Other rental evidence, whether from other lettings outside the definition here or of "tone of the list", may be considered under the heading of "all relevant factors".

26.10.4 Exclusion – "subject to sub-paragraph (3)" – influences on comparables – When considering evidence from the comparable holdings let on similar terms considered by this sub-paragraph, three points in that evidence are to be disregarded:
- scarcity
- marriage value
- the effect of any premium.

The object is to make the evidence of the rents for the comparable holdings comparable to the subject holding by valuing out those factors, even if they are then relevant to the rent of the subject holding when it is considered in its circumstances.

26.10.5 Scarcity – The disregard is of any element of the comparable holdings' rents that is due to

> an appreciable scarcity of comparable holdings available for letting on such terms compared with the number of persons seeking to become tenants of such holdings on such terms.

There may now be few, if any, 1986 Act comparables for which this is still relevant. It could, though, be relevant where an FBT or other letting is found to be such a comparable holding let on similar terms.

26.10.6 There has been little judicial consideration of scarcity under the 1986 Act but it may appear similar in its search for a balance in the overall market to that sought by the fair rent provisions of housing tenancy law (considered by case law such as *Metropolitan Properties v Finegold*) and the equivalent concept in s.13 of the Agricultural Holdings (Scotland) Act 1991 which has been considered several times by the Scottish Land Court and the Court of Session.

26.10.7 The basic point in decisions under those provisions is that, even in a hypothetically balanced market, an attractive holding will still see competition; the assumption of balance does not exclude its qualities from consideration. In the circumstances of the agricultural rent review cases before it, the Scottish Land Court has rarely deducted 30% for scarcity and more often figures closer to 15%.

26.10.8 Marriage value – The disregard is of any element of the rents for comparable holdings due to the fact that the tenant is the occupier of other land in the vicinity of that holding that may conveniently be occupied together with that holding. Apart from taking no account in this of land outside the vicinity of the comparable, this requires the removal of marriage value from the rent for the comparable, so making it more genuinely comparable. As affirmed by the Court of Appeal in *Childers v Anker*, the subject holding may then in its circumstances have its own marriage value opportunities which can be considered.

26.10.9 Premiums – Where the rent for the comparable holding has been affected by its tenant having paid a premium (a capital sum) to take the holding that premium is to be disregarded. This is very rarely met in agriculture.

26.10.10 Use of comparables – Once these three disregards have been applied to the comparable holdings, the valuer should then consider the physical and legal differences between the comparable holdings and the subject holding. Where these differences would affect the rent in practice the comparable rent should be adjusted accordingly. For example, differences in the tenant's fixed equipment and improvements need to be disregarded from the comparable rent.

26.10.11 The fewer adjustments or allowances that are needed in making the comparison, the better the comparable could be as evidence and so the more weight it might be given in determining the new rent.

26.10.12 Component analysis can take available comparables and consider the contribution made to the rent of each by the areas of land, perhaps by quality or type, buildings and dwellings to give unit figures to apply to the subject holding. These will not be market values but instead offer a means of comparison for the elements within entire holdings.

26.10.13 In practice, the discussion of comparables tends to move over the period of the negotiations from general references and an understanding of the "tone of list" at that time to more specific evidence from actual settlements which can be harder to obtain and can rely on the consent of the other tenant for use as formal evidence at arbitration. An expert can rely on his own knowledge of settlements as well as anything the parties put to him.

26.10.14 Where the process becomes more formal, parties are encouraged to agree the comparables to be referred to in an arbitration. Any comparables should be inspected so that the physical nature of the holding is understood access to agreements and relevant paperwork is also desirable in understanding the extent of differences between the comparable and subject holding.

26.10.15 Those factors combined with the reduced number of holdings and the greater differences between them tend to make it harder to find good comparables that can be produced as cross-examinable evidence.

Illustration of a component valuation

The rent is being reviewed for a holding is 270 acres, two thirds of which are assessed as a more useful type A land and one third of a type B land. It has a poorer than average house, a staff cottage and useful landlord's buildings. In its circumstances, it is not considered that marriage value is relevant.

There are three agreed relevant comparables with rents of £10,000, £16,000 and £21,000 set at recent reviews.

The first at £10,000 is a 200-acre bare land unit, equally split between land of the two types identified on the subject holding.

A view based on market knowledge is formed that type A contributes £6,000 to the agreed rent and type B, £4,000 – so in component terms £60/acre and £40/acre respectively. However, as bare land there might be some marriage value in that requiring an adjustment for use with equipped holdings – say that, after review of the other two comparables, these figures are brought back to £50 and £35 respectively.

The second at £16,000 has a farmhouse, two cottages (for staff) and useful landlord's buildings with 240 acres of which 140 are type A and 100 type B. Because the holding is farmed in isolation, it is assumed there is no marriage value in the rent agreed.

The land rents shown by the first comparable before adjustment for marriage value (140 acres at £60 and 100 at £405 totalling £12,400) leave £3,600 for the fixed equipment, which is probably understated at this point given the marriage value in the first holding's rent.

Adjusting the land rents to exclude that marriage value would give a total component rent for the land of £10,500 (140 acres at £50 and 100 acres at £40) and so leave £5,500 out of the agreed £16,000 for fixed equipment. That might initially suggest £2,000 for the house, £1,000 each for the cottages and £1,500 for the buildings – subject to further consideration in the light of other evidence, including the third comparable.

The third at £21,000 has a better house, a staff cottage, a sublet cottage, useful landlord's buildings and 400 acres of type B land. It is assumed there is no marriage value.

At £35/acre, the land component is £14,000, leaving £7,000 for the fixed equipment. The sublet cottage achieves a rent of £400/month; after allowing for voids and greater costs, it is considered that it contributes £1,900 to the overall rent, leaving £5,100 for the rest. The house might contribute £2,600 to that, the cottage £1,000 and the buildings again £1,500.

With a broadly consistent pattern emerging, applying the results of this component analysis to the subject holding would suggest:

– house	£1,800
– cottage	£1,000
– buildings	£1,500
– 180 acres type A	£9,000
– 90 acres type B	£3,150

giving a total rent on this basis of £16,450 to be considered against all the other available evidence and a sense check that a bidder would indeed pay a rent of this order.

The figures used for the house and cottage are their contributions to the component analysis for comparison and **not** a measure of their independent rental value, let alone related to what they might achieve if offered on their own as let dwellings.

Comment – As this illustrative example shows, this approach (drawing heavily on comparables) is very much a matter of practical appraisal, judgment and comparison, supported by an accumulation of evidence in seeking out patterns. In each case, the task is to identify the contribution each component makes to the overall rent agreed, reducing the contribution made by one simply increases those contributions made by the other components. The figures might not in practice always match as neatly as is expected of such a textbook example. Equally, it might have to be recognised that, for example, some of the type A land in one or more cases had been drained with grant aid while the tenant had paid for the useful buildings in another case. Such issues can be taken into account in this analysis. The component approach is a means of analysis to reveal the patterns of behaviour of parties at the time of the rent review.

26.11 The balance between earning capacity and comparables

26.11.1 These two statutory factors have equal standing in the wording of Schedule 2 simply in the point that an arbitrator must have regard to both, whatever weight is then determined should be given to each in the circumstances. Each will fall to be reviewed at an arbitration by the quality of the evidence and argument produced to show their relevance to the determination of the rent of the holding in question. That might mean that one or other (or neither) has particular weight in any case; more simply, each has to be considered.

26.11.2 In terms of valuation principle, comparables can offer direct evidence of rents that are being agreed and so bear more directly on finding the rent "at which the holding might reasonably be expected to be let". Neither the more limited related earning capacity of the Act nor more comprehensive budgets give evidence of behaviour as to rents. However, in a world where evidence may often be imperfect those making a case will argue it as best they can with the tools that they can find.

26.11.3 The hierarchy of evidence set out by the Scottish Court of Session in *Morrison-Low v Paterson*, applying the not dissimilar s.13 of the Scottish 1991 Act is relevant:

- preferring open-market lettings under the Act but accepting that these will not be found with any relevance to the valuation date
- then relevant evidence from other new agricultural tenancies as primary evidence of the market where they fall within the definition of a "agricultural holding" (and potentially of general relevance anyway) with appropriate adjustments made
- the "indirect, and less satisfactory," indication of the market from reviews agreed between landlords and sitting tenants under existing tenancies of comparable holdings
- finally, valuation on the basis of a farm budget "should be a method of last resort". Profitability would be a relevant consideration when

framing an offer of rent, but that offer could, according to the circumstances, be more or less than the budget might suggest while with the assurance of rent reviews:

A farm budget may provide no more than a snapshot of the position as at the valuation date. It is critically dependent on certain key assumptions and on sensitive variables, such as cereal prices. An open market bidder may take a wider view.

26.11.4 That is an agriculturally based expression of the similar judicial view in the English non-agricultural case, *Zubaida v Hargreaves*, of the descending order of weight to be given to rents assessed under different procedures:
  (i)   open market lettings as the better evidence
  (ii) agreements between parties at arm's length on a rent review or lease renewal
  (iii) determination by an independent expert
  (iv) an arbitrator's award as lowest in the hierarchy.

This is a view on the relative potential quality of the evidence where it is available.

26.11.5 It is reasonable that the best evidence as to what would be "expected" to be agreed is drawn from observation and analysis of behaviour in other analogous cases – comparables. The Act gives no direction as to what to do with the assessed "related earning capacity" (now often narrower than the real budget for the overall business) with its treatment in comparable lettings offering a means to address that by reference to actual, rather than theoretical, behaviour.

26.11.6 However, if good comparables are not available for a particular case, it may yet be that the better, if weak, evidence comes from appraising budgets on the way to considering what would be expected to be agreed with the valuer seeking out ways to sustain that final conclusion.

26.12 All relevant factors

26.12.1 While the Schedule requires the arbitrator or expert to have regard to each of the four statutory factors, that is in the broader context of taking into account all relevant factors that help inform the reasonable expectation of what rent would be agreed as at the valuation date between prudent and willing parties. Any factor may be considered as long as it is relevant. It is a matter of the circumstances as to how much or how little weight is accorded to any factor, including the statutory ones.

26.12.2 Examples of matters that might be considered include:
- access to subsidy regimes such as the Basic Payment with their payments and obligations – and then successor post-Brexit regimes

- agri-environment schemes with their opportunities for agreements with their requirements and payments. The opportunity for such an agreement was considered in *Childers v Anker* from which it appears that any actual agreement the tenant might hold is disregarded if personal to him, but the opportunities for the tenant to secure such an agreement can be considered if it would be worthwhile.
- other evidence as to rents outside the definition of those comparables that must be considered
- the opportunities for non-agricultural profit, though noting that the actual tenant may be pursuing an enterprise that a hypothetical tenant might not, but equally the market might see opportunities that the actual tenant has not developed
- the residential aspects of the holding, including the direct benefits and costs to the tenant of housing he may have, accommodation for staff and permitted sublettings
- landlord's dilapidations

26.12.3 Valuation points relevant to the holding can also be considered here, including:
- marriage value where relevant of the subject holding in its context. That flows for the character and situation of the holding and may well vary between farming localities
- its latent value.

26.13 Latent value

26.13.1 Latent value is the value for the opportunities offered by the holding that the tenant can unlock. Potential examples include:
- making an access to an otherwise isolated field as perhaps by the tenant putting a bridge across a river
- the increase in yield for draining a field
- making a cottage legally lettable by adding smoke alarms.

26.13.2 This was considered by the court in *Tummon v Barclays Trust Co Bank Ltd* where a tenant of a north Cornish coastal farm had secured planning permission and developed a field as a caravan park. While some of the value lay in his enterprise and improvements, the field with its location offered that opportunities with that latent value to be reflected in the rent. The court explained this:

> You have to leave out of account the fact that the improvement has been done by the tenant already, but take into account the possibility that the new tenant might himself carry out the improvement and so get much better value out of the farm.

26.13.3 The principle of latent value is that the landlord owns the raw material for the improvement for which hypothetical tenants might bid. That land could be the site of a tenant's building, other improvement or new use and so should be given credit for its contribution to the tenant's benefit.

26.14 Dwellings

26.14.1 While dwellings once played little part in the appraisal of the value of holdings, the role of any farmhouse and other dwellings is now more significant.

26.14.2 They are simply part of the holding as a whole and do not fall to be considered as separate entities. Standalone residential rents should not simply be used as those are for dwellings let on their own with a different pattern of obligations, lacking the larger commitment to the larger holding but usually with fewer liabilities to the dwelling.

26.14.3 The farmhouse offers the tenant accommodation which he might otherwise have to find elsewhere. Many agreements require the tenant to live in the house, whether or not it is precisely suitable to him. Many are larger than families now often want, coming from ages when many farm staff and servants would live in alongside larger families, and so involve extra costs. The house is often beside or amid the working farm buildings, with that location often being convenient or necessary for work but making it problematic if it is not occupied with the farm.

26.14.4 In a market where restructuring of holdings is a major pressure, a residence clause may reduce the value the farmhouse might offer as it could exclude a number of potential tenants who are already well-housed on their own farms. In turn, that can limit the holding's potential for marriage value.

26.14.5 The farmhouse may offer opportunities for bed and breakfast accommodation to be offered. That does depend on the circumstances as well as the tenant's family being interested and able to do this. It may also require improvements to the house to meet the needs of paying guests.

26.14.6 Many holdings have more than one dwelling, having been let when there were many more farmworkers who expected to be housed by the farmer. Indeed, if there are not enough cottages for staff who want to be housed that can be an economic issue. However, it is more usual for the holding to have more cottages than needed by staff, present or retired. The labour requirements of the farming system used for productive capacity may influence this assessment.

26.14.7 Surplus cottages without occupiers can simply be a repairing liability for the tenant. The landlord cannot ordinarily require the surrender of these cottages, though that has often been agreed. More commonly, those cottages have been let out either on tenancies or as furnished holiday lettings, whether by the landlord consenting to the breach of the conventional bar on subletting or by the landlord accepting that the tenant has done that. The economic value of this is relevant to the rent, though allowance should be made for voids, management, tenant's improvements and maintenance and the terms of any landlord's consent for subletting.

26.14.8 Opportunities to let cottages may be limited where they were built subject to an Agricultural Occupancy Condition (AOC) in the planning permission, restricting their occupation to those involved in or retired from agriculture. However, where there is good evidence of ten years continuous breach of that condition it can cease to be enforceable and a certificate of lawful use can be requested.

26.15 Improvements to the holding by landlord and by tenant

26.15.1 The tenant is paying rent for the land and fixed equipment that the landlord has provided. With the length of many 1986 lettings and the changes seen in farming systems, methods and the requirements of the marketplace and the law, many holdings have seen improvements made by both landlord and tenant to their fixed equipment.

26.15.2 It is useful for there to be a proper, accurate and agreed record of improvements, especially of tenant's improvements and fixtures, to clarify matters for both rent reviews and the position at the end of the tenancy.

26.15.3 Improvements are treated differently at rent review according to whether they were made by the landlord or the tenant:
- in general, landlord's improvements are taken into account for the rent except to the extent that they were granted aid
- in general, the effect on the rent of the tenant's improvements is to be disregarded.

These principles (and the exceptions to them) can raise practical problems when determining the rent.

26.15.4 Landlord's improvements – For the purposes of the rent review, a landlord's improvement is simply an addition to the holding, adding to the totality of what is to be valued for the rent.

26.15.5 However, where the landlord received grant aid to make the improvement then the effect of that grant on the rental value is to be disregarded. That money may have come from national government (including the CAP) or possibly local government funds. This is usually considered by disregarding the proportion of the work (such as field drainage or a building) that the grant aid represented of the total cost.

Example – The landlord had paid for the drainage of a field on the holding. He did so at a time when 60% grants were available for this work.

It is established from comparable evidence that, at the time of the rent review, the rent for a drained field is £20/acre more than the rent for an equivalent but undrained field.

Of that £20/acre, £12 falls to be disregarded by applying the 60% grant rate to the uplift in rental value arising from the drainage work. Perhaps the easiest way to apply this disregard in practice is to establish the full rent with the drainage and then deduct that £12/acre.

26.15.6 Under the Agriculture Act 2020, where the landlord has made an investment for which the tenant has agreed to pay a charge, that charge and, so long as the charge lasts, the investment (for example, the building) are to be disregarded. That is to put such landlord's finance on the same footing as from any other funder, rather than being a further issue in finding the rent.

26.15.7 Tenant's improvements and fixed equipment – The tenant may have undertaken works that improve the holding. Examples could include a building or an extension to a building, land drainage, making farm tracks, extending electricity supply or improving rough land. The basic principle is that the tenant is not to pay rent on his investment in the holding. Thus, any increase in the rent that is due to improvements and fixed equipment added by the tenant is to be disregarded.

26.15.8 "Tenant's improvements" are defined by the Schedule as

> any improvements which have been executed on the holding, in so far as they were executed wholly or partly at the expense of the tenant (whether or not that expense has been or will be reimbursed by a grant out of money provided by Parliament or local government funds) without any equivalent allowance or benefit made or given by the landlord in consideration of their execution.

26.15.9 Fixed equipment is defined in s.96 of the 1986 Act as

> any building or structure affixed to land and any works on, in, over or under land, and also includes anything grown on land for a purpose other than use after severance from the land, consumption of the thing grown or of its produce, or amenity.

26.15.10 The disregard for tenant's improvements and fixed equipment at rent review is not limited to those tenant's improvements that are compensatable at the end of the tenancy, nor only to agricultural improvements. They include improvements carried out under any previous tenancy that that tenant had held of the holding (unless the landlord has since paid compensation for them).

26.15.11 The impact on value of tenant's improvements and fixed equipment is to be disregarded unless the improvements or fixed equipment:
- were provided "under an obligation imposed on the tenant by the terms of his contract of tenancy"
- were carried out in contemplation of the tenancy

or the landlord provided an allowance or benefit in respect of them.

26.15.12 Applying the disregard when valuing the rent – The statutory requirement is to disregard any increase in rental value from the tenant's improvement or grant aid to a landlord's improvement: it is not simply to disregard the work itself. This wording poses practical difficulties to which two general approaches have been followed:

- the **"black patch"** approach of disregarding the improvement itself and so considering the holding without it. The difference between the value of the holding with the improvement and without is then the effect on rental value but then potentially subject to allowing for the latent value of the opportunity to make it. A similar exercise on any comparable holding may also assist. This can identify those improvements that may add little or no value as well as those that may add much value.
- **valuing out** the improvement, so looking to find an annual rental value for it within the overall value of the holding including it. Where a budget has been developed, this value may often then, having worked with a partial budget from the profit arising from the improvement, be deducted from the landlord's share of the divisible surplus to give the rent. As this approach does not work directly from rental evidence, it should be used with care as the end result may overemphasise the value of the improvements as part of the whole farm.

In practice, each may serve as a crosscheck on the other.

26.15.13 It has been common for the valuing out approach to work from the current cost of making the improvement, apportioning that over an assumed useful life, to give an annual value as the desired figure. However, that cost-based approach is inappropriate as:

- cost is not the same as value and may have been incurred for reasons other than securing additional value (as, for example, a new slurry store may have been needed simply to stay in business) and may not have been effective in doing that
- the useful value of an improvement is unlikely to decay evenly over its life.

Example – land reclamation by the tenant – The tenant has reclaimed previously rough land to be silage ground. Equivalent rough land would now have a rent of £15/acre. As improved, this land has a rental value of £60/acre.

While the apparent gain in rental value is £45/acre, that work has unlocked the latent value the rough land offered for that improvement – the extra value a tenant would pay knowing he then had that improvement opportunity. If that latent value were found to be £8/acre, the rental value to be disregarded would be £37/acre. That might perhaps also have been identified by comparing rents for poor land capable of drainage with poor land that could not be drained.

Example – a building – A timber framed cattle shed was built by the tenant ten years ago at a cost of £10,000. Its current cost would now be £11,500. Of its expected life of 25 years, 15 years remain. What rental value is added by the building and so should be disregarded?

Method 1 – black patch on the holding – Assume the building is not there and consider the holding and how it would then be farmed. How much less (if at all) would the rent be? In some cases, the improvement may enable an enterprise that would not otherwise be run, so displacing an alternative activity rather than enlarging a current one.

Method 2 – using comparables – One way, as in the land reclamation example described earlier, to answer that basic question of how much less the rent would be without the building would be to take two similar holdings, one with such a building and one without, and identify the rental difference between them. In practice, it may not be easy to find such a simple and direct comparison – save perhaps for really major investments, some non-agricultural improvements or on estates with many tenancies.

For this example, it is fortunate that a nearby holding with similar enterprises has a cattle shed that is larger by the same size as this building. It was reviewed a year ago and after allowing for other issues, analysis of that rent suggests a rental value in the range of £150 to £200 could be identified for it. Rental values and cattle farming returns are both slightly greater than a year ago.

Method 3 – profits method – What would the tenant pay in rent for the economic advantage of the improvement, whether additional income or reduced cost or risk? This typically uses a partial-budget approach to show what is added to the business.

If the building allowed an extra 35 cattle to be fattened at a gross margin of, say, £30 per head, the additional gross margin would be £1,050. If £500 of additional overheads were related to these extra animals, that would give a net margin from the building of £550.

How much of that would the tenant give in rent to secure that advantage? Would that be as a fixed sum or based on a percentage of the £550? If the £550 were equally divided, that would give £275.

One rule of thumb approach might be to apply a typical ratio evidenced from other work. It might be observed that rent is often 10% of a farm's gross output. If the additional gross output attributable to the building were £3,000 a year, that might give comfort to a view of the rent to be disregarded being £300 – but at an arbitration the expert would have to withstand cross examination on that point and be able to show that such an overall figure would also be applied at the margin to an extra facility.

And then? – Compare the answers from the different methods – the other party to the review might use any or all of them. Consider the differences and why they exist.

The output-based approach is pointing to perhaps £300. A simple equal division of the net economic margin (were there evidence that equal split was what was happening in the marketplace) would give £275. Comparison, subject to both any uncertainties in its calculation at this level and a rising market, might now point to £200. The question remains: what would the tenant bid? With the direct evidence of market behaviour to be gleaned from the comparable and perhaps the greater importance of this building to the holding (after all the tenant did choose to spend money on putting it up) than the marginal addition to capacity on the comparable, it might be that £240 is felt right.

Applying a straight-line writing-down basis to a current cost of £11,500 and a return of 10% would give £690, looking adrift whether because the method is inappropriate, the assumptions used are wrong or it was a bad investment for its cost.

One check is whether it is considered that a tenant would think it reasonable to pay any of the figures as rent if he were offered the improvement by the landlord.

Of course, if the building were now surplus to the holding, there might be no increase in rental value to disregard.

26.15.14 Jointly funded improvements – Many tenancies have improvements that have been jointly funded by landlord and tenant, commonly where, as was once the case with field drainage, there was substantial grant aid for the work.

Example – Developing the drainage example of a landlord's improvement used earlier, the scheme cost £1,000 per acre, but, with a 60% grant, the net cost was £400/acre which was then equally funded by landlord and tenant (£200/acre each). The uplift in rental value today is £20/acre

The tenant's share of the work falls to be disregarded (so £10/acre for that half) as does the grant contribution to the landlord's share of the work (60% of that £10/acre for the other half). On those facts, £16/acre of the £20/acre uplift in the rental value would be disregarded.

26.16 "High farming"

26.16.1 In addition to the disregards already noted, the arbitrator or expert is also required to disregard the effect on rent of the tenant's adoption of a special system of farming, commonly referred to as "high farming". This is in practice never met and on close analysis may not even have meaning. The tenant's occupation is anyway disregarded throughout, including the assessment of the suitable farming system when determining the related earning capacity.

26.16.2 The statutory definition is:

> the continuous adoption by the tenant of a system of farming more beneficial to the holding –
> (a) than the system of farming required by the contract of tenancy, or
> (b) in so far as no system is so required, than the system of farming normally practised on comparable agricultural holdings.

Unlike the compensation provisions for "high farming" at s.70, there is no requirement for a record of condition to have been made.

26.17 The valuation

The arbitrator or expert (and so, by extension, the parties) is:
- to take account of the holding as it is
- to allow for the statutory disregards (notably for tenant's improvements and the tenant's occupation)
- consider all relevant factors, giving such weight to each as is right in the circumstances
- in that, to consider the evidence from:
  - comparables, adjusted first for the disregards for scarcity and marriage value and then allowing for differences with the subject holding
  - the related earning capacity (potentially within the extended budget) perhaps using each as cross-check on the other.

The rent is then determined on the basis of this process of assessment.

26.18 Dispute resolution (see chapter 35)

26.18.1 Almost all negotiations over rent reach an agreed conclusion, even if that may require time beyond the review date. This can include the settlement of other issues such as new investment, changes to the tenancy agreement, variations in the holding, a succession or other matters. However, the differences between the parties' positions may, in some cases, require the rent to be settled by a third party. That can be because of genuine and reasonable disagreements over evidence, the law and the realities of the case. It can also be because of misunderstandings, a failure to engage or that other issues between them make it impossible to agree.

26.18.2 The standard method for resolving rental disputes is to use an arbitrator, though, with some constraints, the law now allows the parties to appoint an expert who can make a final and binding determination of the rent. The procedures for appointment have been discussed earlier. The negotiations can continue even where such a person is appointed while the parties may also consider the broader potential of a mediation.

26.19 Some concluding thoughts

26.19.1 Advised parties in active negotiations over rent can be practical in how they deal with the issues that matter in the case, perhaps also settling other matters that may in reality be of equal or greater importance to the future of the tenancy than the rent. However, formal dispute resolution procedures (see Chapter 35) are not only limited to resolving the rent but can see an attritional focus on the analysis of suggested budgets and potential comparables with their detail that can divert attention from that key requirement of Schedule 2 as to the rent, that it be

> the rent at which the holding might reasonably be expected to be let by a prudent and willing landlord to a prudent and willing tenant, taking into account . . . all relevant factors.

26.19.2 The task is to find that expectation. In that, the Scottish Land Court made this useful observation after one lengthy hearing:

> If greater speed, simplicity and economy are going to be achieved in the future – and we believe they can be – close attention will have to be paid to the guidance which now exists as to what evidence is relevant and what is not. It is also for consideration whether matters have to be explored in as much detail as they were in this case. To use what is perhaps an improbable image, a lighter touch with a broader brush might serve equally well. (*Capital Investment Corporation of Montreal Ltd v Elliot*)

Rent Review Factors

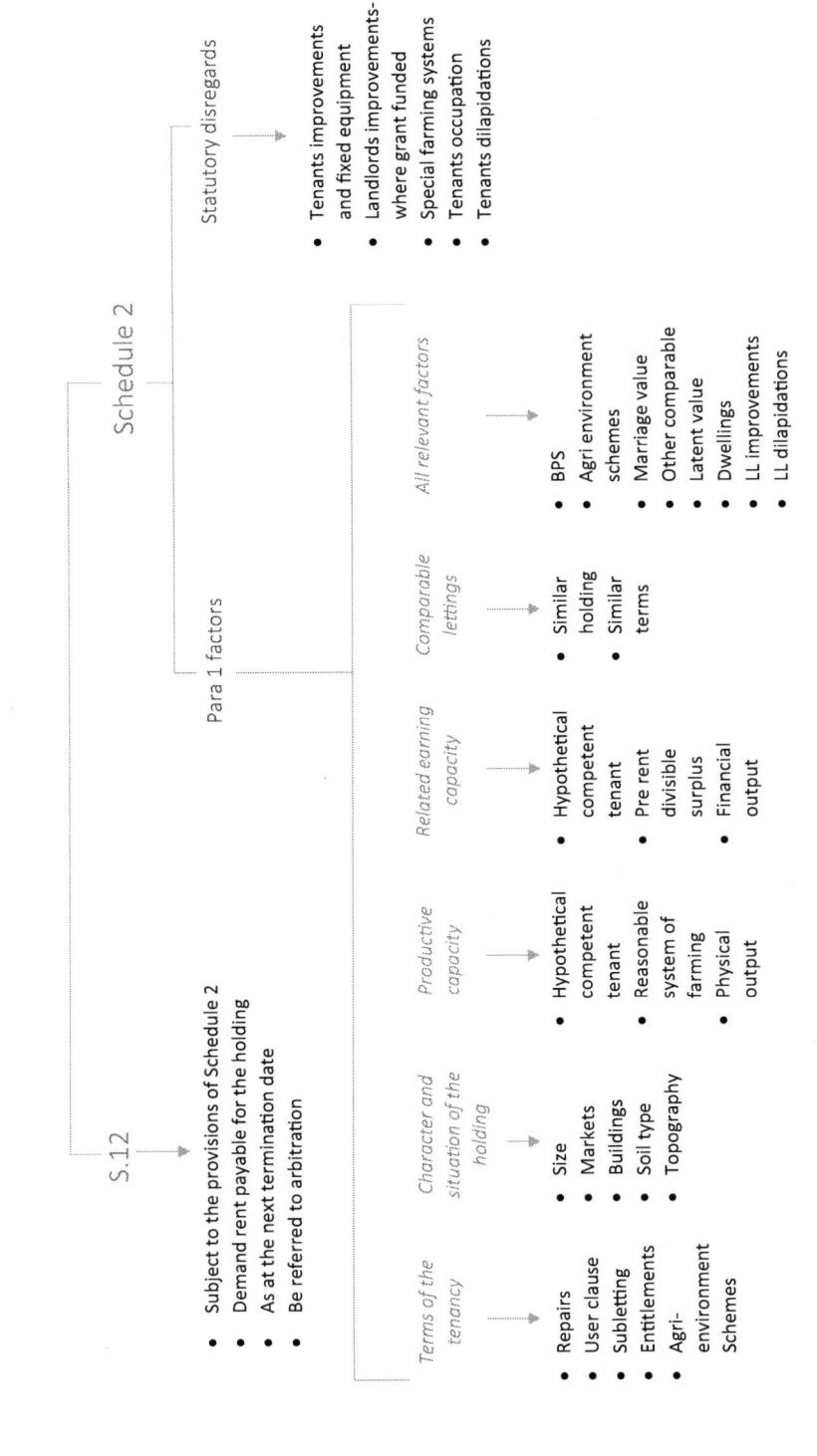

S.12
- Subject to the provisions of Schedule 2
- Demand rent payable for the holding
- As at the next termination date
- Be referred to arbitration

Schedule 2

Para 1 factors

Terms of the tenancy
- Repairs
- User clause
- Subletting
- Entitlements
- Agri-environment Schemes

Character and situation of the holding
- Size
- Markets
- Buildings
- Soil type
- Topography

Productive capacity
- Hypothetical competent tenant
- Reasonable system of farming
- Physical output

Related earning capacity
- Hypothetical competent tenant
- Pre rent divisible surplus
- Financial output

Comparable lettings
- Similar holding
- Similar terms

All relevant factors
- BPS
- Agri environment schemes
- Marriage value
- Other comparable
- Latent value
- Dwellings
- LL improvements
- LL dilapidations

Statutory disregards
- Tenants improvements and fixed equipment
- Landlords improvements- where grant funded
- Special farming systems
- Tenants occupation
- Tenants dilapidations

That is not only about cost but also practicality and effectiveness, two of the hallmarks of an agricultural valuer.

Further reading

Rent Reviews Under the Agricultural Holdings Act 1986 (CAAV, 2008)

27 Rent reviews for farm business tenancies

Note – Rather than repeat text, reference to the preceding Chapter 26 may be useful for some practical valuation points, perhaps particularly concerning the review of comparables and tenant's improvements.

27.1 Introduction

27.1.1 The Agricultural Tenancies Act 1995 provides the framework for new tenancies in England and Wales granted since September 1995, doing so with much greater, but not complete, freedom of contract than allowed by the Agricultural Holdings Act 1986.

27.1.2 Some, often the more substantial and equipped, FBTs are granted for longer terms over which the initial rent may become inappropriate. More tenancies, initially let for shorter terms, may run on indefinitely under the continuity provisions of s.5 of the Act, so making a rent review process desirable if it is not to be achieved by the larger alternative of terminating and replacing the tenancy.

27.1.3 Part II of the Act provides a framework with options for rent variations and rent review, defaulting to a familiar minimum three-year review cycle. In order to pre-empt a reason for lettings to be only for short terms, the Act defaults to an open market basis for rent review but allows other approaches which were broadened in 2006.

27.1.4 The tenancy agreement can expressly exclude any rent review (s.9(1)(a)) or override the default basis with other provisions.

27.1.5 There is an important distinction in those options for changing the rent between:
- bases for *reviewing* the rent, here defaulting to the open market basis but with other possibilities, so that it is determined in the circumstances of the time
- pre-agreed arrangements for *varying* the rent

in neither case precluding a reduction.

27.1.6 This chapter starts with the procedure for a review, then considers the default basis for review and then sets out the alternative approaches.

27.2 Triggering the rent review or variation

27.2.1 Where a rent review or variation has not been excluded under the agreement, s.10 sets out a procedure resembling that under the 1986 Act.

27.2.2 Either landlord or tenant can formally provide for a review or variation by serving a "statutory review notice" on the other at least 12 months but less than 24 months before the potential review date specified in the notice. That does not commit them to a review but is the basis on which either party, not just the server of the notice, can refer a review to arbitration.

27.2.3 Where the tenancy agreement specifies the dates or dates (as, say, every third or fifth anniversary of the commencement of the lease) when the rent can be reviewed or varied, the notice must be for one of those dates. In that case, no other review will be possible.

27.2.4 If no dates are specified, the review date must be an anniversary of the beginning of the tenancy or of another date agreed between landlord and tenant and not less than three years from the latest of:
(i) the beginning of the tenancy,
(ii) any date as from which there took effect a previous direction of an arbitrator as to the amount of the rent,
(iii) any date as from which there took effect a previous determination as to the amount of the rent made, otherwise than as arbitrator, by a person appointed under an agreement between the landlord and the tenant, and
(iv) any date as from which there took effect a previous agreement in writing between the landlord and the tenant, entered into since the grant of the tenancy, as to the amount of the rent.

S.11 protects this minimum review cycle where there has been a simple severance of the reversion with an apportionment of the rent.

27.2.5 It is then for the parties, if one wishes to take the review forward, to negotiate on the rent as well as any other matters that may be between them, doing so in a timely and practical way.

27.2.6 If as the review date approaches the matter is not settled and one party wishes to continue the review then the default position since the Agriculture Act 2020 is that that party will need to require a professional authority (such as the President of the CAAV) to appoint an arbitrator, making that request within the six months before the review date. It is also open to the parties to have agreed on an arbitrator between them, but then that person must be appointed before the review date.

27.3 Default open market basis for review

27.3.1 S.13 prescribes the basis on which the arbitrator is to determine the rent properly payable for the holding at the review date as

> the rent at which the holding might reasonably be expected to be let on the open market by a willing landlord to a willing tenant, taking into

account (subject to subsections (3) and (4) below) all relevant factors, including (in every case) the terms of the tenancy (including those which are relevant for the purposes of section 10(4) to (6) of this Act, but not those which (apart from this section) preclude a reduction in the rent during the tenancy).

27.3.2 That is essentially an open market basis with statutory disregards similar to some of those under Schedule 2 of the 1986 Act:
- *tenant's improvements* (sub-section 3) – The arbitrator is to disregard any increase in the rental value of the holding which is due to any tenant's improvements (as defined in s.15 and so any physical improvement or intangible advantage that remains with the holding, irrespective of whether it is compensatable or not) except:
  - where they had been provided under an obligation which was imposed on the tenant by the terms of the tenancy or any previous tenancy and which arose on or before the beginning of the tenancy in question,
  - to the extent that any allowance or benefit has been made or given by the landlord in consideration of its provision, and
  - to the extent that the tenant has received any compensation from the landlord in respect of it.
- *the tenant's occupation of the holding* (sub-section 4) – The arbitrator is to disregard any effect on the rent of the fact that the tenant who is a party to the arbitration is in occupation of the holding and reduce the rent at a lower amount by reason of any dilapidation or deterioration of, or damage to, buildings or land caused or permitted by the tenant.

Reference should be made to the relevant valuation sections of the previous chapter on these points.

27.3.3 No other disregards, whether or not in the 1986 Act, are applied.

27.3.4 All discussions over rent ultimately reduce to discussion of what other people are paying for similar properties and the economics underpinning that and so, without the formal structure of the 1986 Act, evidence is likely to focus with less complexity on comparables and farming economics. In any direct conflict between the two, good evidence of comparables, showing what people are really doing in markets, will generally be more persuasive than budgets or other economic evidence, on the same logic as advanced by the Scottish Court of Session in *Morrison-Low v Paterson*.

27.3.5 It might therefore be responsive to the plea of the Scottish Land Court in *Capital Investment Corporation of Montreal Ltd v Elliot* for "a lighter touch with a broader brush" in these matters (see 26.19.2 above).

27.4 Alternative approaches to changing the rent

27.4.1 Alternative basis for rent review (s.9(c)) – The 2006 reforms allowed that a tenancy agreement could provide a different basis for a rent review where it:

- also expressly states that Part II of the Act does not apply – so also excluding the statutory trigger notice provisions of s.10, making it sensible for such an agreement to provide for a prior notice procedure
- does not contain any provision which precludes a reduction in the rent during the tenancy and
- provides for the reference of rent reviews to an independent expert whose decision is final.

27.4.2 While this allows the use of any other basis for rent review, the point specifically in mind in 2006 was to facilitate negotiations with an existing 1986 Act tenant about moving to a new FBT. Such a tenant might be reluctant to agree to move to an open market basis from the Schedule 2 basis for rent review of the 1986 Act. Up to that point, some ingenuity had been used to relate variations of the rent to indexes of 1986 Act rents (such clauses will still be found in some leases from before 2006), but this can now be effected more simply and directly.

27.4.3 Thus, this provision is most likely to be found to be used where the parties agree to apply the 1986 Act provisions to a tenancy that is an FBT. In such a case the valuation advice of the previous chapter will be relevant.

27.4.4 As this excludes arbitration, the agreement will need to provide a procedure for the appointment of the expert.

27.4.5 Pre-agreed variations (s.9(b)(i)) – The tenancy agreement can provide that the rent is to change to a specified way on stated dates with the rent otherwise remaining fixed. This might be the case in such situations as where:
- a tenant takes on a farm in run-down condition and is allowed a reduced rent for an agreed period of years to put it right after which a new higher pre-agreed rent takes effect
- the landlord is to make an improvement with a higher pre-agreed rent due once that is done
- a new tenant enters with a series of pre-agreed steps in the rent to assist settling in.

27.4.6 This excludes Part 2 of the Act and so excludes the default review basis. Where the tenancy is for a longer period, some provision for later reviews might be considered.

27.4.7 A rent variation formula (s.9(b)(ii)) – Again excluding Part II of the Act, this allows the tenancy agreement to specify a formula for future rent variations and so substitute completely for the need for reviews. The rent otherwise remains fixed. Such a formula, needing careful drafting, might be:
- a turnover rent (as, say, linked to the milk cheque)
- a link to a measure of product prices or profitability
- an index that had been agreed to be appropriate.

It must not preclude a reduction in rent and must, essentially, be an exercise in arithmetic not requiring judgment. That might suggest the utility of setting out a procedure for a third party to determine the variation.

27.4.8 Such options are vulnerable to the future availability of data, how data is specified and unforeseen changes in circumstances, all of which can have perverse effects or even negate the provision. That can make it prudent to consider a fallback provision for such circumstances.

Further reading

Rent Reviews for Farm Business Tenancies (CAAV, 2008)

28 Rent reviews for Scottish agricultural tenancies

Note – A replacement basis for determining a disputed rent review was legislated for in the Land Reform (Scotland) Act 2016. At the time of writing, it had not been implemented and it is possible that it might not be. This chapter describes the law as it currently is.

Note – Rather than repeat text, reference to the preceding Chapter 26 may be useful for some practical valuation points, perhaps particularly concerning the review of comparables and tenant's improvements.

28.1 General

28.1.1 A rent review is the formal opportunity for tenant and landlord to agree to vary the rent, and the great majority of rents are indeed agreed by negotiation. It can also be a chance to discuss any other issues between them. In conducting a rent review, parties should be aware of the guidance offered by the Tenant Farming Commissioner.

28.1.2 S.13 of the Agricultural Holdings (Scotland) Act 1991 provides a framework of procedure and a valuation basis to determine the rent if the parties cannot agree. This means both applying a legal framework and exercising practical judgment. Both parties should adopt compatible approaches, enabling as many facts as possible to be agreed, so that the remaining differences can be established clearly and focussed upon for resolution.

28.1.3 S.13 is anything but a formula; there are no mechanical routes in finding the proper rent. It is a matter of valuation judgment under the statute with a practical appraisal of the circumstances of each case, considering a wide range of factors in the light of the possible (but often uncertain) interpretations of the law. If the rent review reaches the Scottish Land Court (the Land Court), it is also subject to the limitations of the actual evidence submitted by the parties.

28.1.4 The person wanting to change the rent, whether up or down, should drive the review, explain their position and lead negotiations and, if necessary, make any reference for third party determination.

28.1.5 In pursuing a rent review, ensure that:

- no issues disrupt the intended timing of the review
- the initial s.13 notice is correctly prepared and served well within the time limits.
- time is allowed for the negotiations not to be pressured.
- should it appear that the negotiations will not be resolved before the term date, a reference has been made to the Land Court in good time (or, if agreed, an arbitration).

28.1.6 Either landlord or tenant can initiate the formal rent review process by serving a written notice under s.13(1) on the other party. This notice simply enables the review to be referred to determination in the event of sustained disagreement.

28.1.7 While it had been thought that, as if circumstances changed between the service of the notice and the review date, the party who did not serve the notice could nonetheless rely on it to seek an outcome that was not expected when the notice was served, the Scottish Land Court's decision in *Cawdor Farming No. 1 Partnership v The Cawdor Maintenance Trust* rejected this so that the position is now that only the party that served the notice can use it as a basis for taking the review of the Land Court (or not).

28.1.8 The notice must have been served no less than one year and no more than two years prior to the intended review date, the next termination date. If the tenancy has a fixed term, that chance might not arise until the end of the term. If, as is conventional, it is drafted for a specific date, it will be of no further use for any later date.

28.1.9 It may be prudent to serve the notice in good time to avoid the risk of it being invalid or questioned by leaving service until close to the deadline.

28.2 The minimum three-year rent review cycle

28.2.1 The basic principle is that a rent review cannot be enforced within three years of the commencement of the tenancy (including when there has been a surrender and regrant), the rent last changing or a third-party determination of the rent. Care should be taken to establish when the total rent (not the rent per acre) for the holding last changed and in what circumstances.

28.2.2 S.13 excludes some rent changes from re-triggering the cycle which will only operate in the circumstances defined by statute:
- a variation of the rent by the Land Court when considering the terms of the lease or fixed equipment
- a rent increase under the procedures of s.15 for a landlord's improvement
- a reduction in rent following the landlord's exercise of an early resumption clause or a notice to quit part of the holding under s.31 – but not a surrender of land
- any change in the incidence or rate of VAT on the rent.

28.2.3 Any other variation of the tenancy and the total rent will be by agreement. That would re-trigger the rent review cycle so that no statutory review can then be triggered for three years.

28.3 Serving the s.13 notice

28.3.1 The valuer drafting the notice should ensure that it is correctly prepared and served on the other party. It is often helpful for it to be accompanied by a covering letter, especially as the style effectively required by the law may look heavy-handed where the party receiving it is unfamiliar with the procedure.

28.3.2 The Act sets out specific rules for the service of notices which are to be delivered to the recipient, left at his proper address or sent to him by "registered post" or "recorded delivery". Where the landlord has changed, the tenant can still validly serve notices or other documents on the old landlord until the tenant has received both
- (a) notice that the old landlord has ceased to be entitled to receive the rents and profits of the holding, and
- (b) notice of the name and address of the person now so entitled.

Proof may be required that the notice was served; evidence of service should be kept. While that can be done where the notice is served in person with a witness, it can be harder for fax and email. Posting brings convenience but also risks, another reason for serving in good time.

28.4 Once the notice is served – negotiations

28.4.1 In the period between the service of the s.13 notice and the review date, the party promoting the review should open discussions and pursue negotiations with the opportunity for advised reflection by both parties. Timely progress should usually preclude the need to refer the review to external determination at all.

28.4.2 The valuers should seek to negotiate. The role of the professionals in negotiation is often effectively one of mediation between the parties while honouring their duty to the clients. This may include considering many more issues than the rent. In almost all cases, parties should be able to settle a review without needing to refer to the Land Court or arbitrator. If the review becomes difficult, the valuers should identify the issues, list the areas of agreement and address areas of disagreement.

28.4.3 The review is one aspect within the overall relationship between the landlord and the tenant, which is likely to be a long term one. The professionals involved need to be sensitive to this, with the potentially larger consequences that may flow from actions taken only on an assessment of shorter-term issues.

28.4.4 As general principles and to assist the sensible conclusion of a review:

- the valuer for the party promoting the review should have a timetable in mind to allow for timely resolution of the issues, with neither party having to negotiate against a deadline (although that will, in turn, depend on the behaviour of the other party)
- the parties to a review should put all relevant facts on the table at an early stage rather than produce them piecemeal in the discussions. They should agree on as many facts and points as possible to crystallise any remaining points in dispute.
- the parties should communicate with each other, considering and responding promptly to proposals with a view to a practical and, if more issues than the rent are involved, mutually beneficial outcome.

28.4.5 Perhaps the two most essential points made by the Court of Session in *Morrison-Low v Paterson* are that the task is to find a single figure as to the rent for the holding and that the watchword in doing this is realism. That realism is not only a sensible cross-check at the end of any valuation but also an important approach throughout the process to ensure that the issues in dispute are real ones and that concern over them is proportionate. The better the evidence and the more that can be agreed, the easier it should be to focus efficiently on areas of legitimate and significant difference.

28.4.6 If, as the term date approaches, there is no agreed conclusion to the review, it becomes important to consider reserving the possibility of third-party determination, whether by the Land Court, arbitration or other means, while continuing to negotiate up to and after the review date. This reference is not a reason to stop negotiations. However, a s.13 notice only enables the Land Court to resolve the rent and not any other issues that may be between the parties.

28.5 Valuation date

The valuation date for the rent review is the date for which the s.13 notice is effective, requiring consideration of the actual (or clearly foreseen) conditions at that date which the market would really take into account.

28.6 Define the holding

The subject of the valuation is the holding itself, a physical and a legal entity with opportunities and difficulties. The valuer should identify and appraise its key features. What are the opportunities, restrictions on and obligations of the parties in the context of the holding? It has to be considered as a single unit and not on the basis that it could have been let in separate parts.

28.7 The Valuation basis – the rent properly payable

28.7.1 S.13(3) sets out the statutory basis for assessing the rent:

> the rent properly payable in respect of a holding shall normally be the rent at which, having regard to the terms of the tenancy (other than those relating to rent), the holding might reasonably be expected to be let in the open market by a willing landlord to a willing tenant disregarding –
>
> (a) any effect on rent of the fact that the tenant is in occupation of the holding, and
>
> (b) any distortion in rent due to a scarcity of lets but having regard to the matters referred to in sub-section (4) below.

28.7.2 It is the rent that would reasonably be expected to be agreed between hypothetical willing parties in the open market on specified assumptions. The word "reasonable" is applied to the expectation about the rent, not to the resulting rent itself. Beyond the specific directions of s.13 there are no limits on what factors may influence that expectation. The parties should have evidence on all matters they consider relevant and of potential weight for this assessment.

28.7.3 The rent is such as would be agreed between a hypothetical landlord and a hypothetical tenant, not the actual landlord and actual tenant, both willing to enter into the tenancy but acting reasonably in that.

28.7.4 The open market basis requires evidence of the lettings market affecting the subject holding. Where available, comparables can give direct evidence of this, but budgets on their own do not show what the hypothetical parties would agree. Evidence as to that value in the open market will include what others (rather than one special purchaser) might bid for the holding to run it as a part of their existing businesses – its marriage value.

28.7.5 The **tenant's own occupation** of the holding is to be disregarded so that matters personal to him and not inherent in the land are not relevant. Neither are the costs and potential disruption of having to leave in the event of an unsuccessful bid.

28.7.6 Any increase in rental value due to the certain **improvements**, as defined by the law, whether made by the tenant or to the extent that they were grant-aided, is to be disregarded. The law defines tenant's improvements only to include to include those eligible for compensation at waygo. Other investments and works by the tenant can be taken into account. Proving that status will require either:

- evidence of the relevant procedure having been followed (typically by the tenant having served a notice on the landlord before commencing the work) or

- that the Tenant's Amnesty under the Land Reform (Scotland) Act 2016 was used to agree the improvement's compensatable status.

28.7.7 For the rent review, this requires:
- a schedule of all qualifying improvements and description of each (perhaps with a plan)
- checking whether the lease obliged the tenant to make any of the improvements.
- recognising the grant aid to any landlord's improvements.

28.7.8 Of the various methods of taking this disregard into account the most practical, especially for more significant improvements, is to apply a "black patch" and assume that a relevant improvement is not there. It may then be possible to use comparable evidence to assess the difference in rental value due to the improvement. Where a budgetary approach can be linked to rents, that may assist, offering a crosscheck. Where the improvement has released a latent value in the holding (the value of the opportunity to make the improvement), that is part of the rent and not disregarded.

28.7.9 Any distortion in the rent due to "a **scarcity** of lets" is to be disregarded. Developing the analysis of Land Court decisions finds this to require assuming a "reasonably balanced market", not "a general scarcity of broadly suitable farms overall". This "does not suppose an exact equality of supply and demand" and will still see competition for some farms; the Court of Session observed in *Morrison-Low v Paterson* that

> [t]he value of an agricultural tenancy is not distorted by scarcity because it is highly sought after for the quality of the land or of the buildings or of the convenience of its location.

28.7.10 Two tests should be applied:
- is there scarcity?
- if so, does it distort the rental value?

Each is a matter of evidence and any distortion must be assessed. It suggests forming a view of the larger market for the holding (rather than just its immediate locality). Scarcity in general should be distinguished from any specific limitations of the individual holding. The number, type and distribution of potential bidders may affect the issue, whether found by analysis of the market or from recent patterns of tenders for holdings – recognising that not all bids are likely to be of equal quality – and evidence of the number of farms that have been let. This may be aided by the views of the Land Court in previous cases, perhaps now dated, in which discounts range from 5 to 25%.

28.7.11 Other matters that must be disregarded under s.13 include:
- dilapidations, deterioration or damage by the tenant
- use for a non-agricultural purpose or conservation activity that reduces the rent.

28.8 Regard to comparables and economics

28.8.1 S.13(4) requires the Land Court to have regard to:

(a) information about rents of other agricultural holdings (including when fixed) and any factors affecting those rents (or any of them) except any distortion due to a scarcity of lets; and

(b) the current economic conditions in the relevant sector of agriculture.

28.8.2 The determination of the rent is a matter of evidence and analysis for which the courts will look to expert witnesses. Regard is to be had to evidence as to:
- the **terms of the tenancy** (other than those relating to rent)
- **information about rents** of other agricultural holdings and the factors affecting them. This exposes the holding to evidence of rent settlements on other holdings – decisions taken by real owners and farmers. It is not limited to comparables, but the more closely the evidence can be related to the subject holding and the valuation date, the more effective it is likely to be. "Factors affecting those rents" draws attention to the dynamics in the market.
- **current economic conditions** in the relevant sector of agriculture. As well as prices and margins, this might touch on economic points in the minds of parties, including price volatility and market developments relevant to the holding. Some of this might be revealed by current rent settlements.
- any increase in rental value due to a **use of the land for a purpose that is not an agricultural purpose**.

The force of evidence on these will depend on the quality and weight of the actual evidence submitted. Consideration can also be given to evidence of any other relevant factors except for the statutory disregards.

28.8.3 Farms are individual and information is rarely complete or perfect. Weighing the balance between the evidence of such comparables as are available and the lessons that may be drawn from budgets as to the behaviour of hypothetical parties is a matter of judgement and appraisal. However, good and available comparables are direct evidence of actual market behaviour and budgets are not. Indeed, in *Morrison-Low v Paterson* the Court of Session observed that

> the valuation of an open market rent on the basis of a farm budget should be a method of last resort. The profitability of the holding is obviously a relevant consideration in the framing of an open market offer; but, as the Land Court has rightly held, questions of profitability must invariably be subsidiary to the open market criterion . . . that

budgets should be used for valuation only as a last resort and that a comparison with rent reviews is a preferable, though not an ideal, alternative.

28.8.4 Comparables as evidence – When considering information as to rents, well-presented, relevant and analysed evidence of recently settled comparable farms available for inspection may prove very effective evidence of the decisions taken by real owners and farmers. Any particular reasons affecting the way the rent was settled should be known.

28.8.5 Comparison is an essential valuation skill in appraising real economic decisions in other cases and applying them to the subject holding. A component valuation (as illustrated in Chapter 26), attributing values to any farmhouse, cottages, buildings and land from the evidence of comparables may prove valuable. As an identical comparable is unlikely to be available, judgment and evidence need to be applied to adjusting a comparable's rent to make it directly relevant. The more adjustments there are in the comparison, the less weight it is likely to have as evidence.

28.8.6 The Court of Session has stated in *Morrison-Low v Paterson* that the best evidence would be open market lettings of 1991 Act tenancies. In their probable absence, evidence from the lettings of LDTs and SLDTs can be relevant though needing to be adjusted. The Court of Session was clear that such rents are admissible evidence. Sitting tenant rent reviews rank next.

28.8.7 Economic conditions – This is expressed broadly, more broadly than a matter of a budget, and these conditions might be important in framing the outlook of landlords and would-be tenants in considering the rent to be paid for tenancy. However, it seems that in practice it is largely reduced to the preparation of budgets.

28.8.8 Budgets as evidence – The economics of the holding can be brought together in a budget, with the resulting outcome used as evidence towards determining the reasonable expectation of the rent for the holding. The Court of Session was, however, clear in *Morrison-Low v Paterson* that a rent might, according to the circumstances, reasonably be above the figure indicated by a budget.

28.8.9 The budget is not a valuation method in its own right but perhaps a means to support a valuation, using an approach usually based on standard farming accounts. It is not evidence of actual behaviour in the rental market but of the expectations of profit that may very often be a factor conditioning the offers of hypothetical bidders. It does not shield the tenant from the wider marketplace, though that may reveal how parties are acting in the light of the budgets before them.

28.8.10 However, it may help the preparation for negotiations to draft a budget to understand the economics of the holding, not only in terms of farming but also direct payments, non-agricultural uses, agri-environment agreements, sublettings and other factors. It is an assessment of the holding, not of the actual tenant – and so may not necessarily use the present farming system. Familiarity with farm accounting (see Chapter 6) may make a budget format a means to resolve many important questions about the agricultural performance of the holding – such as the probable farming system and yields.

28.8.11 It will both help reduce costs and assist the Land Court if the valuers involved:
- can agree on the format for the budget that best suits the holding and
- settle as many of the elements of it as they can so that any argument can focus on the areas of genuine and significant contention.

The budget should be consistent with s.13, most obviously in disregarding the benefit of certain improvements. In considering the prices for inputs and outputs to be used in the budget, the underlying requirement is to form a reasonable expectation as to the rent that would be agreed. That may not follow the actual prices and costs at the review date but instead follow the approach that might be taken at the time to prices and costs by parties settling a rent that cannot be reviewed again for three years.

28.8.12 It is conventional for the budget to give a pre-rent surplus available to pay a return to both the landlord (rent) and the tenant (his entrepreneurial return on management investment and labour). There is no prescribed approach for calculating this or as to how that figure is then taken into account in determining the rent. There is no statutory basis for any particular division. History shows this to have moved over time and it may vary between sectors. How might it be linked to the marketplace for holdings? That might be done by considering the equivalent budgets for comparables, linking their calculated pre-rent surpluses to their real rent settlements, subject to this being appropriate to the case in hand and the extent of the dispute warranting the effort and cost of this work.

28.8.13 However, there is no rule that the pre-rent surplus, however calculated, sets a ceiling for the rent, any more than for any particular split.

28.9 Forming a view

The valuer should consider all the relevant evidence available to him and the effects of the statutory disregards on the rent of the subject holding. In practice, both the quantity and quality of the available evidence on each of these matters will vary. Not all evidence potentially admissible at the Land Court will have the same weight: some will be less relevant, more remote or harder to substantiate than other evidence. Inevitably, only limited information will be available in any one case. While evidence of recent real agreements on rents may, in principle, be strong evidence, limitations on the quantity or quality of that evidence may make other factors more persuasive. Strong points of potentially great weight may be advanced by the other party and so should be considered carefully. The weight given to each piece of evidence will be a matter of opinion, on which it will be important for the valuer to give the client his considered professional judgement and advice.

28.10 Rent reviews for LDTs and SLDTs

28.10.1 S.9 of the 2003 Act makes specific default provision, very similar to s.13, for rent reviews under LDTs where no provision is made by the tenancy agreement. While there is the freedom to agree alternative provisions, they may not provide for review to be initiated only by the landlord and upwards-only clauses are barred. Any contractual terms in the lease should be considered carefully.

28.10.2 No statutory provision is made for rent reviews for SLDTs with their shorter term. As a result, if any need is seen for a rent review within, say, a five-year SLDT, it must be provided contractually.

28.11 Third party determination of a rent review

28.11.1 General – Since the 2003 Act, disputes over rent reviews under the 1991 and 2003 Acts between the landlord and the tenant of an agricultural holding can be referred by either party unilaterally to the Land Court (see Chapter 35). Even if a s.13 notice has been served, the parties can agree to use other methods of dispute resolution.

28.11.2 Even if a s.13 notice is served, the parties can exclude the Land Court and agree to refer the rent to arbitration, whether by someone they appoint or nominating someone to make that appointment. The President of the Scottish Agricultural Arbiters and Valuers Association (SAAVA) can undertake this function. The arbitrator must be in place before the termination date in question. The procedure is to be agreed by the parties or (if, as often, they do not agree) as the arbitrator considers appropriate – he may hold a pre-meeting to discuss this. The arbitrator will then issue his directions, confirming the procedure and timetable in writing so that both parties know what is expected of them. They have a duty to comply with those directions. Unless dealt with by written representations, there will be a hearing and the arbitrator will inspect the holding and any comparables that are available.

28.11.3 An appeal may be made against the award on a question of law to the Land Court with appeals over irregularity or jurisdiction to the Court of Session.

28.11.4 Expert determination – As the issues in a rent review may be more for expert knowledge than about contested evidence, the parties might agree to appoint an independent expert to make a final and binding determination of the rent, using his expertise, knowledge and skills, rather than just the evidence of the parties. He is required to act honestly, independently and fairly and has a duty of care to the parties. The expert is still likely to want the parties to make submissions, clarifying facts and issues, but he can proceed to his conclusion and is not bound by the evidence. As suggested in Lord Gill's *Agricultural Tenancies*, there is effectively no appeal against a professionally conducted final and binding determination so this is a means to arrive more swiftly at a definitive answer allowing the parties to move on.

28.11.5 The parties will need to provide the procedure to make it work and to ensure a conclusion, lest they come to disagree on the process.

28.12 Treatment of costs

28.12.1 All involved in a rent review are urged to bear costs in mind, especially as they can escalate towards a hearing. Much can be saved by identifying early those points on which the parties are agreed and clarifying the areas of dispute.

28.12.2 One way to draw attention to and potentially influence costs is to use a settlement offer. If negotiations are not moving to a conclusion, it can be useful to consider carefully with the client whether and when to make a settlement offer and at what figure. Careful consideration should be given to any settlement offer that is received – it is an incentive to realism since all litigation carries risks.

28.12.3 Parties should try to keep an objective view of the issues, their prospects and positions to help manage their liabilities. In particular, their advisers should counsel them as to the risks at stake.

28.13 Mediation (see also Chapter 35)

The parties may also consider mediation. This is a private process, facilitated by a person acting as a mediator, to try to find an agreed resolution of the issues between the parties. Unlike the formal methods for dispute resolution, it is expressly not an adversarial method. It is not limited to the issue in hand (such as the rent) but can look at all the issues troubling each of the parties and so can result in a "package" settlement. It does guarantee a conclusion and cannot be used to impose a settlement, but that can open the door to achieving a positive answer voluntarily. The parties must be willing to enter the process in that light and develop their concerns and solutions. If they do come to an agreement and that agreement is properly recorded, then it is final and binding between them. If they do not come to an agreement then their recourse is to the formal methods already discussed, so far as they are still available, having incurred the costs of mediation.

28.14 Recording the agreement

Once agreement is reached, its terms relating to rent and other matters which have been resolved become part of the terms of the tenancy and should be recorded to avoid subsequent uncertainty or dispute.

Further reading

A Practitioner's Guide to Scottish Agricultural Rent Reviews (CAAV, 2013)

29 End of tenancy
Tenant's claims

29.1 The recognition of tenant's investment on the end of a tenancy

29.1.1 Agricultural tenancies pose the particular problem of the tenant working and investing in the landlord's land, improving it, fertilising it and sowing, growing crops and often adding physical fixtures from buildings to fencing. A basic rule of land law is that what is fixed to land generally becomes part of it. While that is an issue at rent review (assessing the rent to be paid for the landlord's land) and on the end of the tenancy (where the tenant is leaving his work behind), those points can have a larger effect on the functioning of the tenancy if the tenant is to be supported in his care of the land to the end of the tenancy.

29.1.2 At rent review, the effect of tenant's works on rental value are generally disregarded so that the tenant does not ordinarily pay more rent for improving the land – see Chapters 26 to 28.

29.1.3 While longer lettings might defer the issue, the end of tenancy question inevitably arises, casting its shadow on the earlier years, especially when the tenant can see that he is investing for the years after his tenancy will end. Historically, some of this was (and still is) tackled by rights of holdover so the tenant can return and take crops, but that does not answer the problem of longer lasting investments such as buildings, drainage and also improvements to the soil and its fertility. Why should the tenant make those improvements, spending money and effort, if he will not see their full benefit that would justify making them?

29.1.4 Across Great Britain, the practicality of that issue has been answered in two ways:

- **"Tenant's fixtures"** – since 1851, statute law has given a procedure allowing a qualified right for the agricultural tenant to remove many works that he may have made such as buildings that would otherwise, as fixtures, be part of the land and so be left behind. It is more likely to be useful where the work is capable of being removed usefully. The landlord has the pre-emptive right to exclude that power by paying the tenant for the improvement. These items are referred to as Tenant's

Fixtures, effectively excluded from the normal law on fixtures. The law on this is substantively the same for agricultural tenancies throughout the United Kingdom save that the landlord has no statutory pre-emptive powers for FBTs or in Northern Ireland. However, there is no provision for tenant's fixtures under Scottish LDTs, MLDTs and SLDTs.

- **End of tenancy compensation** – the more general answer has been the development of a structure of compensation for the value of the work that is left behind, initially as a matter of local custom from the end of the eighteenth century in County Down and Glamorgan as much as across England and much of Scotland. This is generally handled under the labels of Tenant's Improvements (for more substantive works of fixed equipment) and Tenant Right (for works to the land), though for Farm Business Tenancies in England and Wales, the heading of Routine Improvements. England (in 2015) and Wales (in 2019) have now liberalised the 1986 Act's old provisions on tenant right and short-term improvements. The parallel provisions for improvements in Scotland differ in practical detail and the basis for valuation for improvements. With Northern Ireland's legislative history there are no statutory provisions there.

29.1.5 Compensation first developed as a matter of local practice in various parts of the country from the late eighteenth century, becoming entrenched as local customs, varying from district to district in response to local farming usages and circumstances. First assessed by respected farmers, this work developed into the root of the profession of agricultural valuations, building skills and experience with local groups debating and settling practice, finding answers as problems arose – the origins of the local valuers' associations. The first statutory basis for compensation was provided in 1875. The CAAV was founded by local associations in 1910 with purposes including the promotion of good practice in tenant right valuation. That work included the preparation of analysis, practice and standard figures, the last on an annual basis.

29.1.6 The state stepped in with its legislation for agricultural improvement after the Second World War. Custom was abolished, replaced with statutory rules and, for England and Wales, standard figures for fertilisers and other matters that were only occasionally reviewed and revised, last in 1978–83.

29.1.7 Tenant right valuations (and their corresponding dilapidations valuations), once the staple fare of many valuers' practices, have since become much less common routine in most areas, with landlords and tenants often not seeing the rewards warranting the professional costs, particularly where the tenancy is ending under a negotiated agreement between the parties.

29.1.8 However, those standard figures have now been repealed, leaving values to be found by valuers as before 1948 at their current, usually much greater, levels. That combined with the increased interest in soils can now make significant elements of the outgoing tenant's potential claim more substantial. Those factors, together with a greater awareness by all parties of the need for clarity might now see an increase in the number of end of tenancy valuations, even if not to the numbers involved in the heady days, when the authors first started, when one was faced with six full tenant right valuations on his first Michaelmas in practice. There is value to be recognised.

29.1.9 Even where valuation claims are treated as offset against each other, it is important for the holding to be inspected and agreement reached between valuers as to the condition in which it should be left. That helps both to avoid any disputes at the end of the tenancy and the incomer be clear about what is being taken on.

29.2 Making an end of tenancy claim: the tenant

29.2.1 In Great Britain, statute law provides a procedure and timetable for the tenant to make such a claim on the end of the tenancy and for those claims to be determined.

29.2.2 In **England and Wales**, the timetable is set:
- for claims on the end of a 1986 Act tenancy by its s.83 applying to claims for tenant's compensation (and also dilapidations against the tenant) with:
  - notice of intention to make a claim given to the other party within two months after the end of tenancy, specifying the nature of the claim
  - any appointment of an arbitrator to be made within eight months after the end of the tenancy
- the claim for tenant's compensation for improvements on the end of a farm business tenancy by s.22 of the 1995 Act with:
  - notice of intention to make a claim given to the landlord within two months after the end of tenancy, specifying the nature of the claim
  - the ability to make a unilateral request to the President of the CAAV or other professional authority for the appointment of an arbitrator once four months have passed from the end of the tenancy.

29.2.3 In **Scotland**, the timetable is set for all forms of tenancy by s.62 of the 1991 Act with:
- notice of intention to make a claim given to the other party within two months after the end of tenancy, specifying the nature of the claim
- the parties then to settle matters within four months of the end of the tenancy but able to apply to the Scottish government for up to two successive two-month extensions
- any application to the Land Court or for an arbitrator is to be made within eight months of the end of the tenancy.

29.2.4 In **Northern Ireland**, these issues are entirely matters of contract and common law.

29.2.5 Tenant's fixtures – Where the rules for tenant's fixtures apply, the procedure is covered later on.

29.3 Compensation for value at the valuation date: key concepts

29.3.1 All the varying approaches assess the compensation on the basis of the subject being handed on for subsequent use as an agricultural holding. Thus, the use of the land after the tenancy, whether for farming or housing, does not affect the tenant's compensation.

29.3.2 In their various ways, they look at what the value is to a subsequent farming user, not its value to the outgoing tenant or what it might cost or have cost to create. The two basic approaches are to assess:
- the value of the item to an incoming tenant
- the value added to the holding let as a holding.

29.3.3 Value is not cost – This is one of the areas where it is important not to confuse cost and value. The investment decision in making the work may have been good or bad. The outgoer may have been able to make it very economically or have overspent. The work may once have had value but now be of no use or it may now be of more general use than it was when made. Those are the fortunes of business. It does not matter what the work cost; what matters is its value for a future user. The item may still be of significant value or may be worth nothing. That is to be judged in its circumstances.

29.3.4 Value to an incoming tenant – This is the assessment of the value of the benefit left for the next farming occupier, whether saving cost or offering the potential for profit. This will consider the remaining life of a building, soil improvement or other item and its usefulness to a succeeding tenant.

Example – 20 years ago, the tenant erected a 140ft x 60ft steel portal framed general purpose building with a design life of 40 years and its value to an incoming tenant has to be found.

The building remains suitable for the holding and so the new tenant would take to a useful building with an expected remaining useful life of 20 years – as the value to be found is that to an incoming tenant, it is this 20 year forward prospect that is to be valued. That can be assessed on this basis:
- the current cost of replacing that building is put at £78,000
- with half of its life left that gives a value of £39,000.

That must then be reviewed with the sense check of whether an incoming tenant could actually be expected to pay £39,000 for the building. Does it give that amount of value? One way to approach this could be to consider if a reasonable case could be put to the incomer's bank manager to borrow money for this acquisition. For such a useful general purpose cover of great utility to the holding and in good order, this might well be a valid figure. That value might be lower were the building larger than the holding needed, in poorer condition or compromised in some other way.

29.3.5 Value added to a holding as a holding – This is an assessment of improvement in the value of the landlord's reversion flowing from the improvement, typically in its rental value, and is often more suitable for larger works, such as drainage. Where using an increase in rental value that is then capitalised to give a compensation payment.

29.3.6 Whatever is to happen to the land after the tenancy, the assumption is made that the use will be agricultural so that valuable works are still paid for even if the land is going to a different use, such as development or woodland.

Example – If the underdraining of 100 acres of farmland has increased its rental value by £20/acre compared with its unimproved state and a capitalisation rate of 2% is chosen after reviewing the circumstances, then there could be a claim for £100,000. The validity of that claim would not be affected were the land to be developed for housing.

England and Wales – Agricultural Holdings Act 1986

29.4 Tenant's improvements

29.4.1 Where the tenant has made one of the works of improvement listed in Schedule 7 of the Act (including buildings, drainage and other items of fixed equipment) and has done so with landlord's written consent (or, for many approvals, by the Tribunal) the outgoing tenant can claim compensation for it.

29.4.2 In principle, that compensation is to be assessed as *the value the work has added to the holding let as a holding*. More correctly, that is

> the increase attributable to the improvement in the value of the agricultural holding as a holding, having regard to the character and situation of the holding and the average requirements of tenants reasonably skilled in husbandry (s.66(1)).

29.4.3 Key points in that are:
- the holding is valued as a holding, so that it is the increase in its value as a let unit caused by the improvement that is to be found
- that requirement means that it does not matter whether the holding is indeed re-let. Indeed, the land could go for housing, but with this special assumption the compensation is to be assessed as though it were re-let
- that is likely to be a valuation of the holding with and without the improvement to identify any increase caused by it
- that might commonly be tackled by considering the increase in rental value due to the improvement and then capitalising that over the

remaining life of the improvement. In substance, an investment valuation that may take account of any other factors that appear relevant, as where the fact of the work may be of longer-term value for development control reasons than the work itself
- the reference to the character and situation of the holding takes the holding as it is at the end of the tenancy. Is the item useful or not in that context at that time? It does not matter whether it would be more or less useful in other contexts.
- referring to the average requirements of reasonably skilled tenants points to ordinary uses rather than any special value that the work may offer particular individuals
- both those two points essentially reinforce the view that this is to be judged in the market at the time on conventional market value assumptions
- that would point to a hypothetical rent review for a 1986 Act tenancy with and without the improvement.

29.4.4 Any public grant paid for the making of a Schedule 7 item is to be taken into account, reducing the compensation.

29.4.5 However, and as is commonly found, s.67(2) allows the terms of the landlord's consent to include provisions as to compensation. It has been conventional for landlords to use that to provide that the compensation to be assessed on the basis of the cost of the work being written down over a period of years, often anywhere between 5 and 20, to a nominal sum such as £1 or £5 (but not to nothing because it is compensatable). That sum is only due on the end of the tenancy and, so long as it remains unpaid, the work remains a tenant's improvement. If the parties agree that the payment is made earlier, then the work becomes the landlord's for rent review purposes.

29.4.6 Landlord's consent to the work is most likely to be given in a formal memorandum but may be covered in a letter or even inferred from other evidence such as landlord's counter-signed grant claim form (where the tenant has kept good records).

29.4.7 While landlord's consent is absolutely required for the generally rare and specialist works in Part 1 of Schedule 7 to attract compensation, the tenant can refer to the Tribunal any refusal of consent or conditional consent for works in Part 2. That challenge can be against proposed terms for writing down compensation which may then, in practice, be subject to further negotiation.

29.5 Tenant's fixtures

29.5.1 Where an improvement is made by the tenant without written consent by the landlord, it may still qualify for treatment as a tenant's fixture with qualified power for the tenant to remove it if, after notice, the landlord does not elect to buy it at its value to an incoming tenant.

29.5.2 S.10 applies to both any building erected or acquired by the tenant on the holding and any engine, machinery, fencing or other fixture affixed by the tenant to the holding except where:
- the tenant was required to make the work
- the work replaced a previous landlord's item
- the tenant is entitled to compensation for the building (but not any other fixture).

29.5.3 The tenant can act to remove such items at any point during the tenancy and up to two months after it ends provided that all rent due has been paid and obligations met. The tenant serves on the landlord written notice of his intention to remove the item at least a month before either the intended removal or the end of the tenancy. The landlord then has a month in which to serve notice on the tenant of his election to buy the item at *its fair value to an incoming tenant*. Disputes over value can be unilaterally referred to arbitration or jointly to third-party determination.

29.5.4 If the landlord does not do this, then the tenant can (but does not have to) remove the item whether for his use elsewhere or for sale, as whole or in its components. Concrete may offer little value, but framed buildings and their components may be reused or find a secondhand market. In removing the building or fixture, the tenant is to do no avoidable damage and make any damage good.

29.5.5 The measure of value is *the fair value to an incoming tenant*. What would a hypothetical incoming tenant pay to take on the item on the date of the intended removal? Again, that points to considering the general demand by tenants likely to take the whole holding in question, not the specialist use by a particular individual.

29.5.6 Judging that may typically require appraisal of the item in question, its condition, potential use and prospective life. Would an incoming tenant see value in paying for that or just want to replace or remove it?

29.6 Tenant right and short-term improvements

29.6.1 These items are set out in Schedule 8 to the Act, generally covering shorter-term operational works to the land such as mole drainage, fertilising, growing crops and acts of husbandry but also includes manures in store and severed crops such as silage. The Schedule is set out in two parts:
- Part 1 – Short-Term Improvements
- Part 2 – Tenant Right Matters.

The distinction between the two parts may not now be material save where the law makes a difference. These items do not require landlord's consent to be eligible for compensation, though mole drainage does require the tenant to have served notice at least a month before doing the work.

29.6.2 The list of qualifying items in Part 1 of the Schedule was revised for England in 2015 and Wales in 2019, with an additional focus on soil improvement, the removal of some limitations (notably that it does not now matter whether the manures were purchased or not) and a shift from the value remaining from feedstuffs to the value of manure in store. The tenant right items in Part 2 of the Schedule remain the same.

29.6.3 The outgoing tenant is not entitled to claim compensation for crops or produce, pastures or acts of husbandry that contravene the terms of a written tenancy agreement unless he can show that the term was inconsistent with his responsibility to farm in accordance with the rules of good husbandry as set out in the Agriculture Act 1947.

29.6.4 S.66(2) requires that these be valued at their *value to an incoming tenant* but in accordance with any method that may be prescribed. While some rules (as for sod fertility value) are still prescribed in the Schedule the tables last issued in 1978 to 1983 are now repealed, removing their long-outdated values.

29.6.5 With current values now applying, this head of claim can, in combination with the associated acts of husbandry, amount to a significant figure.

29.6.6 Where the parties have agreed that the tenant has been allowed a benefit for making one of the works in Part 1 of the Schedule that is taken into account in assessing the compensation – the tenant has already had that benefit. Similarly, any public grant paid for the making of a Part 1 item is to be taken into account.

29.6.7 For items in Part 2, any method for assessing compensation that is prescribed in a written tenancy agreement will have effect. However, that situation is rarely, if ever, met.

29.6.8 Approaching a valuation for tenant right – The valuer will need to gather the required information according to the circumstances of the holding. This will cover the items for which a claim might be made, the recent rainfall and waterlogging history and an appraisal of the soils involved.

29.6.9 Where, as often, the valuation is left by the landlord to the ingoing tenant, that tenant's valuer will, like his client, be unlikely to know of his involvement until fairly close to the term date. The outgoing tenant's valuer, who is more likely to have more forewarning, should make good use of that time by negotiating with the landlord's agent over the state in which the holding is to be let and the treatment of the tenant's improvements and fixtures.

29.6.10 In practice, that means that the valuer's task is now to assess what items of tenant's work fall within Schedule 8 and find their value to an incoming tenant. The recent changes not only allow the use of contemporary values rather than the figures from the later 1970s but also a reappraisal of what is now known of soil science. The CAAV has published a review of these issues and produces regular updating guidance as to approaches and figures that might ordinarily be used, including spring and autumn unit values for nitrogen, phosphorus, potash, sulphur, magnesium and sodium.

29.6.11 These are subject to location and site-specific factors, notably:
- the level of rainfall, especially in the winter
- flooding and soil erosion
- the acidity of the soils

as each of these can significantly reduce or remove the value left behind from prior applications.

29.6.12 While it is now understood that **nitrogen** might not be exhausted in a single season, but rather some is mineralised in the soil and more drawn by plants from existing mineralised stocks, that is unlikely to be significant in wetter and more acidic areas (as in much of the south-west). However, where nitrogen has been applied in larger volumes to arable crops, such as potatoes or where residues are incorporated, in drier areas with less acidic, deep medium soils, there may be a carryover to be recognised as of benefit to the incoming tenant provided the nitrogen remaining from the application is more than 50 kg/ha. Equally, all nitrogen is taken as lost after waterlogging.

29.6.13 It is suggested that **sulphur** be treated in a similar way.

29.6.14 Phosphorus and **potash** are written off on the same basis as the old rules (phosphorus written off by halving each season over three growing seasons, potash similarly written off but over two growing seasons) but using current values.

29.6.15 Trace fertilisers and elements (such as manganese, copper, boron, zinc, molybdenum, iron, sodium, cobalt, selenium, nickel, iodine and silicon) are taken to be written off in the growing season if applied as foliar treatments but otherwise might usually be treated like potash if there is no countervailing evidence from the farmer's rotational pattern of application.

29.6.16 Lime is to be written over periods of years varying with acidity of the soil and rainfall levels. In the wettest, most acidic areas that might be over three or four years, but on clay soils in drier areas that might be as long as nine years. A further year can be added for applications to permanent grass or long leys where less than 250 kg/ha of nitrogen applied annually.

29.6.17 The assessment is now of the **manure in store** rather than previous feedstuffs. This requires an assessment of the volume and weight of manure in store, its nature (farmyard manure, slurry, etc) and nutrient composition and how well it has been kept. There will be more value retained in manures protected from the rain rather those in open lagoons.

29.6.18 That has then to be apportioned to exclude the contribution to it from feed (apart from corn) produced on the holding, such as silage and fodder crops. All corn, silage and other feeds from off the holding are taken into account – as imported manurial value. The easiest approach may be to assess the fodder from the holding (excluding that eaten or grazed in the field as manure from that might not be in store) fed to animals and deduct the remaining nutrient values in those feeds from the overall nutrient values in the stored manure. For this, manure from a tonne of grass silage might have typical nutrient values of 3.45 kg of nitrogen, 0.64 kg of phosphorus and 5.28 kg of potash and so, with spring 2020 values of 72p/kg for nitrogen, 63p for phosphorus and 45p for potash, that would give a figure of £5.26/tonne to which might be added something for soil conditioning value – all told, say £5.75/tonne.

29.6.19 Is the manure actually useful to an incomer to the holding? Is it conveniently stored? Is there evidence of market value for manure in that area, remembering that an incoming tenant will not need to consider haulage? If not, then the current values of nutrients should be applied to the analysed manure. A separate value for soil improvement could then be added.

29.6.20 For **manures applied to the land** it no longer matters if they have been purchased. These should be assessed with evidence of their composition on the same basis as other fertilisers, though with an adjustment for the method of application. Swift incorporation is more likely to see nitrogen retained within the season rather than lost into the atmosphere.

29.6.21 With the growing focus on soil health and status, the new Part 1 of Schedule 8 allows claims for soil improvers in their own right. That value can recognise the benefits of soils with more resilience, workability, improved biological activity and cost savings. At the time of writing these benefits may have a value that is less than the cost of making them but that is still a value available to the incoming tenant.

29.6.22 Mole drainage can only be claimed if the tenant served at least one month's notice in writing on the landlord before doing the work. Its costs may then be written off, typically over three to six years according to how soil conditions will preserve the mole.

29.6.23 Protection of fruit trees, if it has a value, is likely to be assessed on it the cost of erecting it apportioned for its remaining useful life.

29.6.24 While **clay burning** has been preserved as a head of claim it is quite possible that none has been done for more than a century.

29.6.25 Growing crops might ordinarily be assessed in their early stages at their cost to date, with CAAV costings available to assist with the typical costs of acts of husbandry. However, if the crop is not in good order, as where flooded or droughted out, then it may have little or no value to an incomer. Once well established, it may be valued more for its potential yield, after allowing for remaining costs and risks, than for its accumulated costs of production. What would an incomer pay to take to the that crop as it stands?

29.6.26 Longer-term crops such as orchards, vineyards or long-term horticulture might be appraised on the basis of their future income.

29.6.27 Severed crops will commonly have market value once their volume and quality has been assessed. However, that is often less true for clamp silage for which the CAAV provides an approach based on values for protein and energy drawn from other feeds with final adjustment as both a reality check as to value and whether the silage is indeed useful to an incoming tenant (see CAAV publication, *Silage: A Valuer's Guide*, Numbered Publication 183).

29.6.28 Acts of husbandry, usually annual in nature, can be assessed using their typical costs, assisted by the CAAV's Costings of Agricultural Operations. Those such as subsoiling which may have longer benefit can be written off over time.

29.6.29 There are various approaches to **tenant's pastures** which are to be assessed, like other items in this Schedule, at their value to an incoming tenant. The object of the pasture is to feed animals effectively, whether by grazing or conserving it as silage or, less commonly now, hay. With the cost of establishing a ley and its potential and continuing value if established and managed well, this is conventionally approached:

- using a costs basis – where no crop has been taken by grazing or mowing, on the basis of the reasonable costs of the materials and operations in its establishment and any further work prior it being grazed or mown. The hypothetical incoming tenant (here most likely in the spring), has at least been saved those costs and has the ley there ready and waiting. If properly established such that there is confidence in a useful crop of grass to be had from it, an extra payment would be paid for that potential yield – the name for that in the now repealed regulations was "enhancement"
- on the "face value" of the pasture – once a crop has been taken, on a more forward-looking basis using the expected unexpired life of the ley, its condition and future productivity with regard to such factors as it being:
  - a good seeds mixture, not dominated by coarser grasses
  - clover will assist a sward by fixing nitrogen but too great a proportion can make it too rich risking bloat in animals
  - clean and free from weeds, including couch grass as well as docks, thistles and dandelions, which would reduce its value and attract a dilapidations claim
  - properly managed since sowing as good management can see a ley improve in quality over time, making a depreciation approach often less apt unless that is the management approach adopted. However, when a sward is worn out, its value might then be negligible. A good and well-managed pasture that is several years old may be more valuable than a three-year-old one with the same seeds mix but which has been poorly managed.

That value will usually be greater if the pasture is grazed rather than taken for silage (especially if mown severely). Some approaches take account of whether it is fenced and has water supplied as well as the location of the field in relation to the buildings, particularly relevant, for example, to a dairy farm.

29.6.30 Whether on enhanced costs with no crop off or at face value, the alternative for the tenant is to go off the farm and take grazing or silage ground managed by someone else with the associated costs of keeping an eye on matters. The value lies in a good sward conveniently located and under the tenant's control. With those two points, one arithmetical approach is to:
- take the cost of a grazing licence across the year
- add the cost of supervising animals that are away – saved by having pasture at home
- deduct the annual husbandry costs of the pasture (fertilisers, sprays, rolling, etc)
- to give a figure for current value
- add for each of the remaining years at fractions of that value to allow for the assessed prospect of that sward deteriorating.

29.6.31 With these issues, valuers may find tenant's pastures a topic ripe for disagreement, making it worth recalling the object of the pasture is to feed animals well and conveniently; its value will lie in its ability to do that. The valuers should assess what is in front of them with its promise for the future. While not universally true, the further west the holding the higher the level of value is likely to be settled for tenant's pastures.

29.6.32 In hill districts, there is a particular head of claim for the **hefting, acclimatisation and settlement of hill sheep on hill land.** This head of claim recognises the value offered by those sheep being accustomed to that hill, knowing their boundaries on the open hill and acquiring a greater immunity to the local ticks and diseases which sheep from elsewhere would not have.

29.6.33 Hill sheep are those that have been reared and managed on a particular hill, have developed an instinct not to stray from it, are able to withstand its typical climatic conditions, and have developed a resistance to the diseases likely to occur in that area. Hill land is land on which only hill sheep will thrive.

29.6.34 What might an incoming tenant pay for those sheep with those attributes without which production might not be possible? The value is likely to be less than the cost of freshly hefting sheep, but it still offers an advantage to a tenant taking such a holding. In practice, the 2001 foot-and-mouth disease outbreak in which many hill sheep were taken saw a hefting addition of 30% of a ewe's value, an approach which links the issue to production values for the enterprise on which the tenancy is based.

29.6.35 While there may have been relatively few claims under the complex rules in Schedule 8 for **residual sod fertility**, the removal of the cash limits imposed by the now repealed 1983 regulations combined with the renewed interest in soil condition points to some significant values being claimable on eligible leys. This head of claim recognises that where a higher than usual fraction of the holding is in longer-term leys it has accumulated a nutrient and soil improvement value in the grass sod which rises as it has more clover and is grazed rather than mown. It may have a particular relevance to land in arable reversion under agri-environment agreements. Its apparently complex provisions are most easily worked through with the flow chart at Sod Fertility Flow Chart.

England and Wales – farm business tenancies

29.7 Tenant's improvements

29.7.1 By contrast to the position under the 1986 Act and in Scotland, the Agricultural Tenancies Act 1995 takes a principles-based approach to identifying compensatable tenant's improvements rather than providing historic lists of qualifying works. If the work in question meets two tests it qualifies for compensation:
- is it on the holding at the end of the tenancy?
- does it have written landlord's consent?

Together with the recognition of intangible advantages, this approach future-proofs the legislation and farm business tenancies against technological and commercial change which may make the other Acts outdated in this respect.

29.7.2 S.15 provides for two types of tenant's improvement:
- "physical improvements" made on the holding by the tenant in whole or part at his expense
- "intangible advantages" which are obtained for holding by the tenant's efforts and/or expense and which become attached to the holding (so as to be useful to a successor). This could include planning permissions, regulatory consents, licences, commercial goodwill and other items.

Those tenant's improvements can then qualify for compensation provided that:
- they remain on or attached to the holding at the end of the tenancy (s.16(2)) – if they have gone there is nothing of value for a successor
- the landlord has given written consent (with or without conditions) for them (s.17(1)).

More specific rules in s.18 apply for a planning permission (as a type of intangible advantage) to qualify for compensation requiring that it be for a physical improvement or a change of use, neither of which has been implemented by the end of the tenancy.

29.7.3 A tenant wishing to challenge a refusal of consent or the conditions proposed by the landlord can refer the matter to arbitration but only before starting to do the work. There is no such right in respect of a planning permission.

29.7.4 The measure of compensation is the value that the tenant's improvement has added to the holding as a holding (s.20). The basic analysis offered previously for tenant's improvements under the 1986

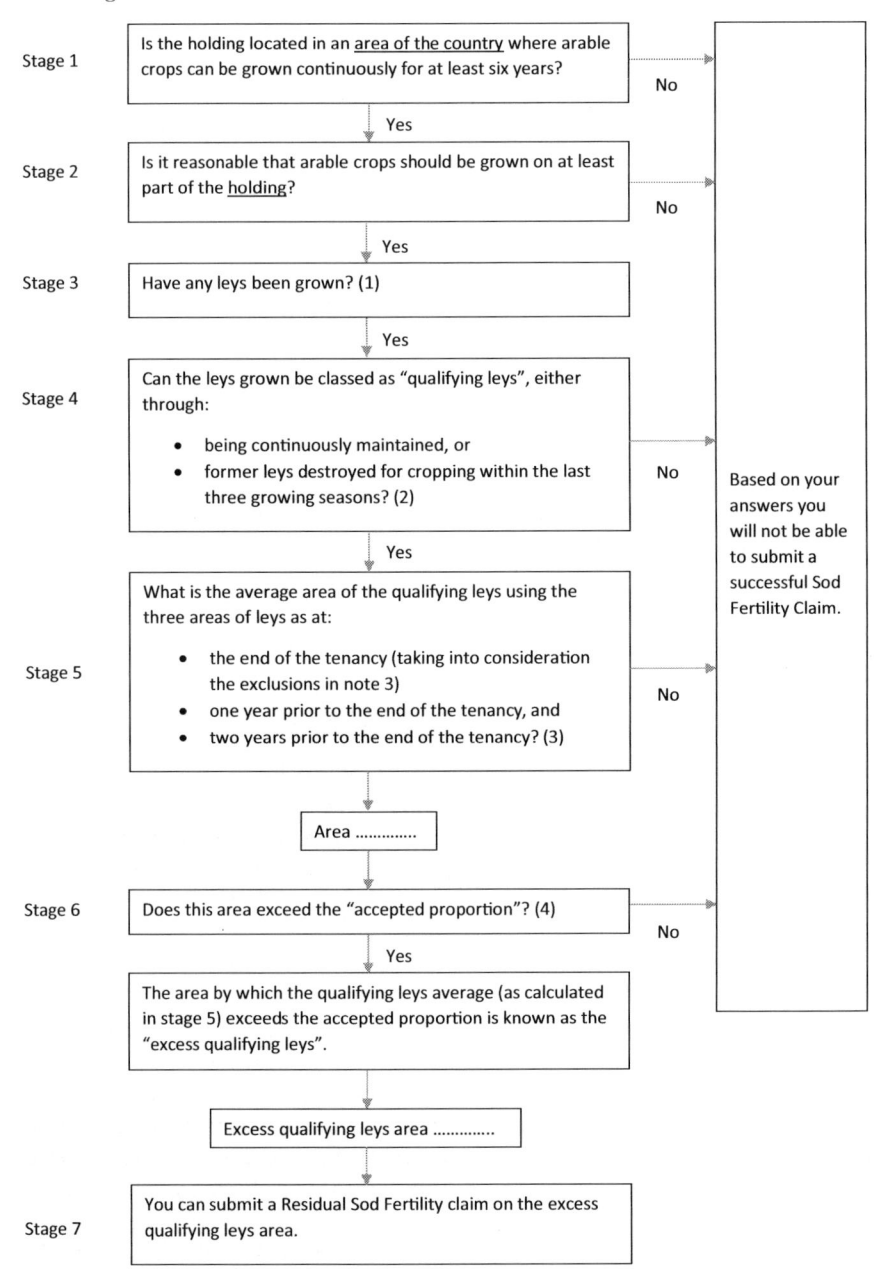

Stage 1	Is the holding located in an <u>area of the country</u> where arable crops can be grown continuously for at least six years? → No	
	↓ Yes	
Stage 2	Is it reasonable that arable crops should be grown on at least part of the <u>holding</u>? → No	
	↓ Yes	
Stage 3	Have any leys been grown? (1)	
	↓ Yes	
Stage 4	Can the leys grown be classed as "qualifying leys", either through: • being continuously maintained, or • former leys destroyed for cropping within the last three growing seasons? (2) → No	Based on your answers you will not be able to submit a successful Sod Fertility Claim.
	↓ Yes	
Stage 5	What is the average area of the qualifying leys using the three areas of leys as at: • the end of the tenancy (taking into consideration the exclusions in note 3) • one year prior to the end of the tenancy, and • two years prior to the end of the tenancy? (3) → No	
	↓ Area	
Stage 6	Does this area exceed the "accepted proportion"? (4) → No	
	↓ Yes	
	The area by which the qualifying leys average (as calculated in stage 5) exceeds the accepted proportion is known as the "excess qualifying leys".	
	↓ Excess qualifying leys area	
Stage 7	You can submit a Residual Sod Fertility claim on the excess qualifying leys area.	

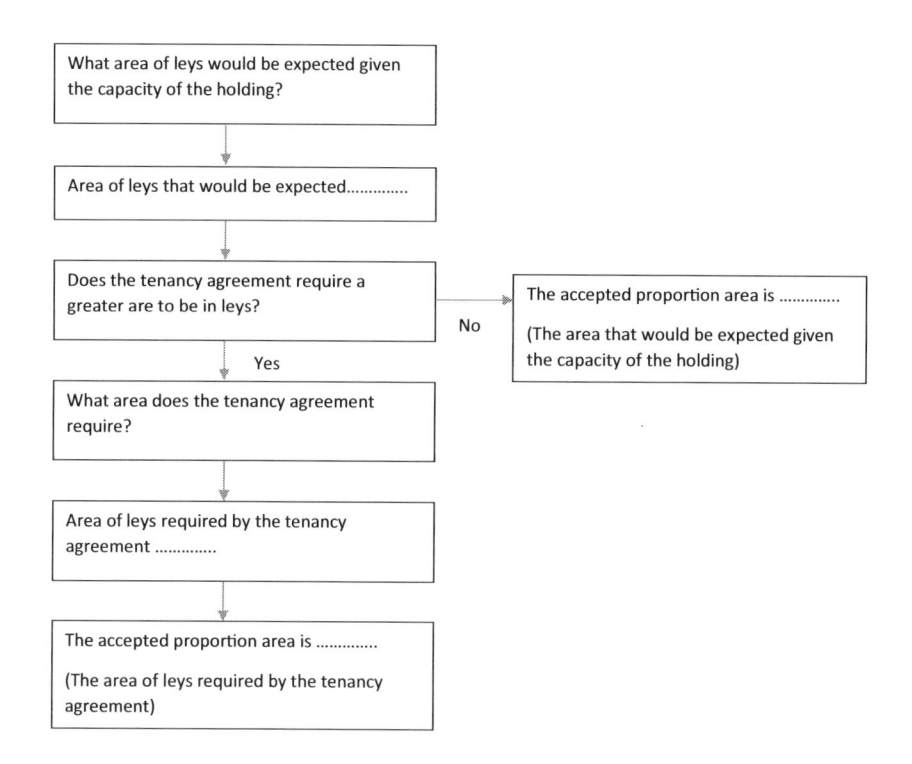

```
┌─────────────────────────────────┐
│ What area of leys would be      │
│ expected given the capacity of  │
│ the holding?                    │
└─────────────────────────────────┘
                │
                ▼
┌─────────────────────────────────┐
│ Area of leys that would be      │
│ expected..............          │
└─────────────────────────────────┘
                │
                ▼
┌─────────────────────────────────┐       ┌──────────────────────────────────┐
│ Does the tenancy agreement      │       │ The accepted proportion area is  │
│ require a greater are to be in  │──No──▶│ ..............                   │
│ leys?                           │       │ (The area that would be expected │
└─────────────────────────────────┘       │ given the capacity of the        │
                │                          │ holding)                         │
               Yes                         └──────────────────────────────────┘
                ▼
┌─────────────────────────────────┐
│ What area does the tenancy      │
│ agreement require?              │
└─────────────────────────────────┘
                │
                ▼
┌─────────────────────────────────┐
│ Area of leys required by the    │
│ tenancy agreement ..............│
└─────────────────────────────────┘
                │
                ▼
┌─────────────────────────────────┐
│ The accepted proportion area is │
│ ..............                  │
│ (The area of leys required by   │
│ the tenancy agreement)          │
└─────────────────────────────────┘
```

Act applies, save that this time the rent review would be that for an FBT. Similarly, where the tenant has been allowed a benefit for making the improvement or a public grant has been made for it, that is to be taken into account.

29.7.5 When the 1995 Act was enacted, it was not possible to contract out of that basis of valuation as with the writing down commonly operated under the 1986 Act. However, amendments in 2006 allow the landlord and tenant to agree a cap on the compensation so that it is the lesser of the statutory basis and a compensation limit agreed between the parties. Where the concept of a limit is agreed but the figure is not settled, the default figure is the cost of the work (s.20(4A) and (4B)). That can be relevant where a landlord wants to limit potential liabilities. As part of agreeing a cap, the tenant might seek other benefits, perhaps extending the tenancy to secure the full economic benefits of the work.

29.7.6 Routine improvements – In the final stages of the 1995 Act, the government accepted that the requirement for prior consent did not properly cater for tenant right type matters and so the concept of routine improvements was provided.

29.7.7 Defined by s.19(10) as a physical improvement that is not fixed equipment and is made in the normal course of farming the holding, routine improvements do not need prior consent, though this can be given and, indeed, can be sensibly provided for in the tenancy agreement.

29.7.8 If there is a dispute about consent that can be referred to arbitration at any time; that may even be in conjunction with the end of tenancy compensation claim itself.

29.7.9 It should be noted that routine improvements do not include items such as straw, silage, hay, or manure in store which as chattels are not fixed to the holding. If these are to be left at the end of the tenancy, this should be covered in the agreement.

29.7.10 The measure of compensation is the same value added to the holding as a holding by the item as for all other improvements under the 1995 Act. In practice, it is considered that the approach now possible in England for tenant right and short-term improvements is appropriate as the best way to identify the value of these items which are of value but in themselves sometimes potentially of relatively little weight compared to the overall value of the holding.

29.8 Tenant's fixtures

S.8 of the 1995 Act makes provisions for tenant's fixtures akin to those in the 1986 Act (see earlier section) though. The points of difference are that:
- all fixtures and buildings that qualify for end of tenancy compensation are excluded provided that the tenant has agreed not to remove them
- it does not require prior notification of the landlord – that interest and any valuation is left entirely for negotiation
- the tenant can remove the item up to six months after the end of the tenancy.

Scotland – Tenancies under the Agricultural Holdings (Scotland) Acts 1991 and 2003 and the Land Reform (Scotland) Act 2016

29.9 Tenant's improvements with tenant right

29.9.1 The 1991 Act sets out the provisions for compensation for tenant's improvements at waygo. For works made since 1948, these are defined by the list in Schedule 5 (revised in January 2019). No other works are capable of compensation under statute. Schedules 3 and 4 cover improvements made earlier with slightly differing rules.

29.9.2 The list in Schedule 5 is divided into three categories for qualification of compensation:
- Part 1 – some specialist and generally uncommon works (though also including the laying down of permanent pasture and the planting of

orchards and fruit bushes) which require landlord's consent which may include conditions
- Part 2 – the generality of works, many involving fixed equipment (including buildings) and land reclamation, for which the tenant has to have served written notice on the landlord at least three months before starting the work, advising the landlord of that intention and the manner in which it is to be carried out. If the landlord wishes to object, he then has one month in which to serve a counter notice saying so, which the tenant can refer to the Land Court
- Part 3 – more operational and tenant right type works which do not need a prior process to qualify for compensation.

29.9.3 The items added to or amended in Schedule 5 in January 2019 are:
- these items in its Part 2 for which notice must be served on the landlord at least three months before starting work:
  - 17 – Installation, provision, distribution or storage of electricity, gas, power, heat or light – so beyond the previous electric light or power
  - 22 – Provision of means of sewage, waste or pollutant disposal, or provision of means of managing water on land – so beyond the previous sewage disposal
  - 23A – Erection of structures for the management or storage of slurries or manures
  - 23B – Provision of means of storing silage
  - 23C – Works to dwellings
- these items in its Part 3 for which no notice is required
  - 29A – Extension to permanent grassland of removal of tree roots, boulders, stones or other similar obstacles to cultivation
  - 30 – Application to land of soil improvers, conditioners, digestates, manure or fertiliser, whether organic or inorganic.

These changes apply to works made after the changes took effect.

29.9.4 With many tenant's works undocumented for this purpose, the Land Reform (Scotland) Act allowed a three-year time window from June 2017 (extended for the Covid-19 pandemic to December 2020) as an Amnesty Period in which tenants can bring forward past works for recognition for end of tenancy compensation. However, all works not completed by 13 June 2017 must follow the statutory rules if they are to be the subject of a compensation claim.

29.9.5 The measure of compensation for all tenant's improvements is "such sum as fairly represents the value of the improvement to an incoming tenant" (s.36(1)). What would a hypothetical incoming tenant pay to take on the item on the end of the tenancy? Again, that points to considering the general demand by tenants likely to take the whole holding in question, not the specialist use by a particular individual. Judging that may typically require appraisal of the item in question, its condition, potential use and prospective life. Would an incoming tenant see value in paying for that or just want to replace or remove it?

29.9.6 The compensation is, though, to take account of and so be reduced for any benefit that the landlord has allowed the tenant for making it. Further, where both landlord and tenant have contributed to the work, any public grant is to be disregarded to the extent of the tenant's share of their combined funding of the work. Thus, in what may well be unintended drafting of the statute, where an improvement has been jointly funded, the tenant will only be due compensation to the extent that his contribution forms of the whole cost and is not enhanced by any grant.

29.9.7 Where the tenant has made an improvement in Parts 2 or 3 and it is work that the landlord should have made under his legal obligations to provide fixed equipment, then the compensation must be assessed on the statutory basis. That overrides any contrary term agreed between landlord and tenant (s.33A).

29.9.8 SLDTs and LDTs – S.45 of the 2003 Act applies Schedule 5 to SLDTs and LDTs with ss.48 and 49 applying the rules on consent and notice for Part 1 and Part 2 improvements to qualify for compensation.

29.9.9 S.47 also applies the measure of value as the fair value to an incoming tenant with account taken of any benefit allowed by the landlord and the same rules on public grants as for 1991 Act tenancies.

29.9.10 MLDTs – Schedule 2 of the Land Reform (Scotland) Act 2016 makes the necessary amendments to the 2003 Act for these provisions to apply to MLDTs.

29.9.11 Repairing tenancies – The rules for improvements for repairing tenancies have yet to be issued.

29.9.12 Other tenant right matters outside Schedule 5 – There are common law rights to claim compensation for these, such as for growing crops.

29.10 Tenant's fixtures

29.10.1 S.18 of the 1991 Act makes provision for tenant's fixtures that are very similar to those of s.10 of the 1986 Act save that the removal may be at any point up to six months after the end of the tenancy, subject to the landlord's right of pre-emption by a notice within one month of the tenant's notice, electing to buy the item at *the fair value to an incoming tenant.*

29.10.2 SLDTS, LDTs and MLDTs – For the first time since 1851, these agricultural tenancies have no statutory provision for tenant's fixtures so that they become part of the landlord's land, with no right to remove or to compensation. That makes it more important for tenants to make correct use of the provisions qualifying an item of work for compensation as an improvement.

29.11 Sheep stock valuations

29.11.1 Where the tenant is required to leave a stock of sheep on the holding at the end of the tenancy, the question of compensation is covered by ss.68 to 72 and Schedules 9 and 10 of the 1991 Act. These provide differing rules according to when the lease was granted and whether the termination date is Whitsunday (28 May) or Martinmas (28 November).

29.11.2 The basic approach is to find the average numbers and values by class of sheep sold in the preceding two or three years, subject to specified adjustments disregarding the lowest values and allowing, within limits, for current condition of the flock.

29.11.3 This is based on external sales and is not an equivalent to the claim in England and Wales for the value of a flock being hefted and acclimatised to its hill.

29.11.4 SLDTS, LDTs and MLDTs – No statutory provisions are made for this and so where a tenant is expected to leave a sheep stock on termination, compensation will rely on contractual provisions in the lease.

29.12 Compensation for tree planting

29.12.1 Where the tenant has planted trees on the holding since 2003 and they are for future cropping, s.45A of the 1991 Act allows a claim can be made between the parties based on:
- the "worth" of the trees to a willing purchaser for future cropping
- any loss of rent by the landlord by retaining the trees until cropping, together with the cost or returning the land to agricultural use afterwards.

If the former exceeds the latter, the tenant can claim the excess and vice versa.

29.12.2 SLDTS and LDTs – S.53(2) of the 2003 Act applies this provision to LDTs but not to SLDTs.

29.13 Diversification

29.13.1 Where part of the holding has been put to a non-agricultural purpose since 2003 under the provisions of ss.40 and 41 of the 2003 Act and that change of use has increased the value of the holding, s.45A of the 1991 Act also allows the tenant to claim from the landlord such compensation as "*as fairly represents the value of the use, change or carrying out of the activities to an incoming tenant*".

29.13.2 Any benefit allowed by the landlord to the tenant for the change and public grants in respect of that change are taken into account in the same way as for a tenant's improvement. Thus, where a diversification use has been jointly funded, the tenant will only be due compensation to the extent that his contribution forms of the whole cost and is not enhanced by any grant.

29.13.3 No compensation is payable if the change of use makes the land unsuitable for agricultural use by an incoming tenant or the use of fixed equipment for the change of use means that the landlord cannot fulfil his obligations at the beginning of a new tenancy.

29.13.4 In this, there is no firm boundary distinguishing use for non-agricultural purposes from Schedule 5 improvements. It is suggested that the same aspect of an improvement should not be compensated twice.

29.13.5 SLDTS and LDTs – S.53(2) of the 2003 Act applies this provision to LDTs but not to SLDTs.

Northern Ireland

29.14 Compensation

29.14.1 Tenant's Fixtures – Even with the general absence of agricultural tenancy law in Northern Ireland, s.17 of the Landlord and Tenant Law Amendment Ireland Act 1860 (Deasy's Act) makes similar provisions for tenant's fixtures in respect of personal chattels, engines, machinery and buildings erected or affixed to the holding at the tenant's sole expense. Such items can be removed at any point up to two months after the end of the tenancy with compensation to the landlord for any damage caused in their removal. The landlord does not have a statutory basis to pre-empt that removal.

29.14.2 Tenant's improvements and tenant right – As the Ulster tenant-right provisions were repealed some years ago, compensation for improvements and tenant right matters is an entirely contractual matter.

29.15 Settling a claim

29.15.1 The approach to settling the valuation varies across the country indeed has changed with the reduction in the number of valuations. However, the traditional practice is for valuers to meet and inspect the holding. They were often accompanied by their clients, offering the outgoer the opportunity to explain some of the peculiarities of the farm to his successor. This approach has the considerable merit of enabling each valuer to understand the issues concerning the other. Very often agreement on matters such as dilapidations to hedges were settled there and then, albeit without the client becoming involved unless there was a point of principle, commonly with the time (number of days) being agreed and costing up left until back in the office.

29.15.2 In other cases, the valuers may inspect separately, prepare their schedules of claim and then meet in the office to settle the claim. This approach demands assiduous note taking and a good memory to recall accurately the point at discussion once away from the holding.

29.15.3 Whichever approach is adopted, this is perhaps the most intense farm inspection that any valuer will undertake, walking each field in detail and worrying, in every sense, at each unhinged gate or overgrown hedge. For those weaned on the exercise, it is also one of the most fun parts of the job, meeting and sparring with one's peers on a variety of highly technical matters. It is also, like a dispersal sale, a significant moment for the clients involved: one quitting, often retiring from a holding on which they may have lived all their life and the other embarking on a new chapter. As with all valuer's work, it requires care and attention to achieve the most satisfactory outcome for the client and professional satisfaction for oneself.

Further reading

Commentary on the Valuation of Improvements Under the Agricultural Tenancies Act 1995 (CAAV, 3rd Edition, 2014)
End of Tenancy Compensation Under the Agricultural Holdings Act 1986 (CAAV, 2006)
Guide to the Amnesty for Tenant's Improvements in Scotland (CAAV, 2017)
A Review of Tenant Right Valuation (CAAV, 2015)
Silage: A Valuer's Guide (CAAV, 2006)
Tenancies, Conacre and Licences: Arrangements for the Occupation of Agricultural Land in Northern Ireland (CAAV, 2015)

30 Compensation for disturbance

30.1 Introduction

While the end of a tenancy is an occasion for the claims against each other by landlord and tenant for compensation for improvements and dilapidations, agricultural tenancy law allows the tenant of a traditional tenancy with extended security of tenure to make further claims for disturbance where the tenancy has been ended in whole or part to the landlord's benefit. Particular provision is made where land is taken back for development. This is not available where the tenant brings a tenancy to an end.

30.2 England and Wales – Agricultural Holdings Act 1986

30.2.1 The opportunity for the tenant to make a claim for compensation for disturbance arises where the tenant quits the holding after:
- the landlord has served an open notice to quit
- the landlord has served a notice to quit under Cases A (retirement for smallholdings), B (permission for a non-agricultural use) or
- the tenant has acted under s.32 to enlarge a landlord's notice to quit part of the holding.

Where this applies, the tenant can claim basic compensation and, in more limited situations, additional compensation.

30.2.2 These payments are calculated on the basis of the rent actually due for the holding at the time of termination. Where the tenant enlarges a notice to quit part of the holding, the relevant rent is only that for the part subject to the original notice unless that part is more than a quarter of the holding and the remaining part would not be viable as a separate holding.

30.2.3 **Basic compensation for disturbance** – This is a minimum of one year's rent with more payable on proof of the tenant's losses and costs directly attributable to leaving the holding, capped at two year's rent. These losses are those

> unavoidably incurred by the tenant upon or in connection with the sale or removal of his household goods, implements of husbandry, fixtures, farm produce or farm stock on or used in connection with the holding,

and includes any expenses reasonably incurred by him in the preparation of his claim for basic compensation.

(s.60(5))

30.2.4 To claim more than the one year's rent, the tenant is to give the landlord:
- at least one month's notice ahead of the end of the tenancy of his intention to make this claim
- the opportunity to make a valuation of any goods, implements, fixtures, produce or stock being sold before it is sold.

30.2.5 Additional compensation – Where basic compensation is due, the main occasion where additional compensation for disturbance can be claimed where the notice to quit is under Case B following a planning permission for non-agricultural use. It will also be available where the notice is under s.27(3)(f) for a non-agricultural use not requiring planning permission and in some more complex situations outside the standard Cases.

30.2.6 The payment is at the rate of four years' rent of the holding.

30.2.7 Compensation for early resumption – While a notice to quit must usually give at least 12 months' notice to the next termination date, it is possible for a tenancy agreement to enable earlier resumption on short notice where all or part of the holding is required for a non-agricultural use.

30.2.8 S.62 allows the tenant in such a case to claim for

> the value of the additional benefit (if any) which would have accrued to the tenant if the tenancy had, instead of being terminated as provided by the notice, been terminated by it on the expiration of twelve months from the end of the year of tenancy current when the notice was given.

Such claims might include lost profits, the costs of replacement housing and other issues but will need to be substantiated.

30.3 England and Wales: farm business tenancies

There is no equivalent provision for disturbance compensation for FBTs which leave this as a contractual matter.

30.4 Scotland – Agricultural Holdings (Scotland) Act 1991

30.4.1 The opportunity for the tenant to make a claim for compensation for disturbance arises under s.43 where the tenant quits the holding after:
- the landlord has served a notice to quit for a non-agricultural use for which planning permission has been obtained
- the tenant has acted under s.30 to enlarge a landlord's notice to quit part of the holding.

Where this applies, the tenant can claim compensation and, in more limited situations, additional compensation.

30.4.2 These payments are calculated on the basis of the rent actually due for the holding at the time of termination (after the deduction, where relevant, of any tax, rate or other charge in respect of the holding that the landlord could set against Income Tax). Where the tenant enlarges a notice to quit part of the holding, the relevant rent is only that for the part subject to the original notice unless that part is more than a quarter of the holding (either by area or by rental value) and the remaining part would not be viable as a separate holding.

30.4.3 Compensation for disturbance – This is a minimum of one year's rent with more payable on proof of the tenant's losses and costs directly attributable to leaving the holding but capped at two year's rent. These losses and expenses are those

> directly attributable to the quitting of the holding which is unavoidably incurred by the tenant upon or in connection with the sale or removal of his household goods, implements of husbandry, fixtures, farm produce or farm stock on or used in connection with the holding, and shall include any expenses reasonably incurred by him in the preparation of his claim.
>
> (s.43(3))

Where claiming more than one year's rent, the tenant must give the landlord at least one month's notice of any sale and the opportunity to value any stock being sold.

30.4.4 Additional compensation – Albeit with some complex provisions, where compensation for disturbance is due, s.54 allows a claim for additional compensation of four years' rent to assist the tenant with the reorganisation of his affairs.

30.4.5 The payment is at the rate of four years' rent of the holding.

30.4.6 Compensation for early resumption – While a notice to quit must usually give at least 12 months' notice to the next termination date, it is possible for a tenancy agreement to enable earlier resumption on short notice where all or part of the holding is required for a non-agricultural use, including building and planting.

30.4.7 S.58 allows the tenant in such a case to claim for

> the value of the additional benefit (if any) which would have accrued to the tenant if the land had, instead of being resumed at the date of resumption, been resumed at the expiry of 12 months from the end of the current year of the tenancy.

Such claims might include lost profits, the costs of replacement housing and other issues but will need to be substantiated.

30.5 Scotland – limited duration tenancies

30.5.1 SLDTs and LDTs – S.52 applies the rules for disturbance compensation to SLDTs and LDTs where resumption is for non-agricultural use under s.17 or the tenant enlarges a notice to resume part.

30.5.2 MLDTs – Schedule 2 of the Land Reform (Scotland) Act 2016 makes the necessary amendments to the 2003 Act for these provisions to apply to MLDTs.

30.6 Northern Ireland

There is no statutory provision. All matters are left for the contract.

31 End of tenancy
Landlord's claims

31.1 Introduction

31.1.1 The landlord of an agricultural holding can also have the power to make a claim for dilapidations or damages against the tenant for falling short on the obligations under the tenancy agreement.

31.1.2 Claims might be made for failures by the tenant to meet obligations on such points as maintaining buildings, keeping pasture in good order and free of weeds, keeping ditches clear and maintaining gates and fences; other weeds, couch and, increasingly, blackgrass are concerns on arable land.

31.1.3 The authority for making a claim for dilapidations varies:
- in England and Wales:
  - under the 1986 Act, there is statutory provision under ss.71 to 73 for deterioration and damage
  - the 1995 Act makes no statutory provision, leaving such matters to the common law and the terms of the tenancy agreement
- in Scotland, the bases of claim under ss.45 (deterioration and damage) and 45A (where arising from diversification and cropping of trees) of the 1991 Act are excluded by s.47(3) where a record of condition has not been made and then only for deficiencies after one is made. In practice, this precludes claims being made in many tenancies.
- in Northern Ireland, it will purely arise under contract, though Deasy's Act makes provision for action for waste.

31.1.4 Under both 1986 and the 1991 Acts, a landlord may make a claim for damages where the tenant's rights of freedom of cropping and disposal of produce have been exercised in a way that has caused injury or deterioration to the holding.

31.1.5 In **England and Wales**, the timetable is set for claims on the end of a 1986 Act tenancy by its s.83 with:
- notice of intention to make a claim given to the other party within two months after the end of tenancy, specifying the nature of the claim

- any appointment of an arbitrator to be made within eight months after the end of the tenancy

31.1.6 In Scotland, the timetable is set for all forms of tenancy by s.62 of the 1991 Act with:
- notice of intention to make a claim given to the other party within two months after the end of tenancy, specifying the nature of the claim
- the parties then to settle matters within four months of the end of the tenancy but able to apply to the Scottish government for up to two successive two-month extensions
- any application to the Land Court or for an arbitrator is to be made within eight months of the end of the tenancy.

31.1.7 In the background, there are common law provisions on waste, but they are very rarely met in practice.

31.2 England and Wales

31.2.1 The 1986 Act – S.71 provides two alternative approaches for a landlord:
- the one normally adopted under s.71(3) making a claim under the terms of the tenancy agreement
- s.71(1) enables the landlord to make an end of tenancy claim for dilapidations or deterioration of the holding or any part of it caused by the tenant's non-fulfilment of the rules of good husbandry as set out in the Agriculture Act 1947. In summary, those require carrying out necessary maintenance and repair work, keeping the land clean and in a good state of cultivation, maintenance of the pastures, keeping the farm properly stocked, maintaining crops and livestock free from disease and protection of harvested crops.

31.2.2 This is then capped by s.71(5) (and the common law) at the diminution in the value of the landlord's reversion in the holding. It has been commonly observed that such a diminution may be hard to prove in the markets of recent years with the strength and thinness of land markets and the opportunity for lettings as a farm business tenancy.

31.2.3 S.72 does allow the landlord to make a further claim, on giving a notice on writing to this effect more than a month before the end of the tenancy, where it can be shown that the "value of holding" has been reduced by dilapidation, deterioration or damage under either the terms of the tenancy or the rules of good husbandry, but that has not been recoverable under s.71. The measure of compensation is

> the decrease attributable to the matter in question in the value of the holding as a holding, having regard to the character and situation of

the holding and the average requirements of tenants reasonably skilled in husbandry.

As with 1995 Act tenant's improvements assessed under the basis of the change in the value of holding, this might typically be assessed by capitalising the reduction in the rent achievable for the holding because of the deterioration. Again, that may limit the recoverable loss.

31.2.4 Farm Business Tenancies – These is no statutory basis for a dilapidations claim which can therefore only arise under the terms of the tenancy agreement and will be aided by including a provision in the agreement framing how such a claim might be made, preferably using a procedure to mirror that for the tenant's claim for improvements.

31.2.5 Any claim would be subject to the common law cap of the diminution in the value of the landlord's reversion.

31.2.6 Making a claim – As with a tenant's claims, the landlord's claim under the 1986 Act must be notified under s.83 within the two months after the end of the tenancy and it is sensible to follow that same basis under the 1995 Act. Ideally, this should be supported by a detailed claim (see 31.4) so that it can be reviewed by the tenant while the facts as at the end of the tenancy are still observable. In practice, a review some months ahead of the tenancy leading to a conversation with the tenant may see more matters remedied by the tenant using his own resources which might often be the more economic answer where that is in a position to do this.

31.3 Scotland

31.3.1 Where the prerequisite of a record of condition has been met, s.45 of the 1991 Act provides for a claim to be made for dilapidations caused by the tenant's failure to fulfil the rules of good husbandry, either for:
- a reduction in the value of the holding, set at that value, or
- the cost of making good any dilapidation, determination or damage to any part of the holding or anything in or on it, set at the cost of making that good, either
  - on the basis of the rules of good husbandry or
  - the terms of the lease.

31.3.2 As part of the 2003 Act's regime for a tenant to take up a non-agricultural diversification or plant trees on the holding, s.45A provides a basis for a dilapidations claim where
- the conversion of the holding or part of it to a non-agricultural use since 2003 has reduced the value of the holding.
- trees have been planted and the value of the trees to a purchaser for future cropping is outweighed by the loss of rent, the landlord is entitled to claim that difference.

The tenant can claim compensation where the reverse of either of these applies.

31.3.3 Making a claim – A landlord's claim under either s.45 or s.45A of the 1991 Act is not enforceable under s.47 unless the landlord served notice on the tenant not less than three months before the end of the tenancy warning of the intention to make a claim.

31.3.4 All end of tenancy claims between the parties are referable to the Land Court for determination. With a common procedure for claims by both parties under s.62, the landlord must serve a notice of intention to make the claim on the tenant within the two months after the end of the tenancy, specifying at least the basis on which it is made (such as which terms of the tenancy will be referred to).

31.3.5 Ideally, this should be supported by a detailed claim (see 31.4) so that it can be reviewed by the tenant while the facts as at the end of the tenancy are still observable. In practice, a review some months ahead of the tenancy leading to a conversation with the tenant may see more matters remedied by the tenant using his own resources which might often be the more economic answer where that is in a position to do this.

31.3.6 The parties then have four months from the end of the tenancy to settle matters, a period that can be extended by two months by Scottish ministers at least twice. If not settled by the end of that period, a claim ceases to be enforceable unless it has been refereed within one month to either the Land Court or an arbiter.

31.4 Preparing a claim

31.4.1 Even where the scale of the enforceable claim is capped, it remains logical to approach the preparation of a dilapidations claim by first listing out and describing the items at issue and considering the cost of putting them right so assessing the ground for any claim and then considering the effect of the relevant cap. Such a list may also assist the practical management of the farm after the tenancy.

31.4.2 However and whenever done, it requires a careful and clear schedule, setting out:
- the field parcel, building or other item affected
- the problem in question
- the remedy
- the unit cost
- the claimed figure

That can now be easily supported by photographs as well as plans.

Example

	Unit Cost	Total

OS No 6531 – 6.8 ha stubble
1. Foul with couch – 6.5 ha
   - Spray with Roundup 2020
   - Spray at half dose 2021
2. High overgrown hedge
   - Cut and lay 224m
3. Pond in NE corner silted up
   - Clean out. 10 hours with drag line

Buildings
4. Covered Yard
   - Clean out all eaves and gutters
   - Replace sheets as marked on the plan

Farmhouse
5. Lounge – Redecorated 11 years ago, overdue 4 years
   - Redecorate including all preparation work, stripping wallpaper, stopping and making good, rubbing down, preparing and applying 2 coats oils to all previously painted wood and iron work, twice whitening ceiling and re-papering walls 4/7 of cost
6. Septic tank
   - Clean out and leave in working order
7. Farm Drive
   - Cut out potholes and fill making good to surface
8. Cattle Grid
   - 5 bent and loose tubular bars
     Take out and renew

31.4.3 Preparing the schedule and the costs requires a knowledge of farming practice and building work and care to link each item of claim to the relevant clause in the tenancy agreement or rule of good husbandry, as appropriate. The CAAV's annual Costings of Agricultural Operations will assist for acts of husbandry and the costs of products used can be ascertained from suppliers. Some may need quotations from a contractor, whether for the work or for the time required on a day works basis.

31.4.4 Past practice, at a time when there were generally more end of tenancy valuations taking place, was to value most field, and some building, works on a day-works basis (the cost of the number of days labour assessed as required by the work). That approach still holds in some parts of the country and among some valuers. There are some cases, particularly clearing long neglected areas, whether land or buildings, where so many days work is still the best way of assessing costs, bearing in mind that most jobs will be more effectively completed with two men rather than one. That said, the vast majority of routine field maintenance is now done by machine. The CAAV costings allow for both approaches.

31.4.5 This is not a matter of improving the farm from what was it supposed to be but of restoring it to the condition required by the agreement or the rules of good husbandry.

31.5 Settling a claim

31.5.1 The approach to settling the valuation varies across the country and indeed has changed with the reduction in the number of valuations. However, the traditional practice is for valuers to meet and inspect the holding. They were often accompanied by their clients, offering the outgoer the opportunity to explain some of the peculiarities of the farm to his successor. This approach has the considerable merit of enabling each valuer to understand the issues concerning the other. Very often agreement on matters such as dilapidations to hedges were settled there and then, albeit without the client becoming involved unless there was a point of principle, commonly with the time (number of days) being agreed and costing up left until back in the office.

31.5.2 In other cases, the valuers may inspect separately, prepare their schedules of claim and then meet in the office to settle the claim. This approach demands assiduous note taking and a good memory to recall accurately the point at discussion once away from the holding.

31.5.3 Whichever approach is adopted, this is perhaps the most intense farm inspection that any valuer will undertake, walking each field in detail and worrying, in every sense, at each unhinged gate or overgrown hedge. For those weaned on the exercise, it is also one of the most fun parts of the job, meeting and sparring with one's peers on a variety of highly technical matters. It is also, like a dispersal sale, a significant moment for the clients involved: one quitting, often retiring from a holding on which they may have lived all their life and the other embarking on a new chapter. As with all valuer's work, it requires care and attention to achieve the most satisfactory outcome for the client and professional satisfaction for oneself.

Further reading

Dilapidations on the End of a Tenancy (CAAV, 2007)

32 Valuing the tenancy

32.1 General

32.1.1 While the conventional agricultural tenancy is not assignable and the tenant is barred from subletting or parting with possession, it can still be an asset with value even if that cannot be directly crystallised in the marketplace unilaterally by the tenant.

32.1.2 The issue of the value of a tenancy arises particularly for the tenancies with extended security protected under the Agricultural Holdings Act 1986 and the Agricultural Holdings (Scotland) Act 1991, where combination of that longer-term security and the statutory rent review provisions are seen to create particular value. For the reasons shown by the illustrations in Chapter 11, shorter fixed-term tenancies with a limited period to their termination and those let at a market rent are much less likely to have a substantive value. While that will apply to the generality of FBTs, LDTs, MLDTs and SLDTs, it will also be so for a tenancy under either 1986 Act or the Scottish 1991 Act which is seen as likely to end in the near future, even if it is not under a current Notice to Quit.

32.1.3 While the most quoted case law on this concerns compulsory purchase and Inheritance Tax, the defining case concerns a partnership and the issue can be relevant to valuations for Capital Gains Tax, divorce and other circumstances. Those differing situations can motivate a client who might ordinarily struggle to perceive a non-assignable tenancy as having a value to want either a very significant value (as typically for compulsory purchase) or a minimal value (as typically for taxation).

32.1.4 For the client struggling to see why a non-assignable tenancy has a value when it cannot be voluntarily sold for value, the answers are:
- first, the real world one that he is choosing to keep the tenancy because it has a value to him – and many 1986 Act tenants, conscious of their scarcity, jealously protect their tenancies
- second and related to that, the rent provisions are seen as valuable
- third, the legal answer founded on the decision in *Crossman* (and applied the cases noted later on), that a value can be found by assuming a single relaxation of the bar on assignment so that the tenancy

can be sold to someone who is then bound by the bar. The question becomes, what would that person pay to stand in the shoes of the present tenant for that tenancy?

32.1.5 The most essential point, beyond seeing that there can be a value in a secure tenancy, is to distinguish two different circumstances in which the value of a tenancy might commonly be assessed:
- the market value of the tenancy to (usually hypothetical) third parties as an asset, usually assessed as a capitalisation of the profit rent advantage that such a tenant has against tenants of similar holdings on open market rents
- the marriage value between the interests of the landlord and tenant and so what each might pay the other merge or release their interests, in either case seeing extinction of the tenancy, also explored in Chapter 11.

32.1.6 Over and above that, Scottish law provides two occasions where these issues are to be considered:
- the 2003 Act provides the opportunity for a tenant to pre-empt a landlord's disposal of the reversion, with a statutory basis for the tenant to buy it
- the 2016 Act provides a mechanism (yet to be implemented) for a 1991 Act tenant to offer first to relinquish the tenancy to the landlord for value on a statutory basis and then, if the landlord declines, to offer the tenancy for assignment and so potentially for value.

32.1.7 The 2019 consultations in England and Wales on tenancy law canvassed a potentially similar approach for a tenant to assign the tenancy in the marketplace subject to the landlord's pre-emption. However, this did not go forward in the Agriculture Act's tenancy measures with DEFRA concluding in its March 2020 response that

> the current proposal is unlikely to achieve the policy aim effectively. . . . As a next step we will work with industry organisations to explore the alternative proposals and options for amending the original proposal to make it more effective.

32.2 The valuation approaches

32.2.1 Value of the asset – While there is a longer history of compulsory purchase and one or two other byways, the exploration of the value of an agricultural tenancy has really been through three cases.

32.2.2 *Baird* was an Inheritance Tax case in Scotland, where the earlier transfer of the tenancy within the family was accounted a chargeable event for Capital Transfer Tax (now Inheritance Tax) on the tenant's death. The tenant's executors asserted there was no value in the tenancy and did not offer a view in the alternative as to what that value might be were it found to exist.

32.2.3 The Tribunal found that the tenancy did indeed have a value. The District Valuer offered evidence of four types that might inform finding that value:
- settlements reached between landlords and tenants for the surrender of a tenancy for which the analysed cases showed the sum had ranged between 18.5% and 50% of the vacant possession premium.
- sale and leaseback transactions, where the price paid by the purchaser showed a discount ranging from 25% to 43% of the vacant possession value as the selling farmer would then retain a 1991 Act tenancy
- compensation paid where tenants' interests were acquired under compulsory purchase and found to be settled at 25% of the vacant possession value, albeit for relatively small areas of land taken for road widening
- the demand for farms offered to let on the open market with tender rents then of the order of £70 per acre compared to the £10 per acre passing rent for the subject farm

The District Valuer drew out of this the conclusion that the value of the tenancy was 25% of the land's vacant value. The tenant's critique was that the surrender and the sale and leaseback cases concerned parties who were motivated to transact; the present landlord had shown no wish to buy the tenant out. The value on the alternative profit rent approach was thought nominal and the compulsory purchase evidence concerned trivial areas.

32.2.4 The Tribunal found there was value in the tenancy and, stressing a sense that the evidence was tenuous, took the district valuer's assessment that it would be a quarter of the vacant value (implicitly half of the vacant possession premium) but downplayed this decision as precedent.

32.2.5 *Walton* was a much more closely argued Inheritance Tax case concerning a half share in a tenancy, setting out the profit rent basis for finding the arm's length value of a tenancy. The Revenue had argued the tenancy should be seen as and valued within the partnership which should be valued as whole and the deceased's share of that valued. The deceased argued that it was only the deceased's half share in the tenancy that fell to be valued – what would someone pay for that? With evidence given that the landlord did not wish to buy the tenancy, the special purchaser/marriage value argument was not relevant. The half share in the tenancy had to be assessed on a profit rent basis – and then take into account the point that the purchaser of that half would be buying into a tenancy with other co-tenants. The Tribunal found for the tenant. While the Revenue appealed, the Court of Appeal upheld the Tribunal in having regard to the actual, rather than a hypothetical, landlord, his inclinations and capacity; that landlord could not be assumed to be a bidder for vacant possession without evidence for that. Further, anyone considering purchasing this part share in the tenancy would have regard to their joint tenant.

Valuing the tenancy

32.2.6 In summary, if the tenancy of the holding in question was at a much lower rent than the market rent of an equivalent holding that created value to be found by capitalising it to show what someone would pay to take that tenancy at its reduced rent compared to one at an open market rent.

32.2.7 *Greenbank v Pickles* – this decision in a partnership dispute affirmed and developed the approach based on profit rent and is seen to stand as the leading case.

32.2.8 While the compulsory purchase case, *Wakerley*, is still sometime cited as authority for a tenancy being worth five times the profit, this part of the Tribunal decision concerned the value of the business, not the tenancy for which, in the absence of evidence, £100 was awarded. Other compulsory purchase cases have, according to their facts and the evidence, generally followed a profit rent route.

32.2.9 In *Layzell v Smith Moreton and Long*, a negligence case, a court awarded damages for a lost succession on the basis of restitution through buying a freehold, granting an equivalent tenancy and then selling the reversion. A cost-based rather than a value-based approach, that has not been widely followed.

32.2.10 The valuation in *Greenbank v Pickles* – The final assessment in this case can be presented as follows in a series of steps, illustrated by the following assessment in *Greenbank v Pickles*:

- what is the difference between the current rent and the rent that would be set at a review? How long might that difference last? That is typically the period to the next date at which a notice served now would see a rent review happen. What is the capital value of that difference in rent?
- what is the difference between that review rent and an open market rent? How long might that last? What is the capital value of that difference? The three-year period assumed here reflected the circumstances of the case; ordinarily a much longer period might be used
- then, a value on the succession that was possible in this case. While this is not directly an asset of the tenant something was included. It could be argued that the tenant has an interest in the potential for succession by a family member, if only to avoid clinging on to the tenancy until death to protect a family home.
- finally, the recognition of end of tenancy compensation claims.

Bottom Slice		
Estimated arbitrated rent	£ 5,000	
Rent Payable	£ 4,720	
Profit rent (bottom slice)	£ 280	
YP @ 5% for 1.5 years	1.4059	
Value of bottom slice profit rent	£ 393.65	£ 393.65
Top Slice		
Estimated tender rent	£10,500	

Bottom Slice		
Estimated arbitrated rent	£ 5,000	
Profit rent (top slice)	£ 5,500	
YP @ 12% for 3 years	2.4018	
Value of top slice profit rent	£13,209	£13,209.90
Statutory compensation to tenant (net of dilapidations)	£ 9,650.00	
Succession		
Top slice profit rent	£ 5,500	
YP of a reversion to perpetuity @ 16% after 33 years (tenant's expected life)	0.04664	
	£ 256.52	£ 256.52
Total		£23,510.07
Say		£23,500.00

32.2.11 Value Between landlord and tenant – This recognises the potential for a deal between landlord and tenant that can unlock value to be shared between them. This special marriage or even fair value is not directly accessible by anyone else and not accessible at all if one party is not interested in such a deal (see also the discussion of marriage value in Chapter 11). However, such transactions have been common whether seeing the landlord pay the tenant to surrender the tenancy or the tenant buying the landlord's reversion and merging the two interests.

32.2.12 The basic principle is that the investment value of agricultural property when let (and so the value of the landlord's interest to any third party) has generally been less, sometimes much less, than the value when vacant. That difference is the vacant possession premium (VPP) and can only achieved by one of these two transactions – or by the landlord waiting for the tenancy to end.

32.2.13 This approach can be simply illustrated:

Value of land vacant	£1,000,000
Value of land as let	£725,000
Vacant Possession Premium	£275,000

32.2.14 If divided equally, then either:
- the tenant could be paid £137,500 to surrender
- the tenant could pay the landlord £862,500 to buy the freehold.

32.2.15 While, as here, conventional illustrations show this value being split equally between the two parties, experience is that that rarely happens in practice. An equal split indicates an equal balance of motivation between the two parties. In practice, more often the tenant is paid less than half the VPP, but there are occasions (as where other value is released for the landlord) that the tenant will get much more. However, the assumption of equality is followed, perhaps necessarily, in the Scottish statutory examples considered later on.

32.2.16 The investment value will itself reflect something of the opportunities for the landlord to secure possession, for such reasons as the tenant's age or health, the presence or lack of a qualifying successor and development opportunities.

32.2.17 The division of the VPP may also reflect the willingness or reluctance of one party to undertake the transaction. A tenant who is ill, aging or under financial pressure may be more motivated and willing to agree a lesser split. Equally, where the landlord has reason to be keen but the tenant sees a long future on the holding, the landlord may have to pay more to attract the tenant's interest and secure vacant possession.

32.3 Scottish statutory cases

32.3.1 The tenant's right to pre-empt a landlord's sale of the holding
– The 2003 Act created the opportunity for 1991 Act tenants, having registered their interest, to have the first chance to buy their holding if the landlord was looking to sell it, setting out procedures and a valuation basis.

32.3.2 The statutory basis for this was to be a payment to the landlord of the investment value plus half the vacant possession premium together with recognition of the waygo compensation and any enforceable landlord's claim for dilapidations.

32.3.3 While it seems that this might never been formally followed on its full statutory basis, this has provided the background to a number of negotiated transactions between landlords and tenants, not only where the landlord might be intending to sell the reversion.

32.3.4 With Scottish investment values for farmland often thought to stand at half of the vacant value (a matter for evidence at the time and markedly lower than values in England), this is often seen as putting a value of 25% of the vacant value on the tenancy for such a transaction.

32.3.5 Relinquishment and assignation – The 2016 Act has provided for this as an opportunity to assist retirement and create openings for new or progressing farmers.

32.3.6 With a structure procedure and timetable, the basic mechanism would be:
- the tenant offering to relinquish the tenancy to the landlord
- with a valuer then to assess the values of the land vacant and let and the waygo claims
- the landlord to choose whether or not to accept the offer, paying the determined value if accepting the relinquishment

- if the landlord does not accept the relinquishment, the tenant then has a period in which to assign the tenancy but can only enforce that on the landlord where the assignation is to new entrant or a progressing farmer, as to be defined.

32.3.7 This sees the potential for use of both the two valuation bases.

32.3.8 For relinquishment, the valuation is that between landlord and tenant, based around the vacant possession premium with:
- the vacant value which would be as ordinarily understood
- the investment value is to be that of the tenancy with the actual tenant in situ and disregarding any prospect of succession. The period to repossession should also allow for the period of a notice to quit. This value will then be lower where the tenant is younger than where the tenant is older. This real world assumption means that there will be no simple standard relationship between vacant and investment value, by contrast to what is often assumed. This actual figure will be a matter for evidence and judgment of each case
- the tenant's claim for waygo compensation as evidenced, including evidence of prior notices to the landlord for improvements or any Amnesty agreement, and quantified
- whether a landlord's claim for dilapidations can be made and, if so that too being quantified with evidence.

32.3.9 The resulting sum to be paid by the landlord for relinquishment is to be half the difference between the vacant and investment values and the balance of the compensation claims.

32.3.10 If the landlord declines to pay that, the tenant can then offer the tenancy to the market for assignation. An assignation can only be enforced on the landlord if the assignee is a new entrant or a "progressing farmer". This will be a matter for the market. Where the assignee is a connected party (though presumably one not able to succeed), a gift by the tenant will be treated as a disposal at market value.

32.3.11 The market value would usually be assessed on the basis of profit rent.

32.3.12 The potential additional factor here though is the limitation on who can be an assignee. That reduces the number of possible bidders. It could appear to limit the bidders to those with the least resources, though it does not rule out those with strong financial backing from family members. The effect of those factors and how they might be managed on what might be paid for an assignation has yet to be seen.

32.4 In practice

32.4.1 Away from such statutory situations, the most common need for the valuation of a tenancy will arise from a family issue or partnership dissolution. In a divorce, the tenancy might even be part of the settlement, as the court can award its transfer between parties, though more often wanting to protect its ability to produce income to be divided.

32.4.2 In a partnership dispute, the terms of the partnership agreement will be critical to the approach to the valuation. In some cases, perhaps more often with the most recent agreements, the terms will be explicit and relevant. In other cases, not always confined to older agreements, matters may not be quite so clear, whether because of poor drafting or that the family has so contrived divisions of family property or partnership assets in ways which significantly affect the value of the property, intentionally or not. While this may also involve vacant property, it becomes more complex with tenancies where there is far less market evidence in the first instance.

32.4.3 In such cases, one of the first jobs of the valuer will be to identify the difficulties that are likely to arise from such arrangements and to propose a practical approach which enables realistic appraisal but still achieves the equitable settlement intended under the original arrangements. This may require constructing an appropriate valuation clause where one may only be partial, made irrelevant by the conduct of the parties or be missing.

32.4.4 The valuer approaching the valuation of an agricultural tenancy has to consider whether there is actual evidence of a special purchaser, most obviously the landlord, for the tenancy or not. If there is not a special purchaser, the profit rent approach is the default basis.

Either answer requires appraisal of the precise nature of the asset in question,

32.4.5 No valuation is necessarily straightforward, and this is particularly the case here. Gwyn Williams, exploring the question of the value of tenancies in the context of compulsory purchase in the fourth edition, summarised the position with his usual practical efficiency:

> It has been stated that valuation is not an exact science. Nowhere is this more true than when valuing and agricultural tenant's unexpired interest.

Further reading

Valuation of Agricultural Tenancies (CAAV, 2010)

33 Valuations for other agreements

33.1 Introduction

33.1.1 Other forms of farming agreement also require practical valuation work. For some like grazing agreements, the licence fee for grazing or the headage fee for tack agreements may simply be knowledge of market rates. In other cases, the figures needed to make contract farming and other business agreements work for both parties, so making them practical and sustainable, will require more insight.

33.1.2 One theme in discussion of the various possible relationships is the possible role for the valuer as adviser to the agreement, helping facilitate it as it continues and mediating on issues and responses to circumstances to avoid or overcome problems. This might apply equally to the longer-term agreements that would facilitate transactions for ecosystem services.

33.2 Contract farming agreements

33.2.1 The principle of contract farming agreements sees the farmer retaining the service of a contractor to help deliver the farmer's business. That may often be motivated by the farmer seeing this as an alternative to the cost of reinvestment in machinery, but the contractor can also bring modern machinery, technology, labour, management skills and buying power to an enterprise. It may be long term or simply to cover a period of ill health or following a death. More generally, if the farmer wants to show that he is in business, he needs to show the activity, the management and the risk of that as well as the hope of return. The contractor should look to see a business return on the services he provides. What is the business deal that does that?

33.2.2 Just as when considering tenancy tenders, so with contract farming tenders a prudent farmer need not take the highest bid. Contractors, as much as would-be tenants, often bid to win, with an optimism that may not be sustainable. Some do not fully recognise the costs that they carry, including depreciation or allowing adequately for the risks of farming life. The strength of the arrangement will ultimately lie in the relationship between farmer and contractor and the ease with which it will work in hard times as well as good.

33.2.3 In the standard model for a whole enterprise agreement:
- the farmer will have a claim on a sum identified as the *basic return*, that is a first call where there is a surplus on the contract account after all income and costs (including the contractor's basic change) have been met. That basic return is only payable to the extent that there is a surplus and so the farmer is at risk for this reward
- the contractor is conventionally able to receive two payments:
  - a *basic charge* for the costs of his services
  - a *share of the calculated divisible surplus*, where there is one after the farmer's basic return.

33.2.4 Those three payments are interlinked:
- the more there is the prospect of a return from a divisible surplus to the contractor, the more he may accept the basic charge as just covering costs
- the less there is the prospect of a return from a divisible surplus to the contractor, the more he will need to see a commercial return on the basic change, leaving less for farmer's basic return
- the higher the farmer's basic return, the less chance there is of a useful divisible surplus.

The art will be to structure the agreement to satisfy both or to recognise that this cannot be done and look to other answers.

33.2.5 These principles can be illustrated by these calculations which are offered only to do just that and are not example accounts. With the divisible amount being split so that the farmer retains 30% and the contractor is paid 70%, these calculations compare the outcomes from:
- a good year in which good yields and prices saw strong receipts and moderate costs saw good returns that could pay the farmer's basic return and give the contractor a share of the divisible surplus
- a bad year with poorer receipts and higher costs, resulting in a balance that cannot fully pay the farmer's basic return (in this illustration only yielding three quarters of the figure used in the arithmetic) and leaves no divisible surplus. Were such an outturn to become seen as the norm for this enterprise, the figures in the agreement or the agreement itself might have to be reconsidered.

	A Good Year	A Bad Year
Receipts	£500,000	£400,000
Costs of the Enterprise (including contractor's costs and basic charge)	£300,000	£325,000
Balance	£200,000	£75,000
Farmer's Basic Return	£100,000	£100,000
Farmer's Shortfall		–£25,000
Divisible Surplus	£100,000	£0
Shares		
• Contractor (70%)	£70,000	£0
• Farmer (30%)	£30,000	£0

33.2.6 More specific issues may need to be considered as where:
- land has to be put into good order, possibly reducing the farmer's basic return for an initial period
- where the contractor is relying on the contract to fund substantial investment in, say, machinery to deliver it
- improvements are to be made
- where improvements are not made and so, for example, without a silage clamp the silage is baled or
- there are other agreements such as for grain storage or sheep hire.

33.3 Joint venture operations

33.3.1 These are individual arrangements built around the circumstances and characters involved.

33.3.2 It might be that the dairy sector has seen more creative use of this approach with the benefit of a regular income and the separate value in the cattle, but there are successful examples in combinable cropping with farmers forming a single business to manage those crops over their land area that allows them:
- to reduce machinery and other overheads and potentially focus on efficiency of operation
- to have time for other enterprises, whether specialist crops or livestock, or businesses on or off the farm.

33.3.3 Typically, these are structured as companies or partnerships. Either model requires decisions about shareholdings for capital input and earnings.

33.3.4 One hill sheep model sees a landowner, a shepherd and an adviser each have shares with the opportunity for the shepherd to build his shareholding for profits were that to be financially feasible.

33.3.5 The valuer's task is to find the structure and the deal that makes for a sustainable business operation.

33.3.6 "Share farming" – While there are many variations of this concept around the world, in the conventional form in which this loose term is used in England, essentially for what might be seen as share land management two separate and independent businesses are on the farm with:
- the landowner responsible for all property matters, including buildings and repairs
- the operator responsible for the farming.

Jointly discussing policy, they then divide what is regarded as a common gross output between them (so distinguishing it from the division of profit and loss under a partnership, where costs are shared).

33.3.7 Other than identifying the deal between the parties and drafting a clear agreement as to what each is to do, the critical decision is the share of the agreed gross output that is allocated to operator who will typically be carrying much larger share of the costs attributable to the two operations. Relatively small adverse movements in this percentage can remove the hope of profit for the operator.

33.3.8 That observation points to:
- the need for initial work on sensitivity testing the basis for the division against a variety of outcomes, and, even then, allowing for outcomes that might not be foreseen. There is not a natural logic based on inputs or other grounds that will drive the outcome objectively; it has to be a pragmatic judgment of what will work.
- the fundamental need for a good and practical relationship between the two businesses so that the landowner has empathy with the operator and the issues the operator can face. If committed to the agreement, they need to be able to rewrite it when circumstances make that what must be done.

Further reading

Arable Farming with Contractors (CAAV, 2020)
Contract Farming for Breeding Livestock Enterprises (CAAV, 2020)

Part 5
Professional practice

34 Farm agency

34.1 Introduction

34.1.1 As discussed in the chapter on freehold property valuations, the amount of land offered for sale has reduced significantly since the Second World War and indeed in the 35 years since the First Edition of this book was published. Whether the economic factors which have influenced the reduction in supply will subsist beyond record low interest rates and Brexit remains to be seen. However, most farming families continue to be driven by the idea of passing property they have inherited down to the next generation and hence the cultural pressure, if nothing else, seems to militate against any significant increase in sales activity.

34.1.2 The situation is by no means uniform around the country. It remains to be seen whether the land reform agenda in Scotland will prompt more owners to sell previously let farms, but aside from that there are parts of the country, northern and western Devon being a classic example, where there are substantial numbers of smaller owner-occupied farms which have been sold to lifestyle buyers who are then far more likely to resell the farm at the end of their tenure than seek to pass it on.

34.1.3 As the number of farms coming forward for sale has reduced so the methods of marketing, or perhaps more properly the media used, have changed. In 1985 the focus would have been on preparing hard copy particulars and advertising and ensuring that local farmers knew of the potential opportunity as well, subject to the nature of the property, as publicising it to the London market. In recent years, the phenomenal reach of e-media, whether through a firm's website, email circulation or social media, has overtaken much traditional media, although not to the extent that many agents have felt able to completely forsake advertising in Farmers Weekly, Country Life or the like and relevant local papers. There are likely to be some prospective purchasers who operate solely on e-media and will not refer to print publications at all; at the same time, there is also a group which may well identify a potential property of interest in print media before downloading digital particulars from the agents' website or, although perhaps less for farm property than smaller residential sales, marketing platforms.

34.1.4 Aside from purchaser activity, the use of print media is still likely to be supported by the other side of the advertising coin; agents marketing property to prospective purchasers are also marketing their services to potential vendors, many of whom may subscribe to the relevant publications for something other than the property advertisements.

34.2 Methods of sale

34.2.1 There remain three principal methods of selling rural property: sales by auction, by tender and by private treaty. The following section considers the position as it applies in England, Wales and Northern Ireland, the position in Scotland being different and more binding on the parties in terms of the making and acceptance of offers.

- Sales by auction – where the property is offered for public bidding at a set time and place; in the past public auction was commonly used for sales of rural land and farms, particularly in the case of sales from a deceased's estate. At the fall of the auctioneer's hammer the purchaser signs and exchanges contracts, also paying a deposit, normally 10%, with completion generally 28 days later.
- Sales by tender – there are two approaches here, formal and informal tender. As with a sale by auction, bids are invited to be submitted prior to a fixed closing date and time, but in writing rather than bidding in person. The vendor will then select the successful bidder, although reserving the right not to accept the highest or indeed any bid. In a formal tender process, the selected bidder will be required to sign and exchange contracts on notification of their success and as with an auction matters will proceed to completion very rapidly thereafter. An informal tender will still reserve the vendor's position in terms of accepting any particular offer but will offer greater flexibility after the tender date.
- Private treaty – in this case, the property is advertised for sale publicly but with no specific dates for bids to be received; rather, potential purchasers submit an offer for consideration at any time by the vendor and, once an offer is accepted, solicitors are instructed and matters progress hopefully to exchange of contracts and ultimately to completion. Matters can become more complicated where there are, as vendors and agents will hope, a number of competing purchasers; in those cases, agents may move to invite purchasers to submit best and final offers a process akin to an informal tender.

34.2.2 There are benefits and challenges with all three approaches. Earlier generations of valuers would have routinely opted to use auction as a method of sale for both land and farms. It offers a degree of certainty not available with other methods, with the sale exchanged on the fall of the hammer. However, that is also a disadvantage to the extent that a prospective purchaser should only bid at auction if they have the funds available to proceed. This tends to exclude prospective purchasers who are relying on the sale of property to support a purchase unless they are willing to fund the purchase in the interim through bridging finance, and those who are unwilling to invest in the cost of a valuation or survey with no guarantee of being able to proceed with a purchase.

34.2.3 In a buoyant market this may not be a problem, but where the market is thin vendors may be happier giving prospective purchasers the opportunity to proceed at their own pace rather than forcing their hand and finding that they decide not to bid. Latterly, auction has been less favoured, albeit during periods when there has been strong demand for rural property. This perhaps reflects in part the increasing influence of urban purchasers who may be less experienced and more nervous of auctions than their rural counterparts who have been weaned on livestock markets and auction sales. That said, auction still has its place particularly where certainty is required and sometimes on occasions where it is necessary to show best value.

34.2.4 Formal tender has similar advantages to auction to the extent that certainty is delivered relatively soon after the tenders have been received. It also has the benefit that tenderers bid without knowledge of the level of bids which one has at a public auction. This can make the approach particularly effective where there is land attractive to a number of special purchasers, paddocks which would enable a number of different neighbouring householders to extend their gardens for example. However, the degree of uncertainty is disliked by prospective purchasers, who must estimate the offer necessary to secure a purchase without being fully aware of the competition. That will dissuade some prospective purchasers from tendering, nervous that they might overbid. Others will face the same problem as at auction; they can only bid if they have the cash readily available or are willing to finance the purchase until they are able to raise funds from a sale.

34.2.5 The informal tender can be attractive where there may be particular complexities or conditions with the sale; consequently, it is often used for sales of development land. Indeed, in its most informal approach it is used to create a shortlist of potential purchasers with whom the vendor then engages in more detailed negotiations, sometimes including a second tender exercise. This has been taken to extremes by public sector bodies both in selling property and in procuring services.

34.2.6 Whilst sometimes warranted for complex properties, provided bidders have been warned in advance, this iterative approach can be vulnerable to abuse with negotiations carrying on long after the original tender to the extent that some bidders and agents can feel the original tender process was academic and indeed a waste of resources. That is particularly the case where bidders are not forewarned that the tender would not, of itself, be decisive. Some purchasers can, very properly, be aggrieved in such circumstances and feel that they have been misled with reputational damage to the agents concerned and the wider profession.

34.2.7 Private treaty is increasingly the favoured approach, offering vendors absolute flexibility over their approach to the market and, in the relatively buoyant market which we have enjoyed for rural properties in recent years, it has become by far the most common method for farm sales. Once again, it is a more instantly recognisable approach to those lifestyle buyers and new entrants to owning farm property who have normally been used to it in residential property transactions.

34.3 The conduct of the sale

34.3.1 Introduction – Whichever approach is adopted much of the work involved in the conduct of the sale will be common across all three methods. Traditionally the agent's role might be seen as falling into four broad, if overlapping, phases:

- initial engagement with the client – often in a competitive situation with other agents and involving a Marketing Report or similar proposal and hopefully culminating in confirmation of instructions
- preparation – involving taking and preparing particulars and marketing
- marketing – launching the property on the market and dealing with potential purchasers and
- negotiation and arranging the sale – drawing prospective purchasers' interest to a conclusion whether through some form of competitive bid or private negotiation, confirming the sale and instructing solicitors.

34.3.2 Increasingly, however, agents are becoming involved in more work after instructing solicitors, particularly in dealing with Enquiries before Contract. In contrast with most residential sales, where these would be dealt with by vendor and solicitor, many farm and particularly estate clients will look to the agent to address the queries raised. This can be particularly time consuming for those who are ill-prepared or who are encountering the task for the first time as explored further later on. Needless to say, this reinforces the need for care and comprehensive data gathering in agency work.

34.3.3 Initial dealings with the client – Sometimes an agent may have the luxury of a sale instruction from a client without the need to compete with other agents. However, vendors are becoming increasingly discriminating and inclined to ask more than one agent to submit proposals for a sale, whether to compare effectiveness, introduce some competition on fees or simply test different opinions of the value of the property.

34.3.4 Consequently, those initial engagements with a potential vendor may be equally akin to a beauty parade as a technical exercise. However, professionalism will remain key with transparency over values and costs and

clarity over approach required, particularly where there are options. The Marketing Report or Appraisal should cover:
- method of sale, or options with recommendation
- lotting, again potentially with options
- any key constraints – whether on the property, access for example, or the particular market
- development potential and how to exploit or protect that
- key influences on demand and anticipated timescales
- anticipated price and guide price
- marketing proposals and budget
- personnel involved
- business terms including fee basis.

34.3.5 This is rather more substantial than used to be the case, but competition and the added pressures of consumer protection regulations and quality assurance demand a more comprehensive approach. It should also include some comment on limitations, both limiting disclosure and particularly being clear that the report is prepared for marketing purposes only, is not a valuation and cannot be used as such.

34.3.6 Where the vendor is a charity in England and Wales the agent should follow the specific procedures of the Charities Act 2011 and further detail will be required including what works, whether of repair or otherwise, are recommended before the property is taken to the market.

34.3.7 Preparation – Assuming that this has not been dealt with in the initial stages, which may be the case where there is no competition, the first step should be to confirm instructions. The point has been laboured elsewhere, but whilst it is tempting, particularly with a sale, to press on pending confirmation of instructions, it is critical that these are secured at the outset to protect against abortive work and confusion but also to meet legal requirements for conducting estate agency business, most particularly the Estate Agency Act 1979 as amended by the Consumer, Estate Agents and Redress Act 2007 and associated regulations.

34.3.8 At the same time agents will need to ensure that they have complied with the current Anti-Money Laundering Regulations (at time of writing, 2019) which are amended almost as often as the Red Book. This can be a somewhat embarrassing conversation with a long-established client asking them to identify themselves, but in practice clients face such regulations in much of their business activity so the antipathy and indeed antagonism anticipated by some agents when the Regulations were first introduced has not really materialised. The *Dreamvar* case, concerning a bogus vendor and the purchase payment disappeared from the country, is but one illustration of the pitfalls of inadequate verification of identity, here leading to an insurance claim against the purchaser's solicitor. In broader terms, the General Data Protection Regulation 2016 has introduced extra disciplines in retaining both clients and potential purchasers' data and thus maintaining lists for circulating property particulars.

34.3.9 Once instructed almost all firms involved in property agency will have established procedures and templates for particulars, advertising, whether traditional or virtual, media and sale boards. These will vary from practice to practice and with different types of property and there will normally be precedents for the new valuer to follow.

34.3.10 Property particulars now need to be available in a wide variety of media and it is important for agents to consider how accessible they are and how easily followed whether being on paper or on laptop, tablet or phone. The latter, in particular, require a far more sophisticated approach than simply uploading a traditional set of particulars which can become almost unintelligible if not formatted for the relevant platform.

34.3.11 Whilst the media may have changed much of the detail required remains the same as was the case at the foundation of the first local valuers' associations in the nineteenth century. All the key elements referred to in earlier valuation chapters need to be covered in some degree, although it is likely that the description of the property will focus very much on the dwellings and buildings with the land covered in a few brief paragraphs, covering land quality, principal cropping and possibly alternative use.

34.3.12 Agents must be careful to keep themselves up to date with the various regulatory requirements, whether statutory or professional, which cover the description of property, whether in advertising or particulars. In particular, the Property Misdescriptions Act 1991 was repealed in 2013, having been replaced by consumer and business protection regulations. The role of the Office of Fair Trading has now migrated to the Competition and Markets Authority (CMA) and Trading Standards are also involved at both a national level, with Powys County Council Trading Standards as the lead enforcement authority for the main legislation of the Estate Agents Act 1979, and locally to the extent that local trading standards teams may be involved where consumers feel that they have suffered at agents' hands. The CMA has become more active recently in investigating agents' practices, although with the focus on the residential more than the farm agency market.

34.3.13 There are professional regulations to consider. The RICS Code of Measurement Practice, whilst still referring to the practice of pacing out a car park, offers guidance on standard definitions and greater accuracy for the measurement of buildings and land.

34.3.14 Other statutes and regulations also apply with the Energy Performance of Buildings (England and Wales) Regulations 2012 requiring particulars to include the first page of the EPC. In Scotland, a Home Report is required for dwellings, with a Single Survey and Valuation, a property questionnaire and and energy report. In planning terms, houses or buildings may be affected by Listing as of Architectural or Historical Interest. Beyond the house and buildings many farms will be affected either by designations (a Nitrate Vulnerable Zone (NVZ) or a Site of Special Scientific Interest (SSSI), for example) or by entering into agri-environmental schemes. This is all information to be captured and included in the particulars, albeit with some discretion over the amount of detail provided.

34.3.15 Alongside the particulars, agents may include video, including drone filming (with its own regulatory issues) and other media offering increased access to an impression of the property for prospective purchasers before they commit time and expense to a physical inspection.

34.3.16 Where sales are by auction or tender with purchasers required to exchange contracts on the fall of the hammer or acceptance of a tender, the particulars will need to be supported by all the necessary legal and technical documentation. This means that the preparation period for these methods of sale will be rather longer than for a sale by private treaty. That begs an interesting question, given the added complexity of transferring property and the far longer timescales associated with the legal processes for farm sales than used to be the case: should the same information be prepared at the outset of a private treaty sale? Whilst that might seem attractive in terms of timescales and accelerating the process, vendors are unlikely to be willing to incur the legal fees involved until they have received an acceptable offer.

34.3.17 Marketing – Once ready and crucially once the particulars and marketing programme have been signed off by the client, ideally but less often with a commitment to advise you of any material changes, the agent will launch the property on the market and hopefully be engaged in a succession of fruitful discussions with prospective purchasers. The old dictum of *caveat emptor*, let the buyer beware, still applies to a considerable extent, but statute effectively holds the agent to a duty of care to the prospective purchaser as well as the vendor, with a particular emphasis on protecting consumers more than businesses.

34.3.18 The National Trading Standards Guidance on Property Sales (2015) sets out summary guidance on the information that should be provided for prospective purchasers alongside guidance on the conduct of the business and in particular ensuring that consumers are able to make informed decisions when dealing with you, referred to as *transactional decisions* in the Consumer Protection and Business Protection Regulations (CPRs and BPRs) applying across the United Kingdom. These transactional decisions and the agents' associated responsibilities are not simply limited to whether or not to buy, or even make an offer, but include whether or not to make further enquiries or view the property. Thus, it is important that agents are careful to give as accurate an impression as possible of the property both in the particulars and other marketing media and in discussions, whether by email, phone or in person, with prospective purchasers.

34.3.19 In making a judgement on an agent's behaviour the CPRs and BPRs refer to misleading information which may be provided:
- over the telephone or in course of discussions
- in details in property particulars of other marketing material and
- visually – in photographs, video clips, floor plans etc.

34.3.20 The test of misleading and reasonableness is not based on a professional's understanding but rather on the *"average consumer"*. This is someone who is "reasonably well-informed, reasonably observant and circumspect" (*Guidance on Property Sales* 2015). In certain circumstances, where a commercial activity is targeted at a particular group of consumers the "average consumer" may be taken to be an average member of that group. Whether or not that might apply to a farm sale and what the impact of that might be may be an interesting conjecture but not something most agents would want to test.

34.3.21 In parallel with the various discussions with prospective purchasers, the agent will need to keep the client vendor updated on progress and occasionally there can be a difficult balance to strike between encouragement and managing expectations. An agent will not wish to admit that their marketing exercise has produced little or no interest and the client will not wish to hear that, whilst there has been some interest, no one has found the property attractive. Whilst much of the focus of the CPRs and BPRs is on the relationship with and behaviour towards prospective purchasers they also cover dealings with vendors, who are also consumers of the agent's services. Agents will be in breach if they give a false impression of interest in the property.

34.3.22 Negotiation and arranging the sale – Where the sale is by auction or tender, the particulars and advertising will notify prospective purchasers of the day of the auction or the closing date of the tender. Auctions may occasionally be postponed, and properties may be withdrawn in advance of the sale date where a particularly attractive offer has been received, but generally the timetable is set in advance.

34.3.23 Where the sale is by private treaty, how long the marketing period continues will turn on the nature of the property and in particular the level of demand. However, at some stage, most vendors and agents would hope that, after between one and two months of marketing, it will be time to bring matters to a conclusion. In some cases, it may be that there is an outstanding offer or, less auspiciously, where only one offer is received, matters can move straight to the legal niceties. In most cases, there will be a number of competing offers and the agent will seek to draw matters to conclusion whether by a series of accelerated negotiations or by inviting 'best and final' offers.

34.3.24 It is important for the agent to agree a strategy with the vendor before embarking on this process. It is also key to keep detailed records of interest during the marketing stage to ensure that no potential bidders are missed from a competitive exercise. Similarly, the agent should make sure that the exercise is transparent for any prospective bidders, in particular ensuring that they explain that the vendor is free to accept any or none of the offers, not simply the highest. The invitation to bid may include many of the requirements that one might see in a tender:

- that the bid should be for a fixed amount
- that it should not be related to any other offer, such as £100 more than the next highest bid
- that it should not be conditional, other than where conditions are specified by the vendor's agent
- where timescales are fixed, such as that exchange of contracts should take place within 28 days of acceptance of an offer, that these are explicit
- that identification information is provided for anti-money laundering purposes
- that funding sources should be clearly explained.

34.3.25 That final point is particularly important; the highest bid may seem attractive but where it is to be funded from the expected sale of another property, not yet on the market when the offer is made, then the inevitable delay to exchange may make a lower cash or fully financed bid more attractive.

34.3.26 In negotiating with prospective purchasers, whether in the context of such a bidding exercise or in one to one negotiations, the agent must be careful to be clear about the terms of any offer that a purchaser might make, ensuring that they have sufficient information to advise the client. Thus, returning to the CPRs and BPRs, alongside blatant falsehoods such as inventing offers, misrepresenting the detail of an offer when reporting to a client, which could include failing to report something through ignorance, is also a breach.

34.3.27 In dealing with prospective purchasers, breaches include:
- failing to report offers to the vendor, unless they have specifically asked you not to report particular types of offer
- failing to inform a purchaser whose offer is accepted subject to contract that the vendor wishes to continue to market the property until exchange
- failing to advise the vendor of any services that you as agent have been asked to provide the buyer or referral fees that might be earnt
- pressuring a potential purchaser through persistent and or aggressive interventions
- pressuring a vendor to accept a lower than reasonable offer.

34.3.28 Whilst most rural valuers would characterise these practices as a problem with residential agency rather than farm sales, there are risks that agents might lapse into over-exuberance in encouraging a bidder or fail to report an offer to a vendor which is below a level which they have previously rejected from another bidder.

34.3.29 Once an offer has been made and accepted, the agent will report that agreement to both vendor and purchaser and their respective solicitors in detail including the price and any conditions or other provisions in respect of the property, the offer or the legal arrangements. Once again, most firms will have a precedent to follow, often referred to as a Memorandum of Sale or Heads of Terms, which will give the new valuer guidance on the detail to be included. Whilst this will focus on the key terms, additional information may be required, particularly where multiple lots are involved and rights of way, fencing between lots and the separation and maintenance of services will all become more complex.

34.3.30 The sale – As mentioned earlier, once the sale had been agreed that was the end of the involvement for most agents other than perhaps valuing any commodities left on site at completion and occasionally liaising between the parties and their respective solicitors. However, land sales have generally become both more complex and far more time consuming.

34.3.31 Theoretically, the fact that far more titles are registered than was previously the case should have simplified at least part of the process, although in common with all of us the Land Registry is far from infallible and the information held is limited. Whilst transfer documents can be shorter than the contracts and conveyances they have replaced, the process of reaching exchange has become far more challenging.

34.3.32 The purchaser's solicitor will want to be sure that they can protect their client against any adverse consequences from the purchase or, failing that, at least draw any potential problems to their attention. The focus of much of this work will be the Enquiries before Contract, whether General Enquiries which, as suggested, apply to most sales, and the specialist Agricultural Enquiries. The latter extend to 36 sections, with multiple questions in each, 32 pages and over 12,000 words before the vendor, their solicitor or agent has attempted a response. Thankfully, there will be some questions, sometimes many, where "not applicable" is a fair answer. However, in many cases considerable research is required and where historic rights of way, private services and long-standing service occupancies are involved much research may be required.

34.3.33 Purchasers' solicitors, who in previous generations might have been content to accept the best guess of the vendor, not least where the purchaser was a neighbour, are now far more concerned about the extension of liabilities to their client, for past environmental issues for example, and by association to their professional indemnity insurance, to take such a cavalier approach. They will be pressing for a comprehensive response and, failing that, for the vendor to accept liability, whether through indemnities or insurance. Increasingly, vendors and their solicitors will look to the agent to research and compile the responses, often under some time pressure.

34.3.34 The amount of detail required is far more than that needed to produce sales particulars and some data, particularly, for example, in respect of services such as water, drainage and electricity, can be gathered prior to the property going to the market. Indeed, it is becoming good practice amongst some agents to produce a summary of the services with a plan and proposals to deal with rights to the service involved, access to equipment and sharing liabilities where services are located off the property to be sold or supply more than one consumer. The same may apply to rights of way where a property is split into multiple lots or a vendor sells some land but retains other property.

34.3.35 Less practical issues may be harder to manage, with long-established tenancies and service occupancies being particularly sensitive especially where they have been dealt with on oral agreements as a matter of trust between the parties.

34.3.36 Lotting – Whether or not to split a property into lots will depend on the nature of the property itself but also the likely interests and strength of the neighbours. This is where the knowledge and expertise of a local agent will become particularly valuable, understanding which of the neighbours would be interested in particular parts of the farm and whether that is sufficient to lot the property to suit that and other competing interests.

34.3.37 A little like auctions, lotting seems at present slightly less prevalent than was previously the case or that may simply be a reflection of the fewer farms which are on offer. There was a time when all but the most commercial of farms would have been lotted to give buyers from outside of farming the chance to buy the house and immediate land and local farmers the chance to bid for the balance in an appropriate number of lots. However, the increased capacity and inclination of city-based purchasers to purchase entire farms, not least to provide privacy around the dwelling, with an intervening period where the financial crisis impacted agents' perceptions of the ability of city purchasers to buy land, means that far more farms and blocks of land seem to be sold either as one lot or "as a whole or in lots" with the agent and vendor wanting to secure the sale of the house and land, where included, before contemplating any other sales.

34.3.38 Timescales – As mentioned earlier, timescales for farm sales seem to have become far more extended with some sales now taking between 6 and 12 months to complete. This leads to frustrations for all the parties, most particularly the principals, and it is important for the agent who will typically be the first port of call, whether or not the matter is in their hands, to manage expectations.

34.3.39 Once again, the position can be improved by gathering data beforehand, although this is unlikely to answer all the questions. Matters can be further improved by the simple expedient of the clients' agents and solicitors meeting to try to deal with issues in bulk rather than the current tendency of an iterative email conversation which, even where that does not lead to errors and omissions, is likely to prolong the process.

34.3.40 Extending this approach and as mentioned previously, there is a strong argument for agent and solicitor preparing the sale together from the outset. This may accelerate some costs for the vendor, but this earlier cost is likely to be save larger costs and more time later that, in extreme cases, may prevent a purchaser losing patience and looking elsewhere.

34.4 Valuation clauses on farm sales

34.4.1 Introduction – The majority of farm sales will include some form of valuation clause to deal with a range of issues including consumable stocks, growing crops, fuel oils and sometimes even fixtures left on the holding when the farm is sold. The extent of the clauses will vary both with individual holdings and the custom of the local area where the farm is being sold. If anything, valuation clauses have become slightly less comprehensive over time, partly as the additional value of items covered by the valuation has paled even further in to insignificance compared with the value of the farm and partly as purchasers, not unreasonably, have become more resistant to paying for other items on top of what they may regard as an already inflated sale price. To that extent, the earlier practice, particularly as observed in previous editions, in the south of England, of wide, all-embracing clauses, very similar to a tenant right valuation, but with the added iniquity for the purchaser of there being no offsetting dilapidations claim, is far less common than was the case.

34.4.2 Basis of valuation – Where valuation clauses are included, for example to cover silage and other fodder in store, care must be taken to ensure that these are clear to avoid subsequent disputes. The basis of the valuation should be explicit, remembering that those purchasing a farm may not be as familiar with these issues. They should not be cut and paste from a tenant right valuation but rather should suit the circumstances of the sale of a farm with produce valued at market value rather than consuming value (unless a sale by a landlord coincides with the end of a tenancy at which consuming value has been paid). As to silage, sometimes fixed sums may be quoted where a clamp has not been opened, although it is more common to see valuations based on CAAV numbered publication 183 (explored in detail in Chapter 29); even with this guidance, the valuation of silage is an area offering plenty of scope for vigorous discussion between valuers.

34.4.3 If growing crops are mature at the date of completion and are to be valued it is advisable to adopt a crop value basis, i.e. the value of the mature crop; again, to avoid dispute, some valuers prefer to quote a fixed price for the crop.

34.4.4 Where crops are recently sown and where cultivations carried out, the basis of valuation should be clearly stated, normally to be taken over at cost of seeds, fertilisers, sprays and cultivations with the latter to be paid for at CAAV rates or contractor's costs (where they have been carried out by a contractor). Enhancement value (now no longer a statutory feature of tenant right valuation but a means to reflect the additional value over its costs of an established crop or pasture) should also be mentioned if it is to be assessed, although mindful of the likely purchase price of the farm that may seem a little excessive.

34.4.5 In the past, prices would be quoted for new leys, but this is one of the areas where valuations have become less common over recent years.

34.4.6 Similarly, it is far less likely for valuation clauses to include equipment or fixtures, except again in a sale after a tenancy or where bespoke grain or other equipment specifically fitted to the buildings are involved. If they are to be taken at valuation, a fixed sum should be specified.

34.4.7 Aside from the arguments cited earlier, one of the arguments against such valuation clauses is that the valuations have taken an increasingly long time to settle, perhaps in an effort to delay payment but equally because purchasers may tend to overlook the valuation clauses until the last moment, although this is no defence. Two valuers acting sensibly with two responsible clients should be able to settle matters relatively swiftly, but the valuation clauses should include some protection for the vendor, whether through a payment on account on completion with a long-stop date for the payment of the balance or interest on delayed payments. That amount on account should be set by the vendor's valuer and, again to avoid dispute over the reasonableness of that judgement, it may be better to specify a sum in advance.

34.4.8 Obvious though it may seem, contracts for sale should include a clause stating that no claim for dilapidations, deterioration or any other offset will be allowed. This is particularly important where a farm or land is sold which was previously occupied by a tenant.

34.4.9 Where Basic Payment Scheme entitlements (or any successors) are to be transferred at the same time as a farm sale, whether as part of the overall sale or as a separate transaction, the sale contract should specify the responsibilities of the parties in terms of effecting the transfer and, subject to the time of year of the sale, who will make that year's application and the responsibilities of the other party not to take actions that would prejudice that claim.

34.4.10 It is important that a valuation clause should be clear, specific and unambiguous, particularly where vendor or agent has felt the urge to ladle lots of extras on top of the purchase price. Otherwise, disputes will surely follow and valuers should not underestimate the capacity of their clients, fellow valuers, the legal profession and when pressured themselves, to find scope for (mis)interpretation of text when arguments begin. As an example, valuers must resist the temptation to refer to Tenant Right even if it seems like a convenient reference point, for fear of bringing issues such as Unexhausted Manurial Values (UMVs) and Residual Manurial Values (RMV) into play when they were not anticipated.

34.4.11 In any event the valuation clause should state that, where applicable, valuers are to be appointed by each party and in the event of disagreement, the matter settled by an arbitrator appointed by them or, failing agreement, by the President of the CAAV.

34.4.12 Previous editions have included a specimen valuation clause; in the same vein and for illustrative purposes only, rather than as recommended practice, this example is offered.

Valuation

The purchaser shall in addition to the purchase price pay the sum of £ for

i) the fitted carpets to the staircase, landing and first floor bedrooms of the farmhouse
ii) the grain drier and associated machinery as described in the valuation clause in the sales particulars

and shall also take over and pay for at valuation on completion, the following:

(a) all hay and straw remaining on the holding, at completion, at market value
(b) all silage remaining on the holding, at completion, valued in accordance with the basis prescribed in numbered publication 183 of the CAAV
(c) any unconsumed grain or other severed crop remaining on the holding at market value
(d) the growing crop of winter barley in NG No 1234 (ha) at the cost of cultivations, seeds, fertiliser and sprays (together with enhancement value of £ (per acre or £ per ha))
(e) all cultivations and acts of husbandry carried out, fertilisers and sprays applied to the unplanted arable land since the last crop was harvested

Note:

i) All acts of husbandry to be assessed at cost based on the CAAV Annual Costings of Agricultural Operations (published annually) or where contractors have been used the actual costs.
ii) No claim will be made for the value of unexhausted value of fertilisers, or lime applied or the residual manurial value feeding stuffs consumed.
iii) There shall be no counterclaim for dilapidations (if any), deterioration or other offset made whatever.

The valuation and payment:

i) The valuation is to be made by two valuers, one appointed by each party, and, failing agreement, shall be referred to a single arbitrator, appointed by agreement between the valuers or, in the event of disagreement, by the President of the Central Association of Agricultural Valuers. Such reference will be under the provisions of the Arbitration Act 1996.
ii) If the amount of the valuation has not been agreed by completion date, then the purchaser shall pay to the vendor's solicitors, as agents for the

vendor on completion date, (either) such sum as shall be certified by the vendor's valuer, as payment on account of the valuation (or perhaps better for certainty) the sum of £15,000 (fifteen thousand pounds). The remaining balance is to be paid within seven days of such valuation being determined or agreed. If full payment is not made upon completion the balance will carry interest at 4% above ...... Bank plc's base rate from the date of completion to the date of payment.

As an alternative, it may be preferable, dependent on the size of the valuation, to substitute the following clause in respect of payment:

If the valuation is not settled and paid on completion date, the purchaser shall pay interest at the rate of 4% above ...... Bank plc's base rate on the sum finally determined due from the date of completion to the date of payment.

34.4.13 However, that does rather assume that, come completion, the vendor's valuer has got on with things and the delay in agreement lies elsewhere. That is by no means a certainty, particularly where the vendor's firm has specialist departments and the agent charged with doing the valuation will receive only a tiny morsel of the fee from the Agency department's table. It should not make any difference, but underneath an acre of tweed agricultural valuers are human too.

34.4.14 It should be emphasised that valuers must give careful thought to the relevance and appropriateness and particularly the contents of such clauses in the sale of a multi-million-pound farm. There is much more to be lost by making the market think the vendor or their agents are too greedy where the farm is concerned than is to be gained from many valuations. Gwyn Williams put things rather more pithily than this in the fourth edition: "*Do not include petty items in a valuation.*" As always, sound advice.

34.5. Farm lettings

34.5.1 As well as offering farms for sale, agricultural valuers are also instructed to offer farms and land to let particularly where they act as managing agents for the owners.

34.5.2 Commonly, lettings involve relatively short-term lettings of accommodation land, and the vast bulk of Farm Business Tenancies are lettings of land only for less than five years. However, farms and larger blocks of land and buildings will also be let.

34.5.3 The smaller short-term lettings may often be dealt with in private negotiation with neighbours and the same agreement may be renewed on many occasions. Otherwise such lettings may come to the open market as, in most cases, will the larger lettings, albeit some larger estates will consider internal promotions; hopefully doing so before going to the market as there is nothing more frustrating for a potential tenant than to go to the time and expense of putting together a bid only to find the farm let to a neighbouring tenant on the landlord's estate.

34.5.4 Where farms do come to the market, they are almost always let by tender with the landlord's agent selecting a shortlist of bidders from those tenders who are then interviewed, commonly on their home farm so that the agent can see some of their farming in practice, before a successful candidate is selected. One of the best tasks for an agent is advising a bidder that they have been successful, particularly a young farmer taking their first farm; conversely, one of the worst is advising the other shortlisted bidders that they have been unsuccessful. There is no prize for coming second, no matter how much praise a sympathetic agent may give your bid and, given the paucity of farms available to let, most potential bidders are setting themselves up for disappointment.

34.5.5 Letting particulars will have many of the attributes of sales particulars but with the added detail on the terms of the tenancy and the ingoing valuation. That added work, together with a desire to defray some of the costs to the landlord, will often see letting agents charging prospective bidders for tender packs although this practice, whilst by no means extinct, is perhaps less commonplace than was the case.

34.5.6 Farm viewings are traditionally dealt with by holding two or three viewing days when prospective bidders can attend and view the farm with the agent in attendance. That may make for travel difficulties for some coming from a distance, but most bidders appreciate the rationale for fixed viewings, not least as the outgoing tenant will normally still be trying to farm the farm and cannot be expected to host multiple visits. The viewing days serve a number of purposes, most obviously letting the prospective tenant view the farm but also enabling the prospective tenant and the landlord's agent to meet and take stock of one another. The importance of that meeting cannot be overemphasised for both parties. Impressions formed then may be a significant influence on the success or otherwise of a bid and, indeed, future relationships with the successful bidder.

34.5.7 The viewing day will also offer an opportunity to understand some of the technical issues with the farm. The letting agent must be careful, as with sales, to be helpful with information where they can be but not to exaggerate the quality of the holding or, more likely with most agents, to stray too far beyond their knowledge base in providing helpful information. The case of *Crown Estate Commissioners v Wakley* was a warning for agents and landlords in which some of the representations made during the letting process over the condition of the holding were very germane to the final judgement.

34.5.8 Traditionally, farms might be offered to let around six months before the commencement date of the tenancy and with the letting process taking two or three months that means the timetables for incoming tenants are very tight. That is particularly the case where the incomer already has a tenancy elsewhere for which, under statute, they would normally need to give between 12 and 24 months' notice to quit. Some landlords are willing to accept short notice where a tenant has been successful in securing another holding, whether out of support for the tenant or interest in securing vacant possession. Others, quixotically including those often most vocal about supporting their tenants, are stricter in holding them to the full notice period which can be particularly difficult where the farms are some distance apart.

34.5.9 In practical terms, the letting agent will need to have settled terms with the outgoer over items such as tenant's fixtures and improvements, tenant right, dilapidations and any holdover, so that these can be made plain to the potential bidders. Similarly, a draft tenancy agreement should be available for inspection. Typically, this will be in a standard form rather than as a final draft agreement as there may be specific changes made to the agreement to accommodate arrangements settled between the landlord and the incoming tenant when the letting is completed.

34.5.10 Where complete farms or larger blocks of land are concerned, the tender process will normally require tenderers to support their bid with detailed personal, farming and financial information, commonly including farm enterprise plans with supporting budgets and cash flows. Landlords are increasingly interested in the extent to which the farm enterprise might be supported by non-faming income and so will want to understand proposals for diversification on the farm or the prospect for off farm income.

34.5.11 Having received the tenders, the letting agent will then need to appraise these to select the shortlist or in unusual cases go straight to the successful bid. This can be a time-consuming task where there are a large number of bids, and it has not been uncommon to get 30 or 40 bids for a holding. It is one that demands a good deal of application and a clear system. It is no longer acceptable simply to tell a bidder that they have been unsuccessful; they will want and indeed deserve some feedback on the attributes or otherwise of their bid and the best agents have been generous with their time in offering constructive feedback to unsuccessful tenderers.

34.5.12 Agents have traditionally used various criteria to make a first 'sift' of bids, whether on:
- experience: how long have they been farming? How long have they held their existing tenancy?
- finance: do they have adequate working capital? How highly geared is the business? (often a tricky test for a tenant farmer). How dependent are they on third party (typically family) capital?
- proposals: do the proposals fit the farm? These are sometimes preconceived – must it be dairy or must it be organic, for example – but they may be a more of a test of realism
- rent level: is the tendered rent attractive? Is it sustainable? It is on this latter point that the best landlords' agents earn their corn, making

sure that anxiety to maximise income for the client is tempered with a degree of realism. The highest bids might not be sustainable ones and there is little benefit for anyone is picking up the pieces where a letting in which so much hope will have been invested fails.

34.5.13 Thereafter, the decision over which bids go through to the shortlist will turn on far less clear-cut issues, including perhaps the quality of the proposal, in both content and presentation, the standing and enthusiasm of the bidders' referees and the reputation of the bidder amongst the landlord's agent's contacts.

34.5.14 Given the demand for greater transparency in most walks of life and in particular the pressures on the public and third sectors where procurement and other decision making is involved, there is merit in combining these various criteria into a "scoring matrix" so that bids can be compared on an even basis and by different agents before a decision is reached. Whilst it is unusual, there can be complaints over the conduct of farm lettings and this can be particularly difficult for public sector clients. Occasionally, for example, an established county council smallholding tenant, disillusioned by their failure to progress, will complain to their division councillor who may not clearly understand the issues involved but will feel obliged or, in certain circumstances, empowered to take the complaint further.

34.5.15 Once the incoming tenant has been selected and the letting agreed, the parties should move to complete the tenancy agreement as soon as possible, not least to ensure that the agreement is concluded whilst the detailed conversations that take place on letting are fresh in the parties' minds and the parties are agreed. The landlord's agent will be involved in the outgoing and incoming valuations to the extent that the landlord is strictly the recipient and payer of valuation monies, although very often this will travel directly between tenants. They will hope not to be involved in the valuation as referee between the parties, although where tenants' agents cannot agree on a valuation they may turn to the landlord's agent for some interpretation of the tenancy agreement or simple advice on the landlord's preferred approach.

34.5.16 The complexity of information required and level of competition amongst bidders is such that some prospective tenants will ask an agent to advise them or indeed to prepare the tender on their behalf. In the past, this might be as much to do with the agent's possession of a word processor and having someone in the office capable of using it as their skill as a bidder. However, with prospective tenants now far better prepared technically, the agent needs to add value in this role, particularly in:

- engaging with the landlord's agent and gaining as much insight as possible into the key objectives of the landlord and their agent
- checking the proposed terms of the tenancy and advising on the inherent risks and liabilities
- inspecting the farm and, where relevant, identifying any likely challenges over ingoing issues

- acting as a critical friend to the client, testing their proposals for the farm, the assumed financial performance and financial conclusion in the light of the agent's wider knowledge of the market.

34.5.17 Given how fierce the competition can be for farms to let, one of the best services an agent might offer is to caution a client to think very carefully before investing in fees and their own time putting together a bid. That said, once the client has decided to bid, the agent must commit fully to the project and should excuse themselves if they cannot do so because they think this is a fool's errand. Once again, acting for a successful bidder is a very gratifying job for an agent whilst seeing one's client just fall short can seem like a personal affront. As in all elements of the rural valuers work a delicate balance of detachment and personal investment in a project is essential to get the best outcome for the client.

35 Dispute resolution and expert witness work

35.1 Introduction

35.1.1 While there are many practical variations, the main ways in which disputes or differences between parties can be settled, other than using courts or tribunals, are:
- *arbitration*, in which the matter referred is determined by an arbitrator only on the basis of the evidence and arguments tested between the parties
- *expert determination*, with the matter at issue referred for a relevant professional to determine using skills and professional knowledge
- *mediation*, providing a framework for accelerated negotiation between the parties, not limited to the issue in question.

35.1.2 Other related tools include *Early Neutral Evaluation* in which a particular key issue or the problem as a whole is put to a third professional for an analysis and a view of what the answer or outcome might be. That external view can help inform subsequent negotiations.

35.1.3 Mediation – The parties may consider mediation. This is a private process, facilitated by a person acting as a mediator, to try to find an agreed resolution of the issues between the parties. While arbitration and expert determination are formal private processes to determine an answer to the question put, mediation is not limited to the issue in hand (such as the rent) but can look at all the issues troubling each of the parties and so can result in a "package" settlement.

35.1.4 Mediation can be seen as enhanced negotiation with a neutral mediator hearing the parties and helping each talk matters through to see if they can reach an agreed solution. That process can embrace any other subject that is at issue, not just the rent, and so potentially reach the root of the disagreement.

35.1.5 It is not guaranteed success – it cannot be used to impose a settlement. The mediator cannot compel agreement but rather manage a process in which agreement can be reached. The parties must be willing to enter the process in that light and develop their concerns and solutions. If they do come to an agreement and that agreement is properly recorded, then it is final and binding between them. If they do not come to an agreement, then their recourse is to the formal methods already discussed, so far as they are still available, having already incurred the costs of mediation.

35.1.6 A mediation might have a style that could be at any point on a spectrum between facilitative to evaluative:
- a facilitative mediation can use a mediator who knows little of the subject but has the skills to help the parties towards their own solution
- an evaluative mediation uses someone more skilled in the area who can be aware of what the relevant issues are likely to be, what the possible answers might be and help the parties tackle them.

The choice between the two may be a mixture of the parties, the issues and the personality of the mediator.

35.1.7 Arbitrator or expert? – These have different approaches to providing an answer:
- an arbitrator operating within the statutory framework of the relevant Arbitration Act determines the answer on the basis of evidence and arguments tested between the parties, whether by their submissions, responses and mutual cross-examination or by the arbitrator putting evidence, points and questions to them for responses, before arriving at an answer based on the evidence and arguments put and tested. The arbitrator should not use evidence that has not been put to the parties for testing and response.
- an expert is able to use his knowledge without that process but with a duty of care to both parties in determining the rent. An expert operates under the terms of his appointment and then under the common law of contract for professional services.

Unless the parties instruct, neither arbitrator nor expert can do more than determine the rent where that is the question statutorily put to them.

35.1.8 Costs – All involved in a dispute are urged to bear costs in mind, especially as they can escalate towards a hearing. Much can be saved by identifying early those points on which the parties are agreed and clarifying the areas of dispute.

35.1.9 Parties should try to keep an objective view of the issues, their prospects and positions to help manage their liabilities. In particular, their advisers should counsel them as to the risks at stake.

35.2 Disputes in agricultural tenancies

35.2.1 Arbitration has been the longstanding means expected for the resolution of disputes in agricultural tenancies and is the default procedure for most disputes under the 1986 Act in England and Wales, with appointment under by the Acts s.84 and governed by the wider code of the Arbitration Act 1996, with more specific provision for rent reviews in s.12 and Schedule 2.

35.2.2 Arbitration is similarly the default approach for farm business tenancies under the 1995 Act, under its ss. 28 to 30 and the general provisions of the Arbitration Act 1996.

35.2.3 However, other methods are coming into use while:
- a number of issues concerning succession and some notices to quit are referred to the First Tier Tribunal, Property Chamber (Agricultural Land and Drainage) in England, still the Agricultural Land Tribunal in Wales
- since 2003, the Scottish Land Court has been the default recourse for Scottish agricultural tenancy disputes. While arbitrations still happen, the awards can still be appealed to the Land Court (and as in *Morrison-Low v Paterson* then potentially to the Court of Session). Where arbitrations under agricultural tenancy law happen with procedure under s.61A of the 1991 Act, they have not yet been brought under the Arbitration (Scotland) Act 2010.

35.2.4 In Northern Ireland, all these issues are simply matters of contract between the parties. Where a tenancy dispute is referred to arbitration, it would simply be under the Arbitration Act 1996.

35.2.5 With concern over cost and procedure, legislative steps have been taken for England and Wales for more use of expert determination, but to date this has been by creating formal opportunities for parties to agree on this *once* there is an issue rather than allowing them to adopt it as an enforceable default mechanism under the tenancy agreement or invoke it unilaterally. The door to this opened with the 1995 Act which allows the use for farm business tenancies of jointly appointed experts save for most reviews, consent for improvements and compensation for improvements. That approach was then extended in 2015 to 1986 tenancies so that all issues save those concerning notices to quit could be referred to a third party, including an expert for decision.

35.2.6 In Scotland Lord Gill's *Agricultural Tenancies* has advised that a final and binding expert determination of a tenancy issue is just that and cannot be referred to the Land Court.

35.3 Arbitration: a statutory framework for and an evidence driven process

35.3.1 The statutory framework – Arbitration is governed by a statutory framework and knowledge of the relevant Arbitration Act, with the duties and powers it gives the arbitrator, is key to making the best use of it. The Arbitration Act 1996 applies to England, Wales and Northern Ireland and covers agricultural disputes.

35.3.2 While the Arbitration (Scotland) Act 2010 operates in an essentially similar way, it does not at the time of writing apply to Scottish agricultural tenancies.

35.3.3 The object of arbitration is

> to obtain the fair resolution of disputes by an impartial tribunal without unnecessary delay or expense. (s.1(a), Arbitration Act 1996)

35.3.4 The arbitrator is to

> act fairly and impartially as between the parties, giving each party a reasonable opportunity of putting his case and dealing with that of his opponent.
> (s.33(1)(a), Arbitration Act 1996)

35.3.5 To achieve that, the arbitrator is to

> adopt procedures suitable to the circumstances of the particular case, avoiding unnecessary delay or expense, so as to provide a fair means for the resolution of the matters falling to be determined.
> (s.33(1)(b), Arbitration Act 1996)

and, unless the parties unite as to how the proceedings should be conducted, the arbitrator has wide discretion in this (s.34).

35.3.6 Moreover, the parties have a duty to co-operate with the process and with the arbitrator's directions:

> The parties shall do all things necessary for the proper and expeditious conduct of the arbitral proceedings. This includes complying without delay with any determination of the [arbitrator] as to procedural or evidential matters.
> (s.40, Arbitration Act 1996)

35.3.7 Appointment – The arbitrator may be appointed:
- by the parties acting jointly, asking an agreed person to do this with the appointment then made. If this is for a tenancy matters, that private appointment must be complete in writing before the statutorily set date.
- by the parties asking a third party, such as a professional body like the CAAV, to nominate a person for them to appoint or to make that appointment. The same rules on timing apply for tenancy matters.
- by one party, acting under the 1986 or 1995 Acts in England or Wales as amended by the Agriculture Act 2020, asking an appointing "professional authority" (such as the President of the CAAV) to appoint an arbitrator. That application with its required fee must be received before the required date, but the appointment itself can be made afterwards.

35.3.8 For Scottish agricultural tenancy matters, the arbitrator's appointment can only be by agreement once the issue has arisen (s.61 of the 1991 Act).

35.3.9 In Northern Ireland, there are no statutory time limits and the parties are free to agree matters in their contract or, if they are able, at the time.

35.3.10 Care over timing is needed where there are statutory time limits for action, as under much agricultural tenancy legislation. Using the common example of a rent review under s.12 of the 1986 Act or s.10 of the 1995 Act:

- a private joint appointment of an arbitrator must be complete in writing before the review date. The parties should allow for the process of appointment taking some time as the first person approached might not be able to act. If the process of private appointment is not completed before the termination date, then the appointment is ineffective and the authority of the s.12 notice for the appointment of an arbitrator lapses
- the same applies where the parties ask a third party, such as a professional body like the CAAV, to nominate or appoint a person
- a unilateral application in England or Wales to an appointing professional authority (such as the President of the CAAV) to appoint an arbitrator must be received before the review date, but the appointment itself can be made afterwards. The authority of the s.12 notice will lapse if an application to an appointing professional authority has not been received (with its required fee) before the termination date.

35.3.11 Once the appointment has been made any dispute over the validity of the appointment should be raised with the arbitrator.

35.3.12 Circumstances can arise where the arbitrator can no longer act. If he becomes incapable or dies, the arbitrator may be replaced under s.84 of the 1996 Act as though he had never been appointed. Resignation is best settled by agreement with the parties. The parties can apply to the court to remove an arbitrator on the grounds set out in s.24 of the 1996 Act.

35.3.13 Procedure – The arbitrator is the servant of the parties whose dispute it is and follow any procedure that they may agree. However, and in practice, not only will it usually fall to the arbitrator to suggest a proper process but the Arbitration Act at s.34 gives the arbitrator substantial powers to determine and drive the process. He has duties of fairness to the parties and to the process of determining the dispute.

35.3.14 The Arbitration Act 1996, applying to England, Wales and Northern Ireland, seeks to deliver the process fairly and effectively. The arbitrator is required by s.33(1) to act fairly and impartially. In turn, the parties to the arbitration are required by s.40 to do all things necessary to the proper and expeditious conduct of the arbitration, with potential sanctions for not doing so. Similar provisions apply to the generality of arbitrations in Scotland.

35.3.15 In Scotland, where the Arbitration (Scotland) Act 2010 does not yet apply to agricultural tenancy disputes, s.61A(4) of the 1991 Act simply provides for tenancy disputes that

> The procedure to be followed at arbitration (including any matters to be taken into account by the arbiter and the matters to be contained in his award) shall, subject to subsection (5) below, be as the parties agree or, in the absence of such agreement, as the arbiter considers appropriate.

The arbitrator here has wide discretion as to how to proceed and, following natural justice, should use this fairly between the parties (who have acted jointly in the appointment) and to achieve the determination.

35.3.16 As agricultural tenancy arbitrations are commonly dealt by a hearing rather than in writing, the conventional procedures have been:
- the arbitrator holding a preliminary meeting with both parties, whether physically or by telephone:
  - to determine the procedure for the arbitration
  - to review which points are in contention so that the process can focus efficiently and cost-effectively on the actual areas of dispute
  - to understand whether negotiations are continuing or have broken down
- the arbitrator would then make directions as to procedure and timing, requiring:
  - early exchange of evidence
  - prior agreement on as much as possible (sometimes as a "Scott schedule")

and making any provision as to costs, such as capping the sum that may be awarded between the parties
- if a hearing proves necessary, it might follow conventional court room procedure with:
  - an opening statement by the Claimant (usually the person wanting the rent to change)
  - evidence taken from the Claimant's witnesses
  - those witnesses are cross-examined by the other party (the Respondent)
  - the arbitrator questions those witnesses with any further points or clarifications
  - the Respondent makes their case
  - the Respondent's witnesses give evidence
  - they are cross-examined by the Claimant
  - the arbitrator questions them with any further processes
  - the Respondent makes a closing statement
  - the Claimant makes a closing statement
- a separate hearing on costs may follow the award as to the issue to be determined.

35.3.17 However, with concerns that arbitration has a reputation for being procedural and expansive, the Arbitration Act 1996 (and it would appear the Scottish 1991 Act's s.61A(4)) allow the arbitrator substantially more control over the proceedings (subject to the joint instruction of the parties) to take a more direct approach to achieve an answer with measures that could involve such points as:
- timetables
- limits on the volume of evidence
- focusing on the key issues
- greater questioning by the arbitrator
- cost capping.

The arbitrator must be fair and impartial between the parties and they must have the opportunity to respond to all points put.

35.3.18 Where the arbitration is to be carried out using written representations, the procedure has often been more flexible. The arbitrator will be provided with written representations from both parties. The parties must then have the opportunity to respond to each other's submissions.

35.3.19 The witnesses may be expert witnesses with their experience and opinions or witnesses as to fact; both owe their duty to the process, not to the party that has retained them. In that, someone who has acted in the rent negotiations might not be able to act as an objective expert. It will be important that they are held to their remits and are not presenting arguments. That role falls to the parties' advocates, a role that a negotiator could adopt.

35.3.20 In preparing their submissions, the parties can ask the arbitrator to exercise his discretion and require the other party to release information – "discovery". Contentious examples include requesting other rents on the landlord's estate or the tenant's accounts. The arbitrator should consider if the request is disproportionate while the other party can object that the documents are privileged. If disclosed, that information is provided to the other party, not to the arbitrator.

35.3.21 The arbitrator is likely to inspect the holding when he should be accompanied by both parties or neither party.

35.3.22 Costs – There are three sets of costs to consider with an arbitration which can amount to a substantial sum:
- the fees and costs of the arbitrator to be met by the parties
- the costs of the landlord for witnesses, advocacy and materials
- the costs of the tenant for witnesses, advocacy and materials.

With the general rule that costs follow the event (the winner has costs paid by the loser), the arbitrator may then direct that one party pays some or all the costs of the other and of the arbitrator.

35.3.23 The arbitrator will confirm his fee to the parties and the basis for charging costs he incurs. In the absence of an agreement on these, the parties are jointly liable for them. The arbitrator can, under s.56 of the 1996 Act, withhold his award until he has been fully paid.

35.3.24 While each party can control its own costs, it has no power over the costs incurred by the other as each is entitled to be represented as they please (but should be aware that the arbitrator can disallow some costs thought unwarranted).

35.3.25 While the costs of arbitration will generally follow the event so that the "loser" may pay all, that can be regulated in several ways:
- the arbitrator, if aware of the potential for costs disproportionate to the case (as perhaps for the rent of a small area of bare land), may direct early in the case that he will not award more than a certain sum to be paid by the "loser" – a cap on the cost award. That may influence the extent to which the parties thereafter incur costs
- the arbitrator may consider that a party has incurred certain costs (such an expert witness) unreasonably or behaved unreasonably, vexatiously or oppressively and so not award that cost against the "loser"
- the arbitrator may occasionally find that each party achieved something in the outcome and divide the award of costs to reflect that
- either or both parties may, in their negotiation, have made a Calderbank offer (see later on) as to the rent which is then made available to the arbitrator when making the award as to costs. If the award as to rent is less generous than was proposed in the Calderbank offer, then costs incurred after that offer are not subject to an award.

35.3.26 Where the parties reach agreement on the rent before the arbitrator had reached an award, it is for the parties to consider the question of costs, particularly those already incurred by the arbitrator.

35.3.27 Calderbank offers – Usually called a settlement offer in Scotland, this is the name for a written offer, expressed to be without prejudice save as to costs, by one party to the other to settle the rent. If made in this way, an offer to settle can not only be effective in encouraging realistic settlement but be relevant when the arbitrator comes to considering how to award costs between the parties.

35.3.28 If the offer is accepted, then the dispute is settled without further contest.

35.3.29 If the offer is not accepted, it is then available for the arbitrator to consider it when making the award as to costs. If when that letter is disclosed, it proves that the offeror had been more generous than the arbitrator, then no costs after the closing date for accepting the offer would be awarded.

35.3.30 This is based on the model of paying money into court when considering the potential outcome of litigation, where it is a means to avoid the need for further dispute when a good settlement is on offer. In practice, it offers a point when the parties can look at what they might really achieve and make an offer to settle accordingly, whatever their formal positions may be.

35.3.31 An example in the context of a rent review might be:
- a holding with a rent of £5,000
- the landlord is seeking £10,000
- the tenant, while arguing that the rent should standstill, nonetheless makes a Calderbank offer to settle at £8,000, perhaps with advice as to where rents were now being settled
- the landlord does not accept that offer
- the arbitrator awards £7,500.

On those facts, the landlord would have been better advised to have accepted the tenant's offer and brought the arbitration process to an end earlier; it could now be seen to have been needless after that point. As he did not, the arbitrator would not ordinarily award the winning landlord's costs against the tenant where they were incurred after the time allowed in the Calderbank offer closed.

35.3.32 To qualify as a Calderbank offer, the letter setting it out should:
- address only the issue that is referred to arbitration (in this case rent)
- clearly offer to settle all parts of the dispute
- set out how all costs are to be handled, including those of the arbitrator
- be clearly expressed to be made "without prejudice save as to costs"; being "without prejudice" means the letter is privileged correspondence and so shielded from the arbitrator, but that having been qualified "save as to costs" it can be shown to the arbitrator once the award of rent is made
- state a time limit for its acceptance that allows the party receiving it time to consider it properly.

35.4 Settling Scottish agricultural tenancy disputes

35.4.1 The Scottish Land Court – Unless agreed otherwise, the Land Court is the body to determine agricultural tenancy disputes that cannot be settled by negotiation. It also deals with a range of other matters from appeals over support payments to crofting issues.

35.4.2 The Court has published its *Plain Guide to Litigation* to assist. As the process of an application unfolds, so the Court will tell the parties, stage by stage, what has to be done and what happens next. The Court makes its decision on the basis of the evidence provided by the parties.

35.4.3 Can it be settled? – The Court will encourage the parties to try and settle the matter by negotiation rather than press ahead with litigation and its costs and uncertainty. The Court can sist (freeze) an application to assist this.

35.4.4 Pleadings – These are to identify the issues in the dispute, so that all parties and the Court can understand the issues, with as many facts as possible agreed. Focusing on the real arguments also helps reduce the costs of the parties and may assist negotiations. The Court expects the parties to be both practical and realistic in their evidence and discourages skeletal applications. Where a party does not take this approach, the Court may impose a penalty for costs incurred as a result.

35.4.5 The application, supported by documents, should set out the issue, giving the other party and the Court clear notice of:
- the basic points of the case
- what outcome is wanted (such as the rent sought)
- the justification for the rent sought

with enough detail for the other party to understand the basis on which the suggested rent is sought.

35.4.6 The Court will ask the other party for a response with the same requirements as for the applicant and confirming any points where the respondent agrees with the applicant.

35.4.7 At the hearing, each side will be able to present its case with witnesses and to question the other's witnesses with the applicant usually presenting his case and evidence first. The Court will inspect the holding and any available comparables and then give a decision and deal with expenses.

35.4.8 The general rule is that the successful party will be found entitled to recover its expenses from the other party, but that is subject to taxation while the Court's award may reflect the parties' conduct. Despite this, a successful party rarely sees full recovery of costs.

35.4.9 One way to draw attention to and potentially influence costs is to use a settlement offer (similar in effect to the Calderbank offer described earlier). If negotiations are not moving to a conclusion, it can be useful to consider carefully with the client whether and when to make a settlement offer and at what figure. Careful consideration should be given to any settlement offer that is received – it is an incentive to realism since all litigation carries risks.

35.4.10 Appeals from the Land Court are to the Court of Session on points of law or misdirection.

35.4.11 The option of arbitration – For an agricultural tenancy dispute, the parties can initially exclude the Land Court and agree to refer the rent to arbitration, whether by someone they appoint or nominating someone to make that appointment. The President of the Scottish Agricultural Arbiters and Valuers Association (SAAVA) can undertake this function. The arbitrator should be in place before the termination date in question. The procedure is to be agreed by the parties or (if they do not agree) as the arbitrator considers appropriate – he may hold a pre-meeting to discuss this. The arbitrator will then issue his directions, confirming the procedure and timetable in writing so that both parties know what is expected of them. They have a duty to comply with those directions. Unless dealt with by written representations, there will be a hearing and the arbitrator will inspect the holding and any comparables that are available.

35.4.12 In a tenancy dispute, an appeal may be made against the award on a question of law to the Land Court with appeals over irregularity or jurisdiction to the Court of Session.

35.4.13 Independent expert determination – As the issues in a rent review may be more for expert knowledge than about contested evidence, the parties might agree to appoint an independent expert to make a final and binding determination of the rent, using his expertise, knowledge and skills, rather than just the evidence of the parties. He is required to act honestly, independently and fairly and has a duty of care to the parties. The expert is still likely to want the parties to make submissions, clarifying facts and issues, but he can proceed to his conclusions and is not bound by the evidence. Lord Gill's *Agricultural Tenancies* observes that there is effectively no appeal against a reasonably conducted determination so that this is a means to arrive more swiftly at a final and binding answer allowing the parties to move on.

35.4.14 The parties will need to provide the procedure to make it work and to ensure a conclusion, lest they come to disagree on the process.

35.5 Expert witnesses

35.5.1 Evidence at any hearing by expert witnesses can offer analysis and insights from their experience. In doing this, they owe their duty to the court, not to any client, and are to fulfil this duty in a professional and objective way. An expert witness is to draw on the full range of his experience, disclose all relevant evidence of which he is aware and advise the court professionally. The expert witness is not to be selective or partial in this and has a duty to explain how he has used that evidence in arriving at his conclusion.

35.5.2 This is a very different role from presenting the client's case. It is outlined in guidance such as the RICS Practice Notes for Surveyors acting as Expert Witnesses (a matter of mandatory professional discipline for RICS members acting as experts). These obligations may mean that someone who has negotiated in the review might not be able to act as an expert witness. On occasion, clients may need to be prepared for this consequence of professional obligations, perhaps sometimes a delicate task.

Further reading

Appropriate Arbitration (CAAV, 2020 – on the CAAV Website)
Means of Dispute Resolution (CAAV, 2015)
Mediation for Agricultural Valuer's (CAAV, 2019)
Rural Arbitration in the United Kingdom (CAAV, 2020)

See other CAAV publications as relevant, including *Rent Reviews under the Agricultural Holdings Act 1986* (2008), *Rent Reviews for Farm Business Tenancies*(2008), and *A Practitioner's Guide to Scottish Agricultural Rent Reviews* (2013).

36 Future skills and services

36.1 Overview

36.1.1 Like all business, agriculture and rural land use respond to long-run economic and social change, but at the time of writing there is the prospect of accelerated change driven by:
- the policy changes and trading circumstances following the UK leaving the European Union
- the impact, mitigation of and adaptation to climate change
- wider environmental concerns from air quality (relevant to slurry management) to biodiversity with the potential flows of money to achieve improvement
- technological change

and whatever changes may flow from policy after the Covid-19 pandemic.

36.1.2 None of the areas of work foreseen here are necessarily new, but they are suggested as being ones that will see development and more emphasis for a profession that has always adapted in response to market circumstances and valuers find where their skills best answers clients' needs. The areas suggested here will overlap; environmental resources will offer business opportunities and technological change will infuse all. While many will have particular specialisms, the profession is ultimately about appraisal and advice to people and businesses in the context of the variety of the agricultural and rural economy, drawing together all the factors that drive decisions and markets.

36.2 Skills in business review and project management

36.2.1 Within agriculture, the withdrawal of Basic Payment area support that is expected in at least England and Wales can be expected to prompt change in those farm businesses, often cereals and grazing livestock that have been accustomed to area payments of one form or another since 1993. These payments typically form some 15% of gross income and appear as a large fraction of profit, but have over time become factored into costs while insulating business structures from change. Some businesses, often grazing livestock or more marginal cereals farms, are more reliant on area support, but it will be of less or negligible importance for some dairy and many horticulture, pigs and poultry business.

36.2.2 While that challenge can be described, the potentially larger economic effects of any changes in trade, whether with the European Union or with other countries, that are being negotiated at the time of writing would again drive economic change with opportunities for import substitution possible for dairy and horticulture but also challenges from more open trade for many livestock producers used to protection by the EU's high tariffs.

36.2.3 Those issues point to many businesses having to review what they do and, indeed their viability, requiring their advisers to have skills in:
- reviewing accounts
- understanding benchmarks
- concluding what steps a business might take
- advising and assisting practical discussion to reach decisions
- implementing decisions

summarised as business review, facilitation and project management. The issue involved might range from generational change in managing the business or withdrawing to let land to others to developing new enterprises and finding ways to add value as well the simple search of a commodity producer for efficiency, cost reduction and profit.

36.2.4 The economic logic in a high cost country confronted with competition is to move up the value chain, searching for where a commercial margin can be won and held, typically by achieving higher value output, more specialist operations and adding value. Essentially, that is what the export-facing manufacturing sector has done. Farming is in the same position but with much smaller businesses, many potentially more adaptable once seized of an opportunity and finding the skills, including marketing, to take it forward.

36.2.5 For food production, it seems noteworthy that, one way or another, it may already be that a third of UK agriculture's output by value comes not from simple broadacre farming but "protected" or higher investment operations, whether indoor dairying, horticulture (from glasshouses and polytunnels to plastic sheeted fields), pigs, poultry and other enterprises. There is increasing experimentation with the technologies for the various forms of controlled environment farming, essentially indoor, rapid turnover, low microbial-load farming, sometimes in locations close to markets.

36.2.6 Other options include seeking out other income streams from off the farm such as outside employment or business to those on the farm such as bed and breakfast, making the family more resilient to keep the farm.

36.2.7 Practical appraisal and professional advice on risk, reward and value tailored to the circumstances and the client with a view to decisions and the management of their effective implementation, marshalling the skills of others, looks likely to be a valuable area of work.

36.2.8 In particular, these are also likely to be skills that will be valuable for those valuers who may wish to seek senior positions in the rural economy, whether managing estates or other larger businesses or those, such as trustees, responsible for them – or, indeed, managing larger professional practices.

36.3 Skills with environmental and other resources

36.3.1 Perhaps really a major aspect of the issues in the business review just discussed, it is still conventional to consider environmental issues separately. The key resources of soil, water and air, the question of biodiversity, the challenge of the impact of, adaptation to and mitigation of climate change are all promising to be even more significant drivers of regulation, opportunity and value in the coming years. Some like soils are fundamentally part of productive agriculture, others are complementary to agricultural use and yet more, such as woodlands, may be alternatives to it.

36.3.2 The agricultural valuer will need an increasing awareness of and knowledge of these issues with some likely to specialise more deeply in these areas. This will include familiarity with the arguments and language of natural capital as a means to rank the expected outcomes of different options and so aid making choices and understanding the transactions involved in agreements for such matters as the delivery of ecosystem services and biodiversity offsetting.

36.3.3 While much discussion focusses on current and prospective governmental environment schemes which will provide work, larger issues will include:
- the role of private sector transactions, as may be driven by the Task Force for Climate-related Financial Disclosures, the needs of insurers, the expectations of investors with the guidance of "green finance" or simple commercial opportunity and self-interest as with purchasers of produce managing risk in their supply chains
- the effect of increasing regulatory standards for land management
- the changing views in markets: to what factors do buyers and sellers become more sensitive in their judgments?

36.3.4 One key area will be the management of land and water to mitigate the flood risks that are the biggest practical challenge to property in the United Kingdom from climate change. That might see pressure for improving soils, some woodland planting, the re-engineering of water courses and drainage and the use of land to hold water. There may be some fundamental issues not only about the development uses of land but even the viability of some existing settlements and how the resulting questions might be practically managed.

36.3.5 Rural land faces the challenge posed by climate change to remove or displace carbon. Especially in England, soils depleted of organic matter may often offer as much or more opportunity to store carbon as tree planting and do so without so substantially excluding future changes of use.

36.3.6 The further development of renewable energy, displacing carbon as the economy moves more heavily to the use of electricity, will be a significant opportunity. A growth area for work over the last decade, valuers will be advising on the arrangements for new developments in generation and storage while the issues of aging existing facilities may also emerge.

36.3.7 However significant, climate change and the net zero target for 2050 do not exhaust the resources issues to be considered as, for example:
- the growing focus on restoring biodiversity through public development and other interests will create new sources of value to be considered, negotiated and advised on
- air quality is one of the government's greatest environmental problems with agriculture being a particular source of the ammonia that poses problems both in nitrifying environmentally protected sites and for air quality, helping generate the very small and damaging PM2.5 particles. As already seen in Northern Ireland and now foreshadowed for England, controls on slurry stores, livestock housing and fertilisers will come to be considered.
- water quality remains an issue with runoff of soil, slurry, phosphates, nitrogen and other inputs from farmland being one of the concerns likely to lead to further intervention. More demanding rules on slurry and silage storage could be an issue for those farms needing to consider expensive changes.

36.4 The competition for land use

36.4.1 With a relatively crowded island and growing population, the competition for the use of rural land sems likely to intensify. Alongside the continuing need for food production, even if much of the substantive value of that becomes done more intensively, competing pressures include those for:
- housing land
- land for employment uses
- land for leisure uses
- land for environmental and biodiversity uses
- woodland and forestry
- the need to respond to climate change with carbon sequestration and displacement
- renewable energy.

Some of these have the potential to be complementary uses, as with soil management for carbon sequestration potentially supporting more productive farming. Wind turbines may take little land, but solar farms are likely to displace farming as will more permanent changes still such as forestry. Some will offer business opportunities and income, others perhaps capital gains but more choices may impose costs and limitations.

36.4.2 Within food production, there will be choices as to who has the farming use of the land and on what terms, especially as current farmers react to the loss of Basic Payment and other economic changes. Improving productivity (the efficiency of profitable production) drives a focus on those who will farm most effectively having access to the land they need.

36.4.3 Scotland, in particular, discusses much greater intervention by public authorities in land use and management decisions. In England, the land value capture debate continues, including proposals for larger housing developments to be taken forward on a new town type basis with compulsory purchase, possibly at less than market value.

36.4.4 With these foreseen new and intensified pressures and the consequences of the choices that could be made, landowners and farmers are going to need careful and considered advice as to how best to navigate these uncertain waters.

36.5 Technology

36.5.1 New technology, where useful, can offer ways to do current work better and more efficiently, but it can also offer ways to add value to clients, making possible work that could not previously be done economically, if at all. Fundamentally it should be seen as augmenting professional work as it changes, not displacing it. It may tend to displace drudgery but call yet more on judgment.

36.5.2 At one level, this is about tools that help work, the laptop with its advancing software and the drone that records fields or inspects gutters from the air. More broadly, agricultural valuation lies where many of the current catch phrases overlap:
- AgriTech, where the great advances in data handling integrated with machinery unlock precision farming, robotics, autonomous machinery, new genetic tools and other developments are all offering efficiencies or gains to those who can use them and aiding land management
- PropTech with application of data to property
- FinTech for ways of providing finance and managing risk.

Some may come together with techniques like BlockChain by which products or transactions can be tracked and verified with data spread across dispersed computers.

36.5.3 Precision farming and precision land management can give information of value through methods such as yield mapping and give a closer understanding of soils, their nature, problems and potential. Mapping systems again aid those looking at what may be needed for the best management of environmental resources or secure the greatest net payments for them.

36.5.4 That comes to questions about the inter-operability of different IT systems and the extent and trustworthiness of data transfer between parties, whether farmer and purchaser, landlord and tenant, owner and contractor or other. The valuer may well come to have a role in such issues, whether directly or being able to marshal the support for them. It is yet to be seen whether data are really to be seen jealously guarded assets of value (albeit their use value rising with volume) or as a free resource with the value lying in how they are used: raw material or application.

36.5.5 Experience to date suggests that even technologically developed businesses can still want external, market-based, independent advice on matters such as stocktaking. While potentially equipped to do it themselves, they may judge their limited time and attention is better focussed on the core issues of their business. Few feel competent on their own in the face of all developing technology; conversation and the benefit of informed comparison offer them value.

36.5.6 Data may lie at the heart of what might often look like digital magic, but the results need to be considered and explained by judgment on such points as:
- whether the data collected and used are right or relevant, and does not include other data that would mislead, to avoid the old computer saying "garbage in, garbage out"
- how that data have been analysed to give an answer, the algorithm
- whether the answer produced is a reasonable one and is not simply given authority just because it has technology behind it.

36.5.7 These points can be illustrated by the growing use in mass residential markets by lenders and others of Automated Valuation Models (AVMs), taking data from housing markets to predict price for an individual dwelling. Aside from noting that that may produce a predicted price, not a market value on a valuation standards definition, that can raise a variety of questions:
- what data have been used? Are they relevant? Are they recent? Are they sufficient? Have other relevant data been excluded by the search definition used? Is it comparing like with like?
- what factors have been taken into account in working from that data to give the result? Are they all relevant? Are they complete?
- does the result look right? If not, that may lead to more questioning of the two previous points
- has the property to be valued been properly understood in its situation?
- have changing economic or market circumstances been picked up? Or have past trends simply been projected?
- is the subject property one for which the AVM can give a useful answer?

36.5.8 Questions beyond that might be about the client's interest. A mortgage lender with large portfolio might in reality be essentially concerned about the security of its overall book and accept a level of individual variation in valuation accuracy within that that someone solely concerned with one property could not accept. Studies point to AVMs misvaluing a significant number of properties by more than 20%. In this, rural housing is frequently more individual in its nature and susceptible to more individual decisions in the marketplace than much of the wider market.

36.5.9 Yet big data can also give new insights. Many property professionals do not really believe that the present Energy Performance Certificates (EPCs) are accurate, particularly relevant or have any market effect. However, big data analysis of housing market transactions in England has shown that there is a positive, if for more properties, small correlation between EPC banding and transaction price (save for detached rural property). Whatever their objective failings, EPCs appear to inform enough of the market to have that effect; once known, it is then hard to disregard that.

36.5.10 Arthur C Clarke's Third Law of Technology says that "Any sufficiently advanced technology is indistinguishable from magic". Yet, in the real world of imperfection the appearance of magic may give the illusion of authority to a spurious, misleading and costly answer. The valuer may be all that stands between a client being misled by the output from a black box technology and mistakes. That will require developing the skills and the confidence to point out where the emperor may have no clothes, as much as where these tools are a genuine aid to judgement and decision.

36.6 Some final thoughts

36.6.1 The view in the round – When other professions, such as law and accountancy, have tended to specialise ever more narrowly, the work of an agricultural valuer's office remains broad and whether looking at valuations in particular or advice in general brings the capacity to appraise, synthesise and apply the wide range of issues that help the individual client in a specific situation find the optimal answer. That will include values and effects of values, family circumstances, financial limitations and resources, taxation, planning, land tenure, available skills and threats and a myriad of other points that make the answer in that case, one that might differ for another client.

36.6.2 That enables valuers to provide a breadth of perspective that can unify the contributions of more specialised advice from other professions from agronomists to accountants, architects to lawyers and so marshal them as a team when required and interpret between them to the benefit of the client. That may help avoid the traps that may be unseen by one specialism and achieve a balance that optimises the outcome.

36.6.3 Ultimately this is about people – After one of the authors had presented a review of the long history of the profession to a multi-disciplinary conference, a former banker pointed to the failings he saw in finance, warning that clients should not be seen as commodities but as continuing personal relationships. It has to be remembered that all the technical skills and specific knowledge reviewed in this text are to be harnessed to the advice being given to a particular individual on which decisions will be taken. That ultimate focus on the person that is the client is the past, present and future truth for the profession.

36.6.4 Appraising and reporting effectively – Synthesising circumstance and advising people requires accurate appraisal and clear reporting. The client will want to be provided with a fully grounded and considered opinion clearly expressed.

37 Advice to young practitioners

37.1 Even though the coming years may be years of enormous change, with the opportunities and work that will bring, the spirit of the advice given in the previous editions of this text holds. Change is not new and previous periods of rapid change show that an adaptable profession can apply the enduring skills of practicality, appraisal, awareness of markets and judgment to new technologies, policies and situations. There is rarely a development that does not have a precedent. However, new work that older practitioners may see as too much of a challenge can be where younger professionals can make their career.

37.2 Work is about people – This book has covered much technical material and professional practice, but we are fortunate to work with and be instructed by individuals with all their varied characteristics, histories, hopes, fears and interactions. The answers will differ between apparently identical situations because of that variety. Gaining experience in the profession is gaining experience about people. Our work has to be attuned to the individuals involved with an empathy not only for the client but also for the other people in any issue, whether other members of the family or the other party. An important lesson drawn from observation of the evolution of other professions that have become more corporate and distant from the client is that the client notices that loss of personal service which leads to best advice. A challenge is to do that and remain commercially efficient.

37.3 A valuer's work is broad – The contents of this book illustrate the range of work that a rural valuer may meet. Few people now cover it all. Valuers may still typically be generalists, gaining a wide knowledge of many things and insight into more, but will not now be universalists, if only because of regional specialisation in farming and, as often, for professional indemnity reasons. Do not only rely on your daily work for this but seek out chances to see other work – whether through CAAV tutorials, going to a livestock or machinery auction or dispersal sale, or asking to be taken on a job by someone else in the firm or even (and there is often goodwill for this) another firm that does different work.

37.4 The CAAV examinations – Preparation for the CAAV's professional examinations with their broad syllabus will give a good grounding for the wider and changing work of the rural valuer, requiring attention to areas of work that may not be part of daily practice. Often now done in conjunction with the RICS APC, tutorials and self-help groups can remedy such gaps as well as strengthen existing areas of experience. Success in the examinations achieves Fellowship of CAAV, not just a label but membership of a professional community of those with recognised specialism in rural work and the CAAV's national and local structures.

37.5 Breadth gives perspective – However, that breadth allows the valuer to provide a more rounded comprehensive advice to a client than can many others. That advice can appreciate and weigh the combined effects of such factors as land tenure, finance, development control policy, succession planning and taxation with the inter-relationships between those and other critical issues in a way that, when many other professionals have become much more specialist, few others can.

37.6 The world is interesting – In starting a rural valuer's career, seek as broad a range of experience as possible. As work may tend to narrow over time, starting on a broad a base as possible will give the breadth of background to inform a lasting career and help find the particular areas of work that are most satisfying and which are likely to be done best. Even then, the in-tray or the in-box is likely each morning to have something new to cause interest and require reflection. There is always something new to learn.

37.7 Ask questions – Do not hesitate to ask questions. Generally, both farmers and practitioners are happy to explain what, why and how something is being done, why it might work or why it might not – there will be reasons, often practical, good and complete but sometimes inadequate. Try to understand why an outcome happened – from the price of something in an auction to a family rearrangement of property.

37.8 Develop and maintain contacts – The rural valuer works in a sociable world, with personal contacts that are not only of benefit in themselves but can be a valuable source of information or a comparable as well as simply a sounding board for a problem when confronted with a new or difficult issue (while recognising client confidentiality). Across the United Kingdom, agricultural valuers are typically of a similar open disposition while also knowing that they too will, in turn, need help.

37.9 Be willing – Our world is one in which the more you give professionally, the greater the return. A client, another party or a principal may put you to the test – or simply need something beyond the immediate job. Meeting someone at an awkward hour, helping round up loose cattle, tackling something new or working to meet the deadline required for an urgent job may appear hard but can earn both respect and self-respect.

37.10 Difficulties – There will be moments of difficulty – with a client, an opposing party, an employer or someone in the office. Try to take a large view of the situation. Not all personalities mesh well together and there can be misunderstandings. Professional and personal networks of contacts can help with this. If the matter is with a job or client, it might usually be right to raise it with your principal, though someone else might help with finding the way forward on a technical issue. If the problem lies within the office, try to keep a sense of perspective as it may appear to loom larger than it might really be, but again it may usually be worth raising it with your principal or someone else senior. Unburdening the problem in conversation is often helpful in itself and assist finding ways forward. Where an issue remains serious, it may require more specific answers, but most situations are more soluble than that.

37.11 The local valuers' association – Within the profession, take an active part in your local valuers' association and, where possible, the CAAV nationally. Not only will this create and strengthen professional contacts but the discussions and opportunities to be involved in matters from technical or legal developments for practice to work on tutorials or the examinations are all personally beneficial as well as contributing to the wider profession. Just as much as you need help, you will be helping someone else.

37.12 Be fair – The rural world is typically one of long and mutual careers, in both farming and professional life, and reputations. You may be acting for people that your firm has worked for over generations. Landlords, tenants and neighbours may go back generations, not only with histories but also the expectations of future dealings. This can be thought as in sharp contrast to the generality of much urban work and requires the adviser to have a longer-term perspective in those dealings. Fully exploiting a temporarily strong negotiating position can leave lasting resentments or grudges that can be repaid when the balance of power is reversed at a later occasion. A little generosity can yield future dividends and it might on occasion be prudent to hold a client back. More than that, fairness, empathy and respect for all involved will build and sustain careers and reputations and the progress of clients.

37.13 Gwyn Williams, a deeply respected and longstanding practitioner, summarised the professional approach of an agricultural valuer in a preceding edition of this book:

> Always act in a straight and honest way and deal with people as you would wish to be dealt with.

37.14 Since the early 1930s, the CAAV has expressed it simply as:

Do what is right, come what may

Appendices

A Calendar of dates for farming agreements and other purposes

Farming agreements, particularly tenancies, typically work to their own series of calendar dates, varying geographically according to historical farming systems. These are the dates when tenancies would end and start, handing over at what were once practical dates, often linked to quarter days or religious festivals, with the autumn dates tending to be more common.

In **England and Wales**, the main dates that will be found are:
- 29 September (Michaelmas)
- 10 October (old Michaelmas)
- 25 March (Lady Day)
- 6 April (old Lady Day).

The difference between the old and new dates stems from the change of calendar in 1752 when it was advanced by 11 days to move to the Gregorian calendar (as well as moving the start of the official year to 1 January). This is also the reason why the tax year starts on 6 April, carrying forward the old start of the year from Lady Day.

Other dates of more local relevance are:
- 2 February (Candlemas), mainly in parts of Wales and the Welsh borders
- 13 May in Northumberland and Durham.

Scotland has statutorily defined its two traditional term dates as:
- Martinmas on 28 November
- Whitsunday on 28 May

In **Northern Ireland**, many seasonal grazing arrangements run from 17 March (St Patrick's day).

Other dates – The quirks of agricultural policy have created their own significant dates:
- since 1993, the main claims for subsidy have had to be submitted by 15 May (if on a weekend, this can slip to the next working day)
- the milk quota year, now no longer relevant, ran from 1 April.

Typical accounting periods for farming end on 31 March and, more in arable areas, 30 September. These set the dates as at which stocktaking is to be assessed for annual accounts purposes.

Other dates and exceptions will be found.

B   Headings for a valuation report for agricultural property: the valuation report

While each practice will have its own style, this offers a checklist of what might ordinarily be expected in a valuation report to a client on freehold agricultural property and so of relevance to other valuation work but always to be adapted to the circumstances with the goal of providing a thorough and explained valuation that can support decision making now and be referred to at a later date.

A. Basis of the instruction

A.1: The property to be valued

1. The property – name (if any)
2. Address of the property
3. Valuer's identification of the property with boundaries on a map
4. Land register reference

A.2: The client

5. Identification of the instructing client (the client's name, details)
6. How the client instructed the valuer and any modification since the date of instruction
7. Third party reliance – where it has been agreed that certain identified third parties will be able to rely on the report, those third parties must be identified
8. Confidentiality clause including any limitations on the report – the valuer must state any limitations on the use of the report as regards publication

A.3: The valuer

9. Identification of the valuer. When a company has been instructed, the individual valuer responsible for the report must be identified

10. The qualifications of the valuer
11. The status of the independent valuer (whether external or internal)
12. Confirmation that the valuer has the experience and market knowledge necessary to value the subject property
13. Confirmation that there are no potential conflicts of interest. Where potential conflicts exist, the report must state that these were brought to the client's attention and detail the measures taken to ensure the valuer's objectivity was not affected
14. Use of specialist valuers or advisers – where the signing valuer has used the services of third party specialists, they must be identified

A.4: The scope of the work

15. The purpose of the valuation
16. The basis of value instructed including full relevant definition (e.g. Market Value with its definition) and the reference to the appropriate valuation standard, law or regulation that defines the basis of the valuation
17. The legal interest of the property that is being valued (freehold, leasehold, etc)
18. Any special assumptions – state if any special assumptions are to be made
19. The investigations carried out
20. The dates of the inspection the completion of the valuation report and to which the valuation applies

A.5: The available information

21. Information received and examined with a list of documents and other information originating from third parties (e.g. land quality, uses, production yields, relevant information about status and history of the property for support schemes, energy performance certificates, building permits, land registry information, current occupancy, leases, etc), including origin of data and supporting evidence (attached as annexes)
22. The valuer must state any important assumptions made as regards documents or information that were not available or about information that could not be verified.
23. If a special assumption is being made, the valuer must state that he has taken it into account
24. Reliance on information obtained from the client and from third parties must be recorded

A.6: The inspection

25. Date of the inspection
26. Confirmation that the inspection was made by the valuer or by a suitably qualified person under the valuer's responsibility

27. The name and qualifications of the person who physically inspected the property, the person's qualifications and the extent of the inspections carried out must be stated. If the inspection has been less complete than usually required for this type of valuation, this must be stated.
28. Responsibility for the inspection: the valuer signing the report (identified earlier under A.3.9)
29. Measurement basis used (e.g. gross area, net farmable area, area eligible for support schemes, etc.)
30. Source of measurement data

B. Description of the property

B.1: The location

31. Description of the area in which the property is situated with factors relevant to potential buyers or tenants
32. Identification and judgment of the relevant market for the property

B.2: The property

33. Property review (with photographs); description of the land:
    - general area, configuration, topography, geology, soil (character, quality, condition, depth, pH, erosion, etc), rainfall, drainage
    - character (including description of fields with sizes, layout and boundaries, permanent pasture, arable, orchard, vineyard, woodland, etc with field identification, areas, current cropping, etc; fencing, water supply, drainage, vehicular access)
    - description of fixed equipment such as buildings and structures, reservoirs, irrigation (nature, dimensions and construction, age and usefulness);
    - description of dwellings (construction, scale and layout, EPC, any property tax liabilities, state of repair, attractiveness and character, etc with photographs as annexes)
    - plant, machinery, livestock, deadstock or contracts passing with the property
    - services/utilities with the parts of the property benefiting from them
    - relevant local processing, storage or marketing facilities
    - relevant observations on the production economics of the property
    - amenity and sporting uses
    - minerals
    - known flood risk, contamination, disease, crop health or other issues or caveats and assumptions made as to these;
    - outgoings to which the property is liable

34. Judgment of the physical characteristics as to quality
35. Identification and judgment of current relevant market conditions

B.3: The legal situation

36. Tenure – including comment on any covenants, third party rights over the property and rights over third party property, public access, restrictions or obligations that could have an effect on value with the identity of the owner and any occupiers
37. Tenancies – information on the main lease terms, the amounts of current rents and any provisions for them to vary during the remaining life of the lease
38. Permissions benefiting the property, such as water abstraction licences
39. Is the property within or subject to any relevant conservation, environmental protection or similar designations, such as Ramsar, Natura 2000, National Parks, Areas of Outstanding Natural Beauty, etc
40. Town and country planning and development control – information about current policies and the relevant development plan(s), allowed uses and development potential, conservation areas and ancient monuments, exposure to compulsory purchase, etc
41. Environmental or other agreements running with the property

C. Valuation

C.1: The methodology

42. Methodology – description of valuation approaches that were considered; which approaches and which methods have been used
43. Key assumptions – as regards capital values, rental values and yields adopted. It is recommended that the choice of these key inputs be explained with reference to the comparables listed
44. Additional assumptions, special assumptions and caveats – if the instruction requires particular additional assumptions or special assumptions and the valuer considers it appropriate to make caveats, details of these must be stated

C.2: The research criteria for relevant market data

45. Comparables, provided to the extent that confidentiality and data protection law permit
46. Complementary relevant evidence; the source of the market data must be provided to the extent that confidentiality and data protection law permit.
47. Justification of the criteria chosen for selection of comparables (market area, size, type, etc)

48. Valuation uncertainty – in those cases where there is a high level of uncertainty about the level of values, the lack of comparables, rents or yields, the valuer must comment on it here
49. List of comparables chosen (redacted as appropriate for confidentiality and privacy)
50. Justification and judgement of each selection

C.3: The analysis of the market data

51. Description of each comparable (photos may be included as annexes, chosen as appropriate in terms of confidentiality and privacy)
52. Adjustment to the property. The valuer must provide appropriate comment reflecting the logic and reasoning for the adjustments made to the comparables provided
53. Adequately supported opinion of market value

D. Conclusion

54. The reported value must be clearly and unambiguously stated, together with confirmation that sufficient investigation has been undertaken to justify the opinion of value reported
55. Confirmation of value
56. Date of valuation
57. A clear statement as to whether transaction costs such as VAT, fees, etc are or are not included in the reported value
58. The valuation report must be signed by the valuer (identified earlier under A.3.9.)

C Schedule of statutes and cases

United Kingdom

Ground Game Act 1880
Partnership Act 1890
Limited Partnership Act 1907
Inheritance Tax Act 1984
Telecommunications Act 1984
Taxation of Chargeable Gains Act 1992
Estate Agents Act 1979
Communications Act 2003
Consumer, Estate Agents and Redress Act 2007
Charities Act 2011 *(NB The restrictions on land disposals apply to England and Wales only)*
Digital Economy Act 2017
European Union (Withdrawal) Act 2018
Limited Liability Partnerships Act 2000
European Union (Withdrawal Agreement) Act 2020
Direct Payments for Farmers (Legislative Continuity) Act 2020
The Business Protection from Misleading Marketing Regulations 2008
The Consumer Protection from Unfair Trading Regulations 2008 (as amended)

England, Wales and Northern Ireland

Arbitration Act 1996
Agriculture Act 2020

England, Wales and Scotland

Pipe-lines Act 1962
Gas Act 1986
Electricity Act 1989
Gas Act 1995

England and Wales

Land Clauses Consolidation Act 1845
Ground Game (Amendment) Act 1906
Agriculture Act 1947
Landlord and Tenant Act 1954
Land Compensation Act 1961
Compulsory Purchase Act 1965
Land Acquisition Act 1961
Agriculture Miscellaneous Provisions Act 1968
Compulsory Purchase (Vesting Declaration) Act 1981
Agricultural Holdings Act 1986
Housing Act 1988
Local Government Finance Act 1988
Town and Country Planning Act 1990
Planning and Compensation Act 1991
Planning and Compensation Act 1994
Water Industry Act 1991
Agricultural Tenancies Act 1995
Planning and Compulsory Purchase Act 2004
Localism Act 2011
Housing and Planning Act 2016
Neighbourhood Planning Act 2017
Regulatory Reform (Business Tenancies) (England and Wales) Order 2003 SI 3096
Regulatory Reform (Agricultural Tenancies) (England and Wales) Order 2006 SI 2805
The Electricity (Necessary Wayleaves and Felling and Lopping of Trees) (Charges) (England and Wales) Regulations 2013 SI 1986
The Electricity (Necessary Wayleaves and Felling and Lopping of Trees) (Hearing Procedures) (England and Wales) Rules 2013 SI 1987

England

Agriculture (Model Clauses for Fixed Equipment) (England) Regulations SI 950
Agricultural Holdings Act 1986 (Variation of Schedule 8) (England) Order 2015 SI 2082

Wales

Renting Homes (Wales) Act 2016
Agriculture (Model Clauses for Fixed Equipment) (Wales) Regulations 2019 SI 1279
Agricultural Holdings Act 1986 (Variation of Schedule 8) (Wales) Order 2019 SI 1404

Scotland

Leases Act 1449
Land Clauses Consolidation (Scotland) Act 1845
Railway Clauses Consolidation (Scotland) Act 1845
Acquisition of Land Authorisation Procedure (Scotland) Act 1963
Land Compensation (Scotland) Act 1963
Land Compensation Act (Scotland) 1973
Agricultural Holdings (Scotland) Act 1991
Town and Country Planning Act (Scotland) 1997
Agricultural Holdings (Scotland) Act 2003
Arbitration (Scotland) Act 2010
Land Reform (Scotland) Act 2016
Private Housing (Tenancies) Act 2016
Agriculture (Retained EU Law and Data) (Scotland) Act 2020

Northern Ireland

Landlord and Tenant Law Amendment (Ireland) Act 1860
Lands Tribunal and Compensation Act (Northern Ireland) 1964
Planning and Land Compensation Act (Northern Ireland) 1971
Land Acquisition and Compensation (Northern Ireland) Order 1973
Planning Blight (Compensation) (Northern Ireland) Order 1981
Land Compensation Order (Northern Ireland) Order 1982
Planning (Northern Ireland) Order 1991
Business Tenancies (Northern Ireland) Order 1996
Private Tenancies (Northern Ireland) Order 2006
Land Acquisition and Compensation (Amendment) Act Northern Ireland 2016

Cases

ADP & E Farmers v Department of Transport [1988] 1 EGLR 209
Allen v Revenue and Customs [2016] UKFTT 342 (TC)
Antrobus 1 and 2 – see Lloyds TSB Personal Representative of Rosemary Antrobus Dec'd
Argylle Motors (Birkenhead) Ltd v Birkenhead Corpn [1974] 1 All ER 201
Arnander, Lloyd and Villers as Executors of David and Lady Cecilia McKenna Dec'd v HMRC [2006] SPC 565
Atkinson – see HMRC v Atkinson
Baird's Executors v IRC [1991] 9 EG 129, 10 EG 153; 1990 SLT (Lands Tribunal) 9
Balfour (see Revenue and Customs Commissioners v Brander)

Batchelor v Kent County Council [1992] 1 EGLR 217
Borwick Development Solutions Ltd v Clear Water Fisheries Ltd [2020] EWCA Civ 578
Brador Properties v British Telecommunications plc [1992] SC 12
Bwllfa and Merthyr Dare Steam Colleries (1891) Ltd v Pontypridd Waterworks Company [1903] AC 426
Camrose (Viscount) v Basingstoke Corpn [1966] 3 All ER 161
Capital Investment Corporation of Montreal v Elliot [2014] SLC 94/09
Cawdor Farming No. 1 Partnership v The Cawdor Maintenance Trust [2018] SLC 151/16
Charkham v IRC [1997] LT DET/3–6/1995; [2000] RVR 7
Charnley v Revenue & Customs [2019] UKFTT 650 (TC)
Trustees of JW Childers Will Trust v Anker [1996] 01 EG 102, [1997 73 P&CR 458
Chivers v Northern Ireland Water Ltd [2015] LTNI R/43/2011 (see also Northern Ireland Water v Chivers)
Clouds Estate Trustees v Southern Electricity Board [1983] 2EGLR 186
Cooper v Northern Ireland Housing Executive [1981] RVR 131
Cornerstone Telecommunications Infrastructure Limited v Compton Beauchamp Estates Ltd [2019] UKUT 107 (LC)
Cornerstone Telecommunications Infrastructure Limited v Ashloch Ltd [2019] UKUT 338 (LC)
Crewe Services and Investment Corporation v Silk [1998] 35 EG 81
Crossman (see IRC v Crossman)
Crown Estate Commissioners v Wakley [2016] – Bristol District Registry
Director of Buildings and Lands v. Shun Fung Ironworks Ltd and Cross-Appeal Co (Hong Kong) [1995] UKPC 7, [1995] 2 AC 111
Donovan v Dwr Cymru [1994] 05 EG 163
Dreamvar v Mishcon de Reya (also P&P v Owen White and Catlin) [2018] EWCA Civ 1082
Duke of Buccleuch v IRC [1967] 1 AC 506
EE Ltd v London Borough of Islington [2019] UKUT 53 (LC)
Elitestone v Morris [1997] UKHL 15
Farmer (Farmer's Executors) v IRC [1999] STC (SCD) 321
Foster v Revenue and Customs [2019] UKUT 251 (LC)
Fyffe v Esslemont [2018] SLC 67/15
Gaze v Holden [1983] 1 EGLR 147
George v South Western Electricity Board [1982] 2EGLR 214
Golding v HMRC [2011] UKFTT 351 (TC)
Gooderam v Department of Transport [1994] 35 RVR 12
Gray (Lady Fox's Executors) v IRC [1994] STC 360
Greenbank v Pickles [2000] EWCA 264
Hanson – see HMRC v Hanson
Executors of John Sidney Higginson v CIR [2002] STC (SCD) 483
Horn v Sunderland Corpn [1941] 2 KB 26

HMRC v Executors of W M Atkinson (dec'd) [2011] UKUT 506 (TCC)
HMRC v Hanson [2013] UKUT 0224 (TCC)
Henke v HMRC [2006] SPC 550
The Ninth Marquess of Hertford, Myddleton, Montagu Scott and Russell, as Executors of the Most Honourable Hugh Edward Conway Seymour Eight Marquess of Hertford (decd) v CIR [2004] SPC 444
Holland v Hodgson [1872] LR 7 CP 328
Hughes v Doncaster Metropolitan Borough Council [1991] 1 A.C. 382
IRC v Crossman [1937] AC 26
Kaufman v Gateshead BC [2012] RVR 128
Kirby v Hunslet Union Assessment Committee [1906] AC 43
Landkreis Bad Dürkheim v Aufsichts- und Dienstleistungsdirektion [2010] ECJ C-61/09
Layzell v Smith Moreton and Long [1992] 13 EG 118
Living Waters Christian Centres v Fetherstonhaugh [1999] 28 EG 121, [1999] 2 EGLR 1
Livingstone v The Rawyards Coal Co. [1879] SLR 16_530
Lloyds TSB Personal Representative of Rosemary Antrobus Dec'd v CIR [2002] STC 468 (Antrobus 1)
Lloyds TSB Personal Representative of Rosemary Antrobus Dec'd v Twiddy [2005] DET/47/2004 (Antrobus 2)
Longson v HMRC [2000] Sp C 238
Lynall v IRC [1972] AC 680
McCall and Keenan (Personal Representatives of Eileen McClean Dec'd) v HMRC [2008] SPC 678
McKenna (see Arnander)
Matthews v Environment Agency [2002] EWLands LCA_192_2000
Metropolitan Properties v Finegold [1975] 1 WLR 349
Minister of Transport v Pettit [1969] EGD 696
Morrison-Low v Paterson [2010] SLC 233/08; [2012 ScotCS CSIH_10
Myers v Milton Keynes Development Corporation [1974] 1WLR 696
MWH Associates Ltd v Wrexham County Borough Council [2011] UKUT 269 (LC)
Naylor v Southern Electricity Board [1996] 1EGLR 195
Newman v Cambridgeshire County Council [2011] UKUT 56 (LC)
Northern Ireland Water Ltd v Chivers & Anor [2016] NICA 22
Oates v HMRC [2014] UKUT 0409 (LC)
Ozanne v Hertfordshire County Council [1992] 2 EGLR 201
Parochial Church Council of the Parish of Aston Cantlow and Wilmcote with Billesley, Warwickshire (Appellants) v Wallbank [2003] UKHL 37
Persimmon Homes (Wales) v Rhonda Cynnon Taff County Borough Council [2005] RVR 59
Phillips v Symes [2004] EWHC 1887 (Ch)
Pointe Gourde Quarrying and Transport Co Ltd v Sub Intendent of Crown Lands [1947] AC 565

Poole v South West Water Ltd [2011] UKUT 84 (LC)

R v Paddington (VO) ex parte Peachey Property Corporation Ltd [1965] RA 177

Revenue and Customs Commissioners v Andrew Michael Brander (As executor of the will of the late fourth Earl of Balfour) [2010] UKUT 300 (TCC)

Rosser v CIR [2003] SPC 368 (Rosser 1)

WSG Russell v HMRC. [2012] UKFTT 623 (TC)

Sawkill v Highways England Company Limited [2020] EWHC 801 (Admin)

St Clair-Ford (As Executor of the estate of Norman Peter Youlden deceased) v Revenue and Customs – Capital Taxes [2006] EW Lands TMA_215_2005

St John's College, Oxford v Thames Water Authority [1990] 1 EGLR 229

Solarin v Wandsworth London Borough [1970] 214 EG 1169

Spath Holme Ltd v Greater Manchester and Lancashire Rent Assessment Committee [1995] 49 EG 128, 2 EGLR 80

Spirerose (see Transport for London)

Starke v CIR [1995] STC 689

Stokes v Cambridge Corporation [1961] EGD 207

Street v Mountford [1985] UKHL 4; 1 AC 209

Stirrat v Whyte [1967] SC 265

Thomas v Sorrell [1673] Vaugh 330

Transport for London (London Underground Limited) v Spirerose Ltd [2009] HL30

Tummon v Barclays Trust Co Bank Ltd [1979] 250 EG 980, [1980] 39 P&CR 300

VodaFone Ltd v Hanover Capital Ltd [2020] EW Misc 18

Wakerley v St Edmundsbury BC [1977] Lands Tribunal 241 EG 921 and 242 EG 49; Court of Appeal [1978] 249 EG 639

Walton's Executors v IRC – Lands Tribunal [1994] 38 EG 161; Court of Appeal [1996] STC 68

Arnold White Estate Ltd v National Grid Electricity Transmission Plc [2014] EWCA CIV 216

Waters v Welsh Development Agency [2004] 2 EGLR 103

White v IRC [1982] 24 EG 935 LT

Williams v Revenue and Customs [2005] UKSPC SPC00500

Youlden (see St Clair-Ford)

Zarraga v Newcastle Upon Tyne Corporation [1968] 19 P&CR 609

Zubaida v Hargreaves [1995] 9 EG 320, [1995] 1 EGLR 127

Index

acclimatisation and settlement of hill sheep 402
accommodation works 288–9
Acquisition of Land Act 1981 250
Acquisition of Land Authorisation Procedure (Scotland) Act 1947 252
active farmer approach 35
adjusted replacement cost 243
ADP&E Farmers v Department of Transport 273
advance payments on compulsory purchase claim 289
Aggregates Levy 182
agricultural buildings 243–5
agricultural holding 362
Agricultural Holdings Act 1986 12, 18, 114–16, 120, 127, 220, 246, 250, 292, 343, 346, 348, 374, 395, 412, 422; additional compensation 413; basic compensation for disturbance 412–13; compensation for early resumption 413; in England and Wales 343, 395–403; farm business tenancies 413, 418; landlord's claim 418; s.12 notice 348–50; s.20 of 346; tenant right and short-term improvements 397–9; tenant's fixtures 396–7; tenant's improvements 395–6; value of tenancy 422
Agricultural Holdings (Scotland) Act 1991 12, 14, 23, 343, 359, 379, 413–14, 422; additional compensation 414; compensation for disturbance 414; compensation for early resumption 414; landlord's claim 419; MLDTs 415; SLDTs and LDTs 415; S.13 of 379; tenancy 40; value of tenancy 422

Agricultural Holdings (Scotland) Act 2003 23, 343, 423
agricultural land 9, 244–5; buildings and fixtures 11–12; character of 9–10; crops 13; development control 11; legal issues 10; location and designations 10–11; tenant's fixtures 12–13
Agricultural Land Classification maps 92
Agricultural Land Tribunal 458
agricultural law 18–19
Agricultural Miscellaneous Provisions Act 1968 250
Agricultural Property Relief (APR) 210, 215, 218–19, 221–4
agricultural support: background 31–3; Basic Payment Scheme and allied payments 33–7; basic policy framework 31–2; BPS and land 40; CAP schemes and valuation 39–41; evolution of EU policy 32–3; future retrospect 47; penalties and reductions 39; prospects after Brexit 41–7; rural development regulation and agri-environment schemes 38
agricultural tenancies 343–4; disputes in 457–8; in England and Wales 18–19, 343; legislation 343–4; renewables and 175–6; in Scotland 23–4, 343–4; valuation of 130
agricultural tenancies, issues requiring valuation: changes to repairing obligations 347; damages and dilapidations 346; exercise of landlord's reservations and covenants 345; game damage 346; landlord's improvements to holding 345

Agricultural Tenancies Act 1995 12, 19, 113–15, 125, 374, 403; in England and Wales 343, 374
Agricultural Tenancies (Gill) 389, 458, 466
Agricultural Value 225–6
Agriculture Act 2020 43, 45–6, 117, 367, 375, 459
agri-environment schemes 38
AgriTech 471
amenity woodlands 135–6
Annual Tax on Enveloped Dwellings (ATED) 28
Anti-Money Laundering Regulations 441
Antrobus (Lloyds TSB [Personal Representatives of Antrobus]) v Inland Revenue Commissioners 2002 224–5
arable land 37
arable stocktaking 234
arbitration 456, 458–64; appointment 459–60; Calderbank offers 463–4, 465; costs 462–3; object of 459; procedure 460–2; statutory framework 458–9
Arbitration Act 1979 69
Arbitration Act 1996 451, 457–62
Arbitration (Scotland) Act 2010 458
arbitrator/expert 457
Areas of Special Scientific Interest (ASSIs) 11
Arnold White Estates Ltd v National Grid Electricity Transmission Plc 332
Atkinson case 223
Automated Valuation Models (AVMs) 472

Baird case 129, 217, 423
Balfour case 226
bases of value 72–3; fair value 73; special value 72; worth or investment value 73
basic land law: in England and Wales 14; in Northern Ireland 14; Scottish law 14
basic loss payment 287–8
Basic Payment Scheme (BPS) 32–7, 205; basics 33–4; claims for young farmers 34; dual use 35–6; eligibility for 34–5; eligible land 35; entitlements 36, 39–40; greening element 33; ineligible claimant 36; issues 41; and land 40; land at disposal 35; livestock support schemes 34; payment rates on entitlements 36–7; redistributive payment 34; use or lose rule 36; Welsh redistributive payment 37
black patch approach 368, 384
BlockChain 471
Brador Properties v British Telecommunications plc 23
Brander 226
Brexit 31
Brexit, prospects after: England 43–5; first steps 43; future policies 43; Northern Ireland 46–7; Scotland 45–6; trade 41–3; Wales 45
Brexit and our Land and *Sustainable Farming and our Land* 45
British Geological Survey 182
buildings and fixtures 11–12
Business Income Manual 233
Business Property Relief (BPR) 218, 221, 226–7
Business Protection Regulations (BPRs) 443
business rates 151, 237–8; agricultural, forestry and sporting exemptions from 243–5; domestic property 238
business tenancies: England and Wales 19–20; Northern Ireland 22; Scotland 24
Bwllfa and Merthyr Dare Steam Collieries v Pontypridd Waterworks Company 273

Calderbank offers 463–4, 465
Capital Gains Tax 39, 60, 63, 160–1, 215, 219; apportionment of value 228–9; part disposals 231; reliefs 229–31; valuation for 227–8
Capital Investment Corporation of Montreal Ltd v Elliot 376
capital spending 50
Capital Transfer Tax (CTT) *see* Inheritance Tax (IHT)
capital valuations 145–50
CAP schemes and valuation 39–41; BPS and land 40; and changes in occupation of land 40–1; entitlements 39
Cawdor Farming No. 1 Partnership v The Cawdor Maintenance Trust 380
Central Association of Agricultural Valuers (CAAV) 2, 21, 61–2, 78, 84; *Costings of Agricultural Operations* 234,

401, 420; *Good Practice in Compulsory Purchase Claims* 248, 260; *Professional Fees in Compensation Claims* 248, 267; Valuation, Compensation and Taxation Committee 151
Charities Act 2011 441
Charkham v CIR 217
Childers v Anker 115, 352, 360, 364
Chivers 314
Civil Justice Council Protocol 84
Civil Procedure Rules (CPR) 84; Part 35 84
clay burning 400
Clean Growth Strategy 202
client and project management 56–8
Clouds Estate Trustees v Southern Electricity Board 329
coarse fishing 142
Code agreement 334
commercial woodlands 136–7
Common Agricultural Policy (CAP) 31–2
Communications Act 2003 333
Community Infrastructure Levy (CIL) 156, 167
company structure 28
comparative method 68
compensation for compulsory purchase 247–9; basis of claim 270–6; claims based on no land taken 294–5; betterment 284–5; blight 285; claims based on value 276–87; development value 274–5; devolution and compulsory purchase 256; devolution and national infrastructure 256; disturbance 287–9; disturbance claim 287; General Vesting Declaration 254, 255; Heads of Claim 268–70; hierarchy of evidence 274; Housing and Planning Act 2016 250–1; injurious affection 278–9; issues during contract 304–12; land taken 276; legal framework 249–53; and let holdings 290–3; by negotiation 254–5; marriage value and development 275–6; no-scheme world 272–3; The Neighbourhood Planning Act 2017 251–2; no land taken 293–5; Notice of Entry 254–5; Notice to Treat 254–5; NSIPs and ancillary development 279–80; severance 277–8; special purchasers 274; statutory routes 253–4; temporary possession 280–4; valuation date 273; valuer's role 256–8
Competition and Markets Authority (CMA) 442
component valuation 361–2, 386
Compulsory Purchase Act 1965 250
Compulsory Purchase and Compensation: The Law in Scotland (Rowan Robinson) 248
Compulsory Purchase and Compensation (Denyer Green) 248
compulsory purchase and let holdings: landlord's interest 290–1; tenant's interest 291–3
Compulsory Purchase Association 249
Compulsory Purchase Orders (CPO) 247, 253
Compulsory Purchase (Vesting Declaration) Act 1981 250
conacre 21
conditional contracts 158
conflicts of interest 58–9
Consumer, Estate Agents and Redress Act 2007 441
Consumer Protection Regulations (CPRs) 443
contingent valuation 208
contract farming agreements 430–2
contractors' method 243
contract price 159–60
Cornerstone Telecommunications Infrastructure Limited v Compton Beauchamp Estates Ltd 336
Corporation Tax 215
cost assessments *see* rebuilding cost assessments; reinstatement cost assessments
Costings of Agricultural Operations 234, 401, 420
cost recovery 159
costs 50; *see also* damage costs, avoided; replacement costs
council tax: and composite hereditaments 245; domestic property 238; and domestic rates 245
Crewe v Silk 346
critical dates 159
crops 13, 188, 400
crops, growing 400; approaches 193; looking at value 194–5; multi-annual crops 195; starting with cost 193–4
Crossman v IRC 216, 217
Crown Estate Commissioners v Wakley 452

CTIL v Ashloch Ltd 337
current assets 51
Current Assets ratio 51
current costs 50

damage costs, avoided 208
data and record keeping 59–60
debtors 51
deer stalking 140–1
DEFRA: Farming Rules for Water 91; *Health and Harmony* 89; 25 Year Environment Plan 44, 202
Denning, Lord 240
deposits 182–3
Depreciated Replacement Cost 110, 179
development 153–4; control 11; and diversification 153; example valuation 164–7; in insurance industry 133; minerals and waste 180–6; net value 154; planning and 154–6; and ransom value 170–1; renewable energy schemes 171–80; of rural land 153; site assembly 160–1; site promotion 156–60; strategic 167–70; valuations 161–71; *see also* Rural Development Regulation; small-scale development; strategic development
development agreements: contract price 159–60; cost recovery 159; critical dates 159; minimum price 159; obligations of parties 159; overage and legacy provisions 160; Promoter's Fee/Developers' Discount 159; tax and CIL 160; term of 158; valuation clauses 160
Development Consent Orders (DCO) 247, 255–6, 260–1
Developments of National Significance (DNS) 256
development valuations 161–71
Digital Economy Act 2017 333
dilapidations 346; end of tenancy claim for 417; making claim 416; preparation claim 419; settling claim 421
Director of Buildings and Land v Shun Fung Ironworks 270
discounted cash flow (DCF) 70, 186; complications 179; valuations 168, 179
dispute resolution 4, 348, 350–1, 371, 456–67; in agricultural tenancies 457–8; arbitrator/expert 457, 458–60; costs 457; mediation 456–7

District Valuer (DV) 218
disturbance claim 287; Basic Loss Payments 288; Home Loss Payment 288; Occupier's Loss Payment 288
diversification 144–5, 409–10; adding value 146; business rates 151; capital valuations 145–50; events 146; example valuation 146–50; farm 144; issue of off-setting 146; letting space 146; lifestyle 146; off-farm working 145; reasons for 144; rental valuations 150–1; sporting 146; tourism 146
domestic property 238
Donovan v Dwr Cymru 264, 325
Dreamvar case 441
Duke of Buccleuch 216

Early Neutral Evaluation 456
earning capacity 362–3; *see also* related earning capacity
easements 16, 106–7
ecosystem services 204; balance of supply and demand 210; costs of alternatives to buyer 210; cultural 204; longer-term value 210–11; operational costs 209; provisioning 204; regulating 204; sellers' opportunity costs 209; start-up and ongoing maintenance costs 209; supporting 204; taxation 209–10; transaction costs 209; transaction prices for 206–7; worth of 207–9
Ecosystem Service Valuation Database 209
effective capital value 243
electricity line wayleaves and claims: basis of claim 329; damage to freehold 328; general 326–7; heads of claim 329–32; losses as result of constructional work 327–8; necessary wayleaves 332
Electronic Communications Code: assessment of compensation 338; basics of 333–5; Code agreement 334; code rights 337–8; consideration (rent) 335–7; further developments 339; new Code 336–8; statutory position 333
Electronic Communications Code rights 333–9; assignment, upgrading and sharing 337; renewal and termination 337–8

Elitestone v Morris 12
Energy Performance Certificates (EPCs) 473
Energy Performance of Buildings (England and Wales) Regulations 2012 442
entitlements 39
Entrepreneurs Relief (Business Assets Disposal Relief in April 2020) 210, 228, 230, 283
Environment Act 2020 156
Environmental Land Management Scheme 44
environmental valuations 202–3; agreeing reward 209–11; natural capital and payments 203–4; payments to change behaviour 205–6; public goods 204–5; transactions prices for agreements 206–7; worth of ecosystem service to a buyer 207–9
Estate Agents Act 1979 441, 442
European Group of Valuers' Associations, The (TEGoVA) *see* TEGoVA (The European Group of Valuers Associations)
European Union 31; Common External Tariff 31, 42; policy 32
European Valuation Standards (EVS) 67
expert determination 456
expert witnesses 466

fair value 73, 235
farm agency 437–8; conduct of sale 440–8; farm lettings 451–5; methods of sale 438–40; valuation clauses on farm sales 448–51
farm: budget 362; diversification 144; lettings 451–4; management accounting 48–50; woodlands 135
farm, conduct of sale on: initial dealings with client 440–1; lotting 447; marketing 443–4; negotiation and arranging sale 444–6; preparation 441–3; sale 446–7; timescales 447
Farm Business Tenancies (FBTs) 2, 113–14, 117–18, 126–7, 129, 374, 403–6, 422, 451; routine improvements 405–6; tenant's fixtures 406; tenant's improvements 403–6; value of tenancy 422
farming agreements 481–2

farming business structures 26; business arrangements between farmers 26–7; company structure 28; farming with contractors 28–9; partnerships 27–8; share farming 30
farmland: licences 20–1; profits of pasturage 20
farm sales method 438–40; private treaty 438, 440; sales by auction 438; sales by tender 438
farmstead: dwellings 93; farm buildings 93–4; farm buildings and value 96–7; modern buildings 94–5; traditional and obsolete buildings 95–6
farm valuations, situation of 97; development value and cottages 105–6; easements, wayleaves, rights of way and other reservations 106–7; elevation and aspect 100–1; example valuation 108–12; farm approaches 99; farming standards 102; farms on flood plains 99–100; fencing, gates, ditches and drains 101–2; general 107; position in relation to roads 98–9; position in relation to towns and infrastructure 97–8; services 103–4; sporting 104–5; woodland 106
fees: on compulsory purchase claim 289; and time management 62–3
financial accounts 50–1; balance sheet 50–1; costs 50; income 50; profit and loss account 50; tax accounts 52; using accounts 51–2
Financial Reporting Standards (FRS 102) 50
FinTech 471
First Tier Tribunal 458
fisheries 142–3
fixed assets 51
fixed costs 48
fixed equipment 367
Forestry Commission 137, 207
Forestry Commission and Natural Resources Wales 137
Forestry Commission in Wales, NRW and Scotland 137
future policies 43
future skills and services 467; business review and project management 467–9; competition for land use 470–1; environmental and other resources 469–70; technology 471–3
Fyffe v Esslemont 19

game fishing 141–2
Gas Act 1986 313–14
Gas Act 1995 314
Gaze v Holden 273
General Data Protection Regulation 2016 441
General Vesting Declaration (GVD) 247, 254–6
George v South Western Electricity Board 329
Gill, L. 458
Good Practice in Compulsory Purchase Claims 248, 260
Grant Thornton Renewable Energy Survey 175
grazing agreements 20, 430
Greenbank v Pickles 425
green finance 469
greening 37–8; arable land 37; and Basic Payment 37; element 33, 37–8
Gross Development Value (GDV) 160, 164
Ground Game Act 1880 346
Ground Game (Amendment) Act 1906 346
Guidance on Property Sales 2015 443

Heads of Claim 268–70; landlord 268; owner/occupier 268; tenant 268
Health and Harmony 44, 89
hedonic pricing 208
hefting 402
Henke v HMRC 231
hereditaments 238–9; composite 245; plant and machinery 239
Hertford case 226
high farming 370
HMRC Help Sheet *232* 233
Holdover (or Gift) Relief 228, 230
Holland v Hodgson 11
Horn v Sunderland Corporation 270
Housing and Planning Act 2016 250–1, 256
HS2 3, 247–8, 252, 260, 269
Hughes v Doncaster Metropolitan District Council 272
husbandry, acts of 401

income 50, 68
income tax, agricultural stocktaking for 232–3; approach 233–5; arable stocktaking 234; fair value 235; livestock 234–5; net realisable value 235; process 235–6

Industrial Revolution 87
Inheritance Tax Act 1984 (IHTA) 216, 220
Inheritance Tax and Stamp Duty Land Tax 216
Inheritance Tax (IHT) 215; Agricultural Property Relief (APR) 221–5; agricultural value 225–6; Business Property Relief (BPR) 221, 226–7; freehold properties 220–1; live and dead farming stock 221; valuation 220; value of tenancy 424
Integrated Administration and Control System (IACS) 33
interest on compulsory purchase claim 289
International Valuation Standards Council (IVSC) 67
International Valuation Standards (IVS) 67
investment value of let property, factors affecting 73, 114–19; nature of letting 114–15; prospect of alternative use 118–19; rent 117–19; reversionary value 119; security of tenure and scope for possession 115–17; terms of letting 118
IRC v Gray 217

joint venture operations 432–3

Kirby v Hunslet Union Assessment Committee 239

land 9; cost budgeting 259; drainage rates 246
Land Acquisition and Compensation (Amendment) Act (Northern Ireland) 2016 253
Land Acquisition and Compensation (Northern Ireland) Order 1973 253
Land and Property Services 239
Land Clauses (Scotland) Act 252
Land Compensation Act 1961 249–51, 270–1
Land Compensation Act 1973 250, 269
Land Compensation Manual 267
Land Compensation (Northern Ireland) Order 1982 253, 271
Land Compensation (Scotland) Act 1963 253, 271
Land Compensation (Scotland) Act 1973 253

500 *Index*

Land Drainage Act 1991 246
Landkreis Bad Dürkheim 35
Landlord and Tenant Act 1954 242
Landlord and Tenant Amendment (Ireland) Act 1860 12
Landlord and Tenant Law Amendment Ireland Act 1860 (Deasy's Act) 14, 21, 410
landlord's claims, end of tenancy 416; in England and Wales 416–18; preparing claim 419–21; in Scotland 417; settling claim 421
Land Parcel Information System (LPIS) 35, 76
Land Reform Act 2016 113
Land Reform (Scotland) Act 2016 14, 24, 253, 343, 384, 406; value of tenancy 423
Lands Tribunal and Compensation Act (Northern Ireland) 1964 253
latent value 364–5
Layzell v Smith Moreton and Long 425
leases 173–5; institutional 173; rent provision 174; user clause 174; *see also* mineral leases
Leases Act 1449 14, 23
Leigh v. Taylor 12
Less Favoured Area (LFA) 10
Less Favoured Areas Support Scheme (LFASS) 10, 38, 46
licences 20–1
lime 399–400
Limited Duration Tenancies (LDTs) 23, 343, 386, 422; compensation for tree planting 409; disturbance compensation 415; diversification 409–10; rent reviews for 388; sheep stock valuations 409; tenant's fixtures 408; tenant's improvements with tenant right 408; value of tenancy 422
Limited Liability Partnerships Act 2000 27
Limited Partnership Act 1907 27
livestock 188–92, 234–5; beef cattle 190; cattle 189; dairy cattle 189–90; high value bulls and rams 190; other livestock 192; pedigree and high value cattle 190; pigs 191–2; sheep 190–1
Livingstone v Rawyards Coal Company 270
Living Waters Christian Centres v Fetherstonhaugh 68–9
Localism Act 2011 255, 274

local planning authorities (LPAs) 155
Longson case 231
lowland shooting 139–40
Lynall v IRC 216

machinery 192–3
manures 400
market risk 74
market value 70–2; assumption 72; date of valuation 71; definition 71; hypothetical parties 72; marketing 72; property 71; result 71; special assumption 72; transaction 71
marriage value 72; and development 275–6; between interests of landlord and tenant 423; and negotiations between parties 127–8; rent reviews 359–60
Matthews v Environment Agency 267
McInerney, J. 144
McKenna 222–4
Means of Dispute Resolution 84
mediation 389, 456–7
Memorandum of Sale/Heads of Terms 446
Mineral and Waste Development Framework 182
mineral leases 183–6
minerals and waste 180–6; deposits 182–3; mineral leases 183–6; mineral rights 181–2; valuation 186
Minerals Local Plan 182
Minimum Energy Efficiency Requirements (MEES) 93
minimum price 159
Modern Limited Duration Tenancies (MLDTs) 24, 343; disturbance compensation 415; sheep stock valuations 409; tenant's fixtures 408; tenant's improvements with tenant right 408; value of tenancy 422
mole drainage 400
Morrison-Low v Paterson 40, 358, 362, 376, 382, 384, 386, 458
MWH Associates v Wrexham CBC 272
Myers v Milton Keynes DC 274

Nationally Significant Infrastructure Projects (NSIPs) 247, 253–6
National Trading Standards Guidance on Property Sales (2015) 443
natural capital and payments for ecosystem services 203–4

Naylor v Southern Electricity Board 332
Neighbourhood Planning Act 2017, The 251–2
net realisable value 235
Net Working Capital Ratio 51
Newman v Cambridgeshire County Council 267
Nicholls, Lord 270
Nitrate Vulnerable Zone (NVZ) 443
nitrogen 399
Non-domestic rating 2017 Valuation List for England 239
Northern Ireland 21–2; agricultural lettings 21; business tenancies 21; compensation 410; conacre 21; farming agreements 481; legal framework 253; prospects after Brexit 46–7; residential tenancies 22; tenant's fixtures 410; tenant's improvements and tenant right 410
Northern Ireland Court of Appeal 315
Northern Ireland Future Agricultural Policy Framework 46
Northern Ireland Lands Tribunal 315
Northern Ireland Water Ltd (NIWL) 314–15
Northern Irish Government Department for Agriculture, Environment and Rural Affairs (DAERA) 34
Northern Irish legislation 315
no-scheme world 256, 272–3
Notice to Treat/Notice of Entry route 253, 254–5, 256

obligations of parties 159
occupier, loss payment of 287–8
option agreements 158
overage and legacy provisions 160
overhead cables 313

Parochial Church Council of the Parish of Aston Cantlow and Wilmcote with Billesley, Warwickshire (Appellants) v Wallbank 106
Parry's Valuation Tables 121
Partnership Act 1890 27
partnerships 26, 27–8
payments to change behaviour 205–6
Phillips v Symes 84
phosphorus 399
pipeline claim: actual crop losses 318; additional cost of maintaining hedges and fences 318; claimant's time 318–19; cost of repairs or reinstatement of drainage 318; cost of replacing lost fertility and improving soil condition 318; cost of replacing topsoil 318; cost of restoration of ground 318; examples 320–6; future crop losses 318; interest on agreed claim 319; livestock losses 319; lost Basic Payment Scheme (and potential successors) and environmental payments 319; professional fees and associated expenses 319; weed control 318
pipelines 313–14; capital payments 314–16; claims 314–16; damage claims 317–19; easement payments 316; easement recognition payment 319; example pipeline claim 320–6; payment 326; project management 316–17
Pipe-lines Act 1962 314
Plain Guide to Litigation 464
Planning Advice for Developments 313
Planning and Compensation Act 1991 250
Planning and Compensation Act 1994 250
planning and development 154–6
Planning and Land Compensation Act (Northern Ireland) 1971 253
Planning Blight (Compensation) (Northern Ireland) Order 1981 253
planning gain 168
Plant and Machinery Regulations 239
Pointe Gourde 251, 271–2
Pointe Gourde principle 272
Poole v South West Water 267
potash 399
Powys County Council Trading Standards 442
Practice Direction and the Civil Justice Council Protocol 84
practitioners, advice to young 475–7
Prag, P. 144
Principal Private Residence Relief (PPR) 219, 228, 230
Principles of Good Professional Practice in Compensation Claims, The 258
Private Housing (Tenancies) Act 2016 24
private treaty 438, 440

produce 188, 195; hay 196; silage 197–201; straw 196–7
productive capacity 356–7
Professional Fees in Compensation Claims 248, 267
professional issues 55; clarity and understanding 55–6; client and project management 56–8; conflicts of interest 58–9; data and record keeping 59–60; fees and time management 62–3; standardisation and systemisation 60–2; utility and advice 63–4
profit: margin 52; method 242–3; of pasturage 20
Promoter's Fee/Developers' Discount 159
promotion agreements 158
Promotion Agreements, Option Agreements, and Conditional Contracts 157
Property Chamber (Agricultural Land and Drainage) 458
Property Misdescriptions Act 1991 442
PropTech 471
Protocol for the Instruction of Experts to give Evidence in the Civil Courts 84
public goods 204–5

Quick Assets ratio 51

Railways Clauses Consolidation (Scotland) Act 252
rateable value 239–41; assumptions 240
rebuilding cost assessments 131
Red Book 56, 63, 78, 85, 162, 176, 215–16, 220
reinstatement cost assessments 131
related earning capacity 357, 363
renewable energy schemes 171–80; and agricultural tenancies 175–6; growth in 171; industrial renewables 173–4; leases 173–5; natural resource renewables 173–4; site finding 171–3; stages of development 171; valuation 176–80
rental: comparison 242; valuations 150–1
rent review 375, 379
rent reviews for farm business tenancies 374–5; alternative basis for rent review 376–7; open market basis 375–6; pre-agreed variations 377; rent review or variation 375; rent variation formula 377–8
rent reviews for Scottish agricultural tenancies: budgets as evidence 386–7; comparables and economics 285–7; comparables as evidence 386; economic conditions 386; forming view 387–8; general 379–80; holding 382; LDTs and SLDTs 388; mediation 389; minimum three-year rent review cycle 380–1; negotiations 381–2; recording agreement 390; serving s.13 notice 381; third party determination of 388–9; treatment of costs 389; valuation basis (rent properly payable) 383–4; valuation date 382
rent reviews under Agricultural Holdings Act 1986 348; balance between earning capacity and comparables 362–3; black patch 368; character and situation of holding 355; current level of rents for comparable lettings 358–9; dispute resolution 371; disregard in rental value 367–8; dwellings 365–6; high farming 370; jointly funded improvements 370; landlord's improvements to holding 366; latent value 364–5; marriage value 359–60; negotiating rent 350; premiums 360; productive capacity of holding 355–6; profits method 369; recourse to dispute resolution 350–1; related earning capacity 357; relevant factors 363–4; rent review factors 372; scarcity 359; serving a s.12 notice 348–50; statutory factors 354; tenant's improvements and fixed equipment 367; terms of tenancy 354–5; use of comparables 360, 369; valuation 370; valuation of rent 351–4; valuing out 368
Repairing Tenancy 24
replacement costs 208
residential property: council tax and composite hereditaments 245; council tax and domestic rates 245
residential tenancies: England and Wales 20; Northern Ireland 22; Scotland 24
Residual Manurial Values (RMV) 449
residual sod fertility 402

Return on Assets ratio 51
revealed preferences 208
RICS: Code of Measurement Practice 442; Guidance 177, 179, 186, 249, 266–7; Practice Notes for Surveyors acting as Expert Witnesses 466; Property Measurement 76; Red Book 63, 215; Registered Valuers 78; UK Guidance Note 3 (UKGN 3) 216
Rollover Relief 228–30
Rural Arbitration in the United Kingdom 84
Rural Development Regulation 38
Rural Diversification (Estates Gazette 2002) 144
Rural Payments Agency (RPA) 34
Rural Payments Wales (RPW) 34
rural valuers 162, 172

"s.12 notice" (Agricultural Holdings Act 1986) 348–50
S.13 of the Agricultural Holdings (Scotland) Act 1991 379
sales: by auction 438; by tender 438
scoring matrix 454
Scotland 22; Agricultural Holdings (Scotland) Acts 1991 and 2003 406–8; agricultural tenancies 23–4; business tenancies 24; England and Wales: making claim 419; farming agreements 481; Land Reform (Scotland) Act 2016 406–8; legal framework 249, 253; other arrangements 24; prospects after Brexit 45–6; repairing tenancies 408; residential tenancies 24; servitudes 24; SLDTS, LDTs and MLDTs 408; tenancies 23; tenant's fixtures 408; tenant's improvements with tenant right 406–8
Scottish 1991 Act 19, 347, 362, 422
Scottish Agricultural Arbiters and Valuers Association (SAAVA) 465
Scottish agricultural tenancy disputes: independent expert determination 466; option of arbitration 465–6; pleadings 464–5; Scottish Land Court 458, 464; settling 464
Scottish Government Rural Payments and Inspections Department (SGRPID) 34
Scottish Land Court 458, 464
Scottish law 14, 24, 423

Scottish statutory cases: relinquishment and assignation 427–8; tenant's right to pre-empt landlord's sale of holding 427
servitudes 24
Severely Disadvantaged Area (SDA) 10
share farming 17, 30, 433
sheep stock valuations 409
Short Limited Duration Tenancies (SLDTs) 23, 343; compensation for tree planting 409; disturbance compensation 415; diversification 409–10; rent reviews for 388; sheep stock valuations 409; tenant's fixtures 408; tenant's improvements with tenant right 408; value of tenancy 422
Silage: A Valuer's Guide 401
Single Payment Scheme (SPS) 32–3
Sites of Special Scientific Interest (SSSIs) 11, 443
Small Landholders (Scotland) Acts 1886 to 1931 24
small-scale development 163; location 163; planning applications/promotions 164; planning permission 163; planning status 163; practical development 163
soil quality 89–93; challenges of assessing 91; clay soils 89; indicators of 91; sandy soils 90; stony soils 90
Solarin v Wandsworth Borough Council 274
sole trader 26
Spatholme v Greater Manchester Rent Assessment Committee 358
Special Purpose Vehicle (SPV) 172, 175
special value 72
sporting property 138–9; coarse fishing 142; deer stalking 140–1; fisheries 142–3; game fishing 141–2; lowland shooting 139–40; rights 139; upland shooting 140; valuation of 104–5
Stability and Simplicity 45
Stamp Duty Land Tax 40, 123
standardisation and systemisation 60–2
stated preferences 208
statutory review notice 375
St John's College case 314, 315
stocks 51
Stokes v Cambridge Corporation 170, 271
strategic development 167–70; market capacity and demand 169; phasing and timescales 169; proposed development 169

Street v Mountford 17, 23
stubble-to-stubble contracting 29
sulphur 399
surrogate market valuation 208
Surveyors acting as Expert Witnesses 84
Sustainable Farming Scheme (SFS) 45, 205

Task Force for Climate Change Disclosure 469
tax: accounts 52; and CIL 160
taxation 14–15, 209
tax valuations 216–17; basis of value 216–17; case management 218–19; positive husbandry 219; untradeable assets 217–18
TEGoVA (The European Group of Valuers Associations) 67
TEGoVA Blue Book 78, 176
TEGoVA's European Code of Measurement 76
Telecommunications Act 1984 333
tenancies: and valuation 15; in Scotland 23
tenancies in England and Wales 17–20; agricultural 18–19; business 19; residential 20
tenants: fixtures 12–13; improvements 367; pastures of 401
tenant's claims, end of tenancy: Agricultural Holdings Act 1986 395–6; compensation 392–3; compensation for tree planting 409; compensation for value at valuation date 394–5; diversification 409–10; in England and Wales 393; farm business tenancies 403–6; in Northern Ireland 394; recognition of tenant's investment on 391; in Scotland 393; settling claim 410–11; sheep stock valuations 409; tenant right and short-term improvements 397–403; tenant's claims 393–4; tenant's fixtures 391–2, 394, 396–7; tenant's improvements 395–6; value added to holding as holding 395; value to incoming tenant 394–5
terms of agreements 158
Third Law of Technology 473
Thomas v Sorrell 20
Total Income from Farming (TIFF) 44
Town and Country Planning (Scotland) Act 1997 253

trace fertilisers and elements 399
trade 42–3
transactional decisions 443
transactions prices for agreements 206–7
tree planting, compensation for 409
Tummon v Barclays Trust Co Bank Ltd 364

UK Soil Observatory 92
underground pipes and cables 313
Unexhausted Manurial Values (UMVs) 449
United Kingdom, devolution in 14
upland shooting 140
Upper Tribunal (Lands Chamber) 332, 334
US Appraisal Institute's Uniform Standards of Professional Appraisal Practice (USPAP) 67
utility and advice 63–4

valuation 65–6; of agricultural tenancies 130; bases of value 72–3; for capital taxes 215–16; comparative method 68; development 161–71; difficult markets 73–4; evidence and analysis 66; for financial accounts 68; income method 68; legal nature 65; longer-term value 74–6; market value 70–2; measurement 76; mineral deposit 186; physical nature of 65; property as comparable 69–70; renewable energy schemes 176–80; sense check 73; stages 177; standards 66–70; uncertainty 73
valuation clauses 160, 448; basis of valuation 448–50; valuation 450–1
valuation methods 241–2; contractor's methods 243; profits method 242–3; rental comparison 242; use of other rating assessments 243
valuation of farm property with vacant possession 87–8; example valuation 108–12; farmstead 93–7; quality of soil 89–93; situation 97–8
valuation of let property 113; of agricultural tenancies 130; Agricultural Tenancies Act tenancy 125–6; AHA tenancy 120–1; basic principles 113–14; comparative values 126–7; example valuations 119–26; factors affecting investment value 114–19; marriage value and negotiations between parties 127–30

valuation report: basis of instruction 483–5; description of property 485–6; valuation 486–7
Valuations for Capital Gains Tax, Inheritance Tax and Stamp Duty UKGN3 216
valuations for insurance 131; assessing the cost 132; developments in insurance industry 133; process 131–3; *see also* rebuilding cost assessments; reinstatement cost assessments
valuation undertaking 77–8; assessing value 81–2; for court proceedings 83–5; information gathering 80–1; instructions 78–80; report 82–3; 5 "w's" 77
valuer, role of (compulsory purchase) 256–8; acquiring authority 256–61; advance payments 261–2; agent 257; claimant 256–60; before construction 261; during construction 262–4; early stages 258–61; end of construction 265; fees 266–8; landowner 257; preparing, submitting and negotiating the claim 265–6
values 65; assessing 81–2; of asset 68, 423–5; bases of 72–3; comparative 126–7; longer-term 74–6; of market risk 74
valuing tenancy 422–3; *Greenbank v Pickles* 425; market value of tenancy 423; marriage value between interests of landlord and tenant 423; in practice 428–9; value between landlord and tenant 426–7; value of asset 423–5
variable costs 48
Viscount Camrose v Basingstoke 274

Wakerley case 425
Wales: legal framework 252; prospects after Brexit 45; *see also* England and Wales
Walton case 216–17, 424
waste *see* minerals and waste
Water and Sewages Service Order 1973 315
Water Industry Act 1991 313, 320, 325, 334
Waters v Welsh Development Agency 272
Welsh Parliament 14, 252
Wight v Moss 217–18
Williams v HMRC 2005 222
woodland 106, 134–5; amenity 135–6; commercial 136–7; farm 135; leisure 137–8; markets for 134–5
World Trade Organisation 32
worth/investment value 73, 207

Zarraga v City of Newcastle 274
Zubaida v Hargreaves 363

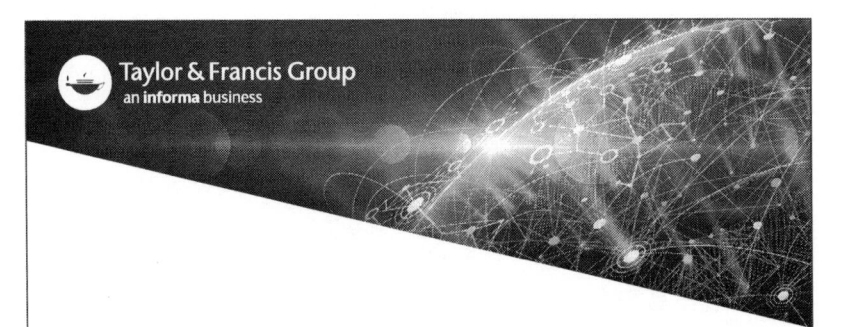

Taylor & Francis eBooks

www.taylorfrancis.com

A single destination for eBooks from Taylor & Francis with increased functionality and an improved user experience to meet the needs of our customers.

90,000+ eBooks of award-winning academic content in Humanities, Social Science, Science, Technology, Engineering, and Medical written by a global network of editors and authors.

TAYLOR & FRANCIS EBOOKS OFFERS:

- A streamlined experience for our library customers
- A single point of discovery for all of our eBook content
- Improved search and discovery of content at both book and chapter level

REQUEST A FREE TRIAL
support@taylorfrancis.com

Printed in Great Britain
by Amazon